Subcellular Biochemistry

Volume 17
Plant Genetic Engineering

SUBCELLULAR BIOCHEMISTRY

SERIES EDITOR

J. R. HARRIS, Institute for Cell and Tumor Biology,
German Cancer Research Center, Heidelberg, Germany

ASSISTANT EDITOR

H. J. HILDERSON, University of Antwerp, Antwerp, Belgium

Subcellular Biochemistry

Volume 17
Plant Genetic Engineering

Edited by

B. B. Biswas
Bose Institute
Calcutta, India

and

J. R. Harris
Institute for Cell and Tumor Biology
German Cancer Research Center
Heidelberg, Germany

PLENUM PRESS • NEW YORK AND LONDON

The Library of Congress cataloged the first volume of this title as follows:

Sub-cellular biochemistry.

London, New York, Plenum Press.
v. illus. 23 cm. quarterly.
Began with Sept. 1971 issue. Cf. New serial titles.
1. Cytochemistry — Periodicals. 2. Cell organelles — Periodicals.
QH611.S84   574.8'76   73-643479

ISBN-13:978-1-4613-9367-2   e-ISBN-13:978-1-4613-9365-8
DOI: 10.1007/978-1-4613-9365-8

This series is a continuation of the journal *Sub-Cellular Biochemistry,*
Volumes 1 to 4 of which were published quarterly from 1972 to 1975

A Division of Plenum Publishing Corporation
233 Spring Street, New York, N.Y. 10013

Contributors

Giorgio Binelli Department of Genetics and Microbiology, University of Milan, 20133 Milan, Italy

B. B. Biswas Bose Institute, Centenary Building, Calcutta 700 054, India

Paula P. Chee Molecular Biology, The Upjohn Company, Kalamazoo, Michigan 49007

Abhaya M. Dandekar Davis Crown Gall Group, Department of Pomology, University of California, Davis, California 95616-8630

Andrew Davidson Max-Planck Institut für Züchtungsforschung, Abt. Genetische Grundlagen der Pflanzenzüchtung, 5000 Köln-30, Germany

Tristan A. Dyer Molecular Genetics Department, Cambridge Laboratory, John Innes Centre for Plant Science Research, Norwich, NR4 7UJ, United Kingdom

David A. Evans DNA Plant Technology Corporation, Cinnaminson, New Jersey 08077

Zhegong Fan DNA Plant Technology Corporation, Cinnaminson, New Jersey 08077

Hans-Jörg Jacobsen FB Biologie, Lehrgebiet Molekulargenetik, Universität Hannover, D-3000 Hannover, Germany

Dieter Jahn Department of Molecular Biophysics and Biochemistry, Yale University, New Haven, Connecticut 06511

David I. Jokhadze Institute of Plant Biochemistry, Georgian Academy of Sciences, 380059 Tbilisi USSR

Gary Kochert Department of Botany, University of Georgia, Athens, Georgia 30602

Robert A. Morrison   DNA Plant Technology Corporation, Cinnaminson, New Jersey 08077

Richard J. Mural   Biology Division, Oak Ridge National Laboratory, Oak Ridge, Tennessee 37831-8077

Nihal K. Notani   Biomedical Group, Bhabha Atomic Research Centre, Bombay 400085, India

Gary P. O'Neill   Department of Molecular Biophysics and Biochemistry, Yale University, New Haven, Connecticut 06511

Ercole Ottaviano[†]   Department of Genetics and Microbiology, University of Milan, 20133 Milan, Italy

M. Enrico Pè   Department of Genetics and Microbiology, University of Milan, 20133 Milan, Italy

Peter L. Schuerman   Davis Crown Gall Group, Department of Pomology, University of California, Davis, California 95616-8630

Jerry L. Slightom   Molecular Biology, The Upjohn Company, Kalamazoo, Michigan 49007

Dieter Söll   Department of Molecular Biophysics and Biochemistry, Yale University, New Haven, Connecticut 06511

Hans-Henning Steinbiss   Max-Planck Institut für Züchtungsforschung, Abt. Genetische Grundlagen der Pflanzenzüchtung, 5000 Köln-30, Germany

Dominique Van Der Straeten   Laboratorium voor Genetica, Universiteit Gent, B-9000 Gent, Belgium

Marc Van Montagu   Laboratorium voor Genetica, Universiteit Gent, B-9000 Gent, Belgium

[†]Deceased

Preface

Plant genetic engineering and the techniques of plant gene manipulation at large have gained importance because of their application in the improvement of plants. The ability to introduce foreign genes into a wide variety of plants has created a revolution in plant biology. A tremendous information explosion in this area of research, both basic and applied, has taken place during the last decade. Despite this advancement of knowledge, however, transformation, regeneration, and stabilization of genes inserted by direct or indirect method, particularly in the case of cereals, legumes, and trees, are still difficult to achieve. No general prescription can be given for producing fertile transgenic crop plants. It appears that each plant will have to be studied individually for proper manipulation of genes. This volume of *Subcellular Biochemistry* gives an overall coverage of the achievements in plant gene manipulation, since technological advances in plant cell transformation and molecular biology have led to enormous progress in understanding how plant cells function.

In the opening chapter, an overview of prospects, perspectives, and problems of plant genetic engineering is presented by Biswas in order to describe the progress made so far in gene manipulation in plants. Different problems and methods available for gene transfer in plants are well documented. In the second chapter, Chee and Slightom discuss the expression of storage protein, particularly vicilin gene from legumes in tobacco. This is one of the model plants in which regulation of expression of plant genes to find *cis* and *trans* elements has been studied. Expression of high sulfur-containing seed storage protein genes in tobacco suggests there is potential for legumes to be enriched with vicilin and phaseolin in sulfur-containing amino acids. The technology of producing haploid plants, particularly for the improvement of crop plants, has gained attention recently for its simplicity and transfer of specific traits via hybrid sorting. This is described in Chapter 3 by Morrison, Evans, and Fan.

Plant transposable elements are extremely valuable for gene tagging, mutation, and monitoring of putative gene function. A succinct discussion of the role of these elements in diversification of plants is presented by Notani in Chapter 4. Woody plants are important and difficult to engineer. Recombinant DNA technology is now applied to forest trees to contain the spread of diseases. Some aspects of this possibility are discussed in Chapter 5 by Schuerman and Dan-dekar. Pollen transformation and gene transfer by transformed pollen is an important technique for generation of engineered plants. This aspect also has implications in manipulating self-incompatibility in plants. This is stressed by Ottaviano, Pè, and Binelli in Chapter 6. Chimeric gene expression in plant cells is important in monitoring the transformation of cells and also in designing the expression vectors for plant cells. Steinbiss and Davidson project the problems and possibilities connected with it in Chapter 7.

Restriction fragment length polymorphism (RFLP) mapping has emerged as an important tool in plant breeding programs. RFLP mapping is being used to expedite the acquisition of important genes from wild species, which may increase yield, resistance, and adaptability to extreme environments, since these traits are found to be polygenic. Thus the uses of RFLP and RFLP mapping in DNA fingerprinting, plant breeding, and cloning of genes are elaborately presented by Kochert in Chapter 8. In the next three chapters, light-regulated gene expression in plants is discussed. Specific promoters for transcription of ribulose bisphosphate carboxylase responsive to light and other effects of light, such as translation and protein stability, are discussed by Mural. Light-mediated reactions are involved in photosynthesis. Although the process is very complex, the question arises as to how this process can be improved upon by available genetic engineering techniques. Selection of photosynthetic components for manipulation is essential. Dyer covers this aspect of the photosynthetic process and O'Neill *et al.* discuss the interesting process of involvement of glutamyl-tRNA in chlorophyll biosynthesis.

Finally, in the last three chapters, the molecular basis of the action of auxins, ethylene, and gibberellin in plants in eliciting the signal transaction, as well as in modulating gene expression, is discussed. Jacobsen discusses the role of auxins in somatic embryogenesis as well as auxin-induced genes in plants, including what is known about biochemical markers for developmental processes and the possible relevance of specific markers for improving regeneration systems. Van Der Straeten and Van Montagu review the molecular basis of ethylene biosynthesis along with gene manipulation for endogenous ethylene formation and stress production. Ethylene-induced genes have been implicated in plant development. Jokhadze discusses the gibberellin binding protein and transcription process in nuclei and chloroplasts in order to elucidate the genetic system of plants under the influence of this plant growth substance.

It is hoped that this volume of *Subcellular Biochemistry*, containing a spectrum of information on diverse aspects ranging from the action of plant growth substances to genetic manipulation of plant genes to produce transgenic plants, will generate much interest among research workers in this field.

B. B. Biswas

Calcutta, India

J. R. Harris

Heidelberg, Germany

Contents

Chapter 3
Haploid Plants from Tissue Culture: Application in Crop Improvement

Robert A. Morrison, David A. Evans, and Zhegong Fan

Chapter 4
Plant Transposable Elements

Nihal K. Notani

Chapter 5
Potentials of Woody Plant Transformation

Peter L. Schuerman and Abhaya M. Dandekar

Chapter 6
Genetic Manipulation of Male Gametophytic Generation in Higher Plants

Ercole Ottaviano, Enrico Pè, and Giorgio Binelli

Chapter 7
Transient Gene Expression of Chimeric Genes in Cells and Tissues of Crops

Hans-Henning Steinbiss and Andrew Davidson

Chapter 8
**Restriction Fragment Length Polymorphism in Plants
and Its Implications**
Gary Kochert

Chapter 9
Fundamentals of Light-Regulated Gene Expression in Plants
Richard J. Mural

Chapter 10
Genetic Manipulation of Photosynthetic Processes in Plants
Tristan A. Dyer

Chapter 11
Transfer RNA Involvement in Chlorophyll Biosynthesis
Gary P. O'Neill, Dieter Jahn, and Dieter Söll

Chapter 12
Biochemical and Molecular Studies on Plant Development *In Vitro*
Hans-Jörg Jacobsen

Chapter 13
The Molecular Basis of Ethylene Biosynthesis, Mode of Action, and Effects in Higher Plants
Dominique Van Der Straeten and Marc Van Montagu

Chapter 14
Gibberellin-Binding Proteins and Hormonal Regulation of Transcription in Cell Nuclei and Chloroplasts of Higher Plants
David I. Jokhadze

Contents   xxi

Chapter 1

Prospects, Perspectives, and Problems of Plant Genetic Engineering

B. B. Biswas

1. INTRODUCTION

Plant genetic engineering has already gained importance because of its implications for crop improvement and productivity. The transgenic plant was first reported in 1983 (Herrera-Estaella *et al.*, 1983a). During the last 6–7 years, a tremendous informational explosion in plant molecular biology and genetic engineering had taken place. It is expected that transgenic plants may yield a second green revolution in agriculture. At present a number of plant genes have been isolated, and not only genes from plants but also genes from bacteria and animals have been introduced into plants to study their expression and functions. Several reviews have already been published on the implications of transformation, regeneration, and expression of genes of agroeconomic importance (Eckes *et al.*, 1987; Weising *et al.*, 1988; Gasser and Fraley, 1989; Benfey and Chua, 1989). This chapter presents an overview of the most important aspects of plant genetic engineering. It is not our purpose to catalogue all the publications in the field or to examine critically the results of the work thus far reported. In this review the aim is to cover broad perspectives and problems so that an overall

B. B. Biswas   Bose Institute, Centenary Building, Calcutta 700 054, India.

Subcellular Biochemistry, Volume 17: Plant Genetic Engineering, edited by B. B. Biswas and J. R. Harris. Plenum Press, New York, 1991.

understanding of different aspects of research in plant genetic engineering can be obtained.

1.1. Special Features in Plants

The principles of genetics were worked out first in plants and the first enzyme to be crystallized was bean urease. Transposon was discovered first in plants. Yet, molecular biology developed independently of plant sciences. This was perhaps due to difficulties associated with working with the plant systems as well as the prolonged time required for such work. Now the potential of plant systems for molecular biological work has been realized and the systems have also been developed to work out such problems with ease. What is there in plants to be attracted to? The plant is a unique system wherein photons can be captured and channeled to fix atmospheric CO_2; the plant cell is totipotent; i.e., a whole plant can be regenerated from a single cell; plants can synthesize an extraordinary range of specific chemicals, including secondary metabolites and cell wall components, many used in defense against bacterial and fungal pathogens. The plant cell does not migrate during morphogenesis; its division is unlike that in other eukaryotes. Plant cells are connected to their neighboring cells by unique structures, plasmodesmata, and are encased in a singular type of matrix, i.e., the cell wall. In response to their environment, plants utilize unique pathways of light, temperature, and gravity detection; as well as plasticity of genomes; in both environmental and developmental responses they use hormones, quite unlike those of other organisms. Not only are these hormones chemically different from those in animals, but their use does not involve specific endocrine organs. Many of the most perplexing aspects of plant physiology, development, and biochemistry are now being studied by the methods of classical and molecular genetics. During the past few years much information has been generated on gene expression in plant systems (Schell, 1987; Collinge and Slusarenko, 1987; Weising *et al.*, 1988; Benfey and Chua, 1989; Gasser and Fraley, 1989). Such information is centered around seed protein genes, light-regulated genes, modulation of specific genes induced by hormones, environmental stresses, wounding, and pathogens, including atmospheric nitrogen fixation genes. These gene systems have been studied because they are involved in processes manifested in plant systems. In addition, plants have unique features entailing chloroplasts and mitochondria giving feedback to unique molecular events such as *trans* splicing of RNA, involvement of aminoacyl tRNA in δ-aminolevulinic acid synthesis, and male sterility, respectively. The cell wall is an important component in the structure of the plant cell. The best characterized proteins of dicot cell wall are the extensins, a family of hydroxyproline-rich glycoproteins having repeating peptides of specific sequences. Some cell wall proteins are rich in glycine and others in cysteine, having antifungal activity. The implication of oligosaccharide messengers (oligosaccharins) generated from both fungal and higher plant cell wall in eliciting a variety of responses from plants has already been stressed. Therefore, it is apparent from the above scenario that plants constitute an interesting and unique system for exploration of molecular events with applied and basic implications.

1.2. Model System in Plants Amenable to Genetic Engineering Investigations

It is difficult to work with each plant system as far as transformation and regeneration are concerned. Tobacco, tomato, potato, and petunia are commonly used. However, *Arabidopsis thaliana,* like the *Drosophila* in animal sciences, belonging to the family of Cruciferae, has long been used as a subject for classical genetic experiments in plant sciences. It is easy to cultivate this plant in the laboratory and several mutations and linkage maps are already available. The generation time is 5–6 weeks. A number of laboratories have turned their attention to use this plant to solve problems in plant physiology, biochemistry, development, and genetics (Meyerowitz, 1987, 1989). The haploid genome consists of five chromosomes. The recombinational length of the genome is 500 centi-Morgan (cM) and physical length is 70,000 kb. The small genome of *Arabidopsis* permits rapid screening of genomic libraries in lambda or cosmid vectors. Many mutants are available, including those affecting lipid biosynthesis (Browse *et al.,* 1987), starch biosynthesis (Lin *et al.,* 1988), nucleotide synthesis (Moffatt and Somerville, 1988), phototropism, (Khurana and Poff, 1988), photosynthesis (Estelle and Somerville, 1986), amino acid synthesis (Klee *et al.,* 1987), nitrate reductase, growth hormone synthesis, and flower formation, (Braaksma and Feenstra, 1982; Karssen *et al.,* 1987; Bleeker *et al.,* 1988; Meyerowitz, 1989). It is also possible to isolate a number of genes from *Arabidopsis thaliana.* In particular, the small genome size and low abundance of interspersed repetitive DNA makes it technically feasible to isolate genes by chromosome walking. RFLP maps for *Arabidopsis* along with the linkage map make it possible to analyze the genome in details (Chang *et al.,* 1988). A library of the whole genome containing an insert of about 130 kb of *Arabidopsis* DNA in yeast artificial chromosome (YAC) has already been obtained (Somerville and Grill, 1989).

2. PERSPECTIVES

2.1. Gene Cloning for Plant Improvements

Genetic engineering of plants is rapidly becoming a reality and plant gene transfer is now a fertile field. Gene transfer into dicot plants can generate varieties with herbicide tolerance or disease resistance and can also be used to study plant gene control. Genetic manipulation of the monocot plants has certain difficulties. However, these difficulties are now being resolved at a quicker pace by development of methods such as electroportation or high-velocity micropro-jectiles for introduction of foreign genes into monocot cells through a tailored vector. Nevertheless, the long-standing inability to regenerate whole plants from transformable cells still remains an obstacle to the successful genetic engineering of monocots and legumes. A model genetic engineering of a plant consists of the following general steps: (1) selection of a gene whose introduction in the plant will have positive agricultural value; (2) identification and isolation of such genes; (3) transfer of the isolated genes to the plant cell; (4) regeneration of complete plants from transformed cells/tissues; and (5) expression of that gene in transgenic plant.

2.2. Introduction of Foreign Genes into Plants

Various attempts have been made by different laboratories to introduce foreign genes, from bacteria, plants, or animals, into plants. Foreign genes can be introduced into plants by several methods (Weising *et al.*, 1988). The ultimate purpose of inserting a gene is stable integration as well as having it in the right place so that it can be expressed and can impart the characteristic that was hitherto not present in the plant. Natural transformation of plant via the *Ti* plasmid of *Agrobacterium tumefaciens* has, however, long been known.

2.2.1. Ti and Ri Plasmids

By taking advantage of the Ti plasmid system it was possible as early as 1980 to transfer foreign DNA to plant cells (Hernalsteens *et al.*, 1980). Early genetic analyses showed that a large plasmid present in most *Agrobacterium tumefaciens*, called the Ti plasmid, was responsible for tumor formation in many species of plants (Zaenen *et al.*, 1974; Van Larebeke *et al.*, 1974; Watson *et al.*, 1975). This was followed by the discovery that part of the Ti plasmid, called T-DNA, is integrated into the genome (Chilton *et al.*, 1982). T-DNA encodes at least two classes of compounds: (1) those involved in opine synthesis (Murai and Kemp, 1982; Joos *et al.*, 1983) and (2) those involved in biosynthesis of plant growth substances (Akiyoshi *et al.*, 1984; Thomashow *et al.*, 1986). The only classes of mutants of the T-DNA that abolish DNA transfer are those that remove terminal repeats of the right and left border (Zambryski *et al.*, 1980; Yadav *et al.*, 1982). *Vir* genes have also been found to have a role in transfer and integration (Garfinkel and Nester, 1980; Douglas *et al.*, 1985). *Vir* gene products and their induction by acetosyringone have also been worked out and the products of *vir* D gene are found to be responsible for nicking at the right border and formation of the single T-DNA strand and are bound to its 5' end during its entry into the plant cells (Yanofsky *et al.*, 1986). The tentative involvement of other *vir* genes is also exemplified (Zambryski *et al.*, 1989).

The concept of two types of vector has already emerged: (1) In integrative vectors advantage is taken of the fact that the tumor-inducing genes are not required for the infection and that any DNA sequence that is inserted between the two borders is to be transmitted to the plant. In these vectors, the tumor genes of T-DNA are replaced by sequences of the *Escherichia coli* vector pBR322. The foreign DNA that is to be integrated into the plant genome is cloned in a pBR vector and then inserted by homologous recombination, via these pBR sequences, into Ti plasmid, containing the *vir* genes. (2) In binary vectors *vir* genes and the borders are localized on two separate plasmids. Different types of constructions of binary vectors are already available.

Integrative Vectors. The first nononcogenic vector designed to cointegrate into the T region of Ti plasmid was pGV 3850 (Zambryski *et al.*, 1983), in which the entire T-DNA of nopaline plasmid pTiC58Tra except the border sequences and nopaline synthase gene was replaced by pBR322 sequences. The gene to be transferred in pBR322 (the intermediate vector) can be mobilized into pGV 3850 where the gene is integrated into Ti plasmid by recombination. The cointegrative vectors have also been designed in the form of split-end vectors (Fraley *et al.*, 1985). In these, T-DNA borders are present on separate plasmids. The intermediate vector, pLGV 2382, frequently used is a pBR322 derivative containing chimeric genes, *nos-npt*II, which confers kanamycin resistance to the transformed plants (Herrera-Estrella *et al.*, 1983b).

Binary Vectors. Most binary vectors are based on wide host range replicon pRK2 (Ditta *et al.*, 1980), but others are pVS1 derived from *Pseudomonas* plasmid and an *Agrobacterium* plasmid pArA4a (Vilaine and Casse-Delbert, 1987). In many vectors using RK2 replicon, the plasmids use Ori T and Ori V so that they can be efficiently mobilized from *E. coli* to *Agrobacterium*, generally by the helper strains HB101/pRK2013 that provide RK2 helper mobilization functions (Van Haute *et al.*, 1983). However, the efficiency of integrative or binary vector is somehow dependent on the plant systems to work with. Both *A. tumefaciens* and *A. rhizogenes* transform plant cells in a similar way as far as T-DNA transfer and synthesis of opines are concerned (Chilton *et al.*, 1982; Tepfer, 1984; Weising *et al.*, 1988). Infection by *A. rhizogenes* results in hairy root formation, and the genes are located in Ri T-DNA. Attempts have also been made to construct a mini Ri plasmid for transformation of plants (Vilaine and Casse-Delbert, 1987). Transformation of *Brassica napus* by Ri T-DNA has been obtained (Guerche *et al.*, 1987). The use of Ti plasmid in transformation of plants far exceeds that of Ri plasmid. It has also been reported that the Ri T-DNA genes are rather complex and involved in modulation of signal transduction mechanisms underlying growth hormone activity. The roots produced by *A. rhizogenes* can be regenerated into shoots and as such may prove useful in the transformation of many recalcitrant species. Plants have been regenerated containing TL of Ri plasmids containing *rol* A, B, and C genes from roots of soybean, tobacco, and tomato (see Zambryski *et al.*, 1989).

2.2.2. Cauliflower Mosaic Virus as Vector

Cauliflower mosaic virus (CaMV) contains double-stranded DNA molecule of 8000 bp and *per se* it is not integrated into plant genome. Thus, in general, their use would be as expression vectors. The infectivity of CaMV shows that all CaMV ORFs except ORFII and VII are indispensable for systematic virus spreading (Howell *et al.*, 1981). The Rationale for construction of CaMV vector is as follows: (1) Some space should be available in the CaMV genome to allow the insertion of foreign sequences without packaging limitation problems. This could be accomplished by deleting a dispensable region of the viral genome. (2) Since the tight arrangement of the CaMV ORFs appears to be a prerequisite for the virus translation mechanism, the distances between termination and initiation of the coding sequences should be kept to a minimum. (3) The foreign gene sequence should be maintained during the propagation of the chimeric viral DNA in plants and expressed under the control of the viral promoter (Brisson and Hohn, 1986). Expression of a bacterial gene in plants by using CaMV vector was first reported in 1984 (Brisson *et al.*, 1984). The dihydrofolate reductase gene in CaMV-infected turnip plants has been also found to be expressed. Later, transforming capability of Ti plasmid and viral vectors were combined under the term "agroinfection" (Grimsley *et al.*, 1986), and viral DNA or cDNA was introduced into plants through disarmed Ti plasmids, which resulted in systemic viral infections of transgenic plants. The 35S region of CaMV is now being used as a strong promoter in the various constructs of vectors for transformation of plant cells (see Weising *et al.*, 1988). However, there is a limitation in using CaMV *per se* as a vector in that (1) CaMV cannot carry more than 0.8 kb inserts, (2) viral disease might be lethal to host plants, (3) there might be incorrect expression of inserted gene, and (4) there might be a problem in transfer of inserted gene to the offspring.

2.2.3. Gemini Viruses as Vectors

The gemini viruses constitute a group of plant viruses having circular single-stranded DNA of about 2.7 kb. These viruses are subdivided on the basis of host range. All monocot-infecting members are transmitted by leaf hoppers, and this type of genome has been found in maize streak virus (MSV), wheat dwarf virus (WDV), digitaria streak virus (DSV), and chlorosis striate mosaic virus (CSMV). Most of the members infecting dicot plants are transmitted by white fly and have a bipartite genome comprising similar-sized DNAs, generally designated A and

B. This type of molecule has been found in cassava latent virus (CLV), tomato golden mosaic virus (TGMV), and bean golden mosaic virus (BGMV). Both genomic components are essential for infection. DNAs A and B show homology only for 200 bases involved perhaps in DNA replication. Beet curly top virus (BCTV), having a single genomic component equivalent to DNA, occupies an intermediate position between the above two subgroups (Stanley *et al.*, 1986). The fact that coat protein deletion mutants remain infectious implies that gemini virus DNA may be used to construct gene vectors. The ability to express high levels of reporter genes, e.g., CAT gene from coat protein promoter of CLV, gives transient expression in the tobacco plant (Ward *et al.*, 1988). Although replacement genes that restore the size of DNA A appear to be relatively stable (Etessami *et al.*, 1988), it has also been demonstrated that CLV DNA A and B become inherently unstable in plants when they contain a large nongeminiviral insert (Stanley and Townsend, 1986). The application of agroinfection to monocot plants has allowed cereals and grasses to be infected with cloned multimers of geminivirus DNA, not only providing an extremely sensitive assay for T-DNA transfer to such plants, but also opening up a means of testing mutants and possibly constructing gene replacement vectors, as is already being achieved with CLV and TGMV (see Davies and Stanley, 1989).

RNA virus may also be used as a vector for plant transformation. Replacement of the viral coat protein gene with a *CAT* reporter gene results in *CAT* gene expression in transformed protoplasts (French *et al.*, 1986).

2.2.4. Direct Gene Transfer

Protoplasts are versatile recipient systems for genetically engineered vectors based on both Ti and other bacterial plasmids. Methods of DNA delivery to protoplasts include (1) cocultivation of protoplasts with intact Agrobacteria, (2) chemically stimulated uptake of isolated DNA into protoplasts, (3) fusion of bacterial spheroplasts containing the recombinant plasmid with protoplasts, and (4) fusion and/or uptake of liposomes carrying DNA into protoplasts.

The use of polyethylene glycol (PEG) to mediate uptake of DNA by protoplasts has been explored. Protoplasts from tobacco were found to take up Ti plasmid DNA in presence of PEG (Krens *et al.*, 1982) although the frequency of transformation was very low. A plasmid containing the *nptII* gene with *nos* promoter and octopine synthetase polyadenylation signals was used for PEG-mediated transformation of protoplasts from *Triticum monococcum*, and *npt*II activity was demonstrated in cell lines selected for kanamycin resistance (Lorz *et al.*, 1985). This method is also successful in transfoming rice plants expressing β-glucuronidase activity (Zhang and Wu, 1988).

When the electroporation method, employing electrical impulses of high field strength (1–1.5 kv/cm) for a short period, was used with the protoplast preparation, a higher efficiency of transformation was obtained. Maize protoplasts have been transformed by plasmid DNA containing *nptII* gene linked with CaMV 35S promoter using electroporation (Rhodes *et al.*, 1988).

Though DNA transfer has been achieved using the electroporation method, the regeneration of the protoplast is not always successful. For that purpose a direct method to introduce DNA into the plant cell was devised. This entails the use of high-velocity microprojectiles to deliver the nucleic acid. Chloramphenicol acetyltransferase gene containing the 35S promoter was introduced by bombardment with microprojectiles exhibited a high level of CAT activity in the epidermal tissue after transformation (Klein *et al.*, 1987). Similarly particle acceleration has been employed to introduce DNA-coated gold particles into meristematic tissues of soybean (McCabe *et al.*, 1988).

Further, it has also been indicated that DNA can even be taken up by dry germinating embryos and transiently expressed. Transient expression has been reported in many species, including wheat, rice, pea, and maize. However, it remains to be demonstrated whether transgenic plants containing stably integrated foreign DNA can be generated by this approach (Topfer *et al.*, 1989).

A relatively simple method of transformation is afforded by use of the pollen tube pathway for introducing DNA. This method has been successfully used to generate transgenic plants (Ohta, 1986).

2.3. Approaches to Making the Plants Disease and Pest Resistant

It appears from the vast literature that a number of genes are selectively expressed in plants following infection by bacterial, fungal, and viral pathogens (Collinge and Slusarenko, 1987; Lamb *et al.*, 1989). Furthermore, similar genes are also expressed in response to other stimuli, such as wounding, chemical treatment, and UV light. The induction of a number of genes each by several stimuli does not imply that a single regulatory mechanism is involved. Broadly, what can be formulated is that there are at least three steps, such as (1) recognition, (2) signal transduction, and (3) response. Plants possess a number of defense mechanisms against infections, such as (1) production of phytoalexins and enhanced enzyme activities involved in secondary metabolite synthesis, (2) increased lignin formation at the cell wall and thus greater protection against invaders, (3) formation of hydrolases, i.e., proteases, β-1,3-glucanase, and chitinases, (4) accumulation of hydroxyproline-rich proteins in the cell wall, and (5) induction of pathogenesis-related proteins.

Defense responses involving secondary plant metabolites have long been known, and the biochemistry of phenolic substances and their relevance in plant diseases have been extensively reviewed (Friend, 1981; Legrand, 1983; Dixon, 1986; Smith and Banks, 1986). Chemically phytoalexins are grouped into three categories: (1) phenylpropanoids, (2) terpenoids, and (3) fatty acid derivatives (polyacetylene). This aspect has been reviewed by Darvill and Albersheim (1984). Phytoalexins have been identified in different plant systems after challenging with pathogens or elicitors, and a putative correlation of enhanced activity of certain enzymes with developed resistance has been observed (Table I). Enhanced cell wall lignification has been observed in different plants when challenged by various plant pathogens, such as fungi, viruses, and nematodes (Vance et al., 1980; Friend, 1985). Lignification was found to be associated with resistance to pathogen (Grand and Rossignol, 1982). Cinnamyl alcohol dehydrogenase (CAD) activity involved in lignification is increased in *Phaseolus vulgaris* cell suspension culture after challenging with *Cladosporium sp.* (Collinge and Slusarenko, 1987). Several other enzymes such as peroxidases and polyphenoloxidases were induced in different plant systems after infection with pathogens but the role of these enzymes in lignification is not clear (Collinge and Slusarenko, 1987).

Hydrolytic enzymes such as chitinase and β-1,3-glucanase have been found to be activated after infection of plant pathogens for treatment with elicitors (Mauch et al., 1984; Roby et al., 1986). However, the data so far accumulated indicate that different control mechanisms exist for the induction of chitinase in different plants, since chitinase is encoded by a multigene family (Broglie et al., 1986). Plant chitanases are potent inhibitors of fungal growth (Schlumbaum et al., 1986) and show lysozyme activity that might be effective against bacterial attack (Boller et al., 1983). Both chitinase and β-1,3-glucanase are induced in *Fusarium*-inoculated pea (Nichols et al., 1980).

There are several reports of increase in hydroxyproline-rich glycoprotein

Table I
Phytoalexins and Enzyme Activities after Infections or Treatment of Plants

Plant/tissues	Enzyme activity increased[a]	Phytoalexins produced	Pathogen or elicitor treatment	Reference
1. Parsley cells	PAL/4CL/CHS	Furanocumarins	Elicotor for *Phytophthora megasperma*	Hahlbrock et al., 1981
2. Parsley cells	PAL/4CL	Furanocumarins	UV	Kuhn et al., 1984
3. French bean	PAL/CHS	Isoflavonoids	*Colletotrichum linde muthianum*	Lawton et al., 1983
4. Soybean	PAL/4CL/CHS	Glyceollins	*Phytophthora megasperma*	Hille et al., 1982
5. Castor been	Casbene synthetase	Casbene	*Rhizopus stolonifer*	Lee and West, 1981

[a]PAL, phenylalanine ammonia lyase; CHS, 6'-hydroxychalcone synthase; 4CL, 4-coumarate:COA ligase.

(HRGP) in wounded or pathogen-challenged tissues of *Phaseolus vulgaris,
Cucumis sativus, Cucumis melo, Nicotiana tabaccum, Solanum tuberosum, Hordeum vulgare, Oriza sativa,* and *Triticum aestivum* (Mazan and Esquerre-Tygaye, 1986). Hydroxyproline-rich proteins consist of extensin interlinking the
cellulose mesh in the cell wall (Wilson and Fry, 1986). When melon seedlings
were infected with *Celletotrichum lagenarium,* an increase in HRGP was correlated with the resistance to the pathogen (Esquerre-Tugaye and Lamport, 1979).
HRGP increase has also been recorded from soybean treated with elicitor from
Phytophthora megasperma (Roby *et al.,* 1985). Some cell wall proteins rich in
cysteine are also synthesized after infection with fungi (Bohlmann *et al.,* 1988).

Pathogenesis-related proteins (PR proteins) are induced in a number of plant
species following infections (Van Loon, 1985). These proteins are produced in
the leaves when infected with bacteria, fungi, and viruses or treated with chemical elicitors, UV light, and ethephon (Collinge and Slusarenko, 1987). The role
of these proteins in disease resistance is supported by the observed resistance to
TMV following treatment with the elicitor salicylic acid, which can protect
tobacco from TMV infection (White, 1979). About 10 proteins, soluble in acid
(pH 3) are induced; one of them has been found to be similar to thaumatin and
some of them are hydrolytic enzymes. However, the function of all these proteins
in disease resistance remains unknown (Van Loon, 1985). Other proteins, such as
protease inhibitors, are induced after wounding of *Phytophtora* infection
(Sanchez-Serrano *et al.,* 1986). The levels of protease inhibitors increase in
melon when the plants are treated with elicitor prepared from *Colletotrichum
lagenarium* (Esquerre-Tugaye *et al.,* 1984).

In most cases, there is no direct evidence that observed pathogen-induced
responses constitute effective defense mechanisms. The difficulty lies in identifying the putative disease resistance genes in a particular plant with respect to a
specific disease. The question arises as to how the link between recognition and
expression of the defense responsive genes could be established. The hypothesis
proposed indicates that the avirulent gene product of pathogen might interact
with host resistance gene product to dictate recognition of the first step of
infection. This interaction might trigger the production of elecitor/second messengers such as, cyclic AMP, myoinositol-1,4,5-tris phosphate, and Ca^{2+} so
that the signal transduction pathway is followed. Phytoalexins, PR proteins,
hydrolases, and other enzymes of secondary metabolism are indirectly involved
in the responsive reaction or there might be a direct pathway to trigger the
resistance genes by those second messengers (Lamb *et al.,* 1989).

However, by inserting genes from different sources into the plants, it is
observed that the susceptible plant can be made resistant toward pathogens or
pests. This gives direct correlation with the gene product and the resistance in the
lack of knowledge of genetics of resistance. Resistance to specific viruses such
as TMV, AIMV, and potatovirus X in transgenic plants as a result of expression

of inserted coat protein or satellite RNA sequences has been reported (Cuozzo *et al.*, 1988; Davies and Stanley, 1989). Transgenic tobacco plants to which the gene encoding the coat protein of TMV was inserted showed protection toward infection (Abel *et al.*, 1986). Similarly, when antisense sequence of TMV coat protein was inserted, a protection was discernible. However, plants that expressed RNA complementary to coat protein coding sequence alone are not protected, indicating that sequences complementary to the terminal 117 bp of TMV, which include the putative replicase binding site, are responsible for protection (Powell *et al.*, 1989). The detailed reports of such antisense RNA in protection of viral diseases will be mentioned (see Section 2.7). The tobacco, tomato, and cotton transformed plants containing *Bacillus thuringiensis* toxin gene (Bt toxin) show resistance to the insects (Vaeck *et al.*, 1987; Fischhop *et al.*, 1987). When the cow pea trypsin inhibitor gene was transferred to tobacco plants, similar protection against the insect *Callosobruchus maculatus* was observed (Hilder *et al.*, 1987). Only very limited success in engineering resistance to fungal diseases has thus far been reported (Jones *et al.*, 1988; and see Gasser and Fraley, 1989). A novel lectinlike protein, arcelin, was identified in the resistant plants (*Phaseolus vulgaris*) and demonstrated to confer toxicity to bean bruchid pests (Osborn *et al.*, 1988). The sequence of arcelin 2 cDNA has also been reported (John and Long, 1990). Another alpha amylase inhibitor that inhibits insect and mammalian alpha amylases has been reported in the seeds of *P. vulgaris* (Moreno and Chrispeels, 1989). This gene is also different from, but closely related to, the genes that code for phytohemagglutinin of bean. It is possible that the genes of those lectinlike proteins, when inserted into the susceptible crop plants (legumes), might confer resistance to bruchid pests.

2.4. Engineering Herbicide Resistance

It is relatively easy to produce herbicides selective for plants. Atrazine and diuron interfere with photosynthesis and glyphosate, and the sulfonylureas and imidazolinones block the synthesis of amino acids. But crop plants share these processes with weeds, so crop plants must be protected through differential uptake and herbicide inhibition of metabolism. It has been found that atrazine is an effective herbicide for use with maize, which can detoxify the compounds, but soybean is sensitive. Herbicide resistance can be achieved by at least three different mechanisms: (1) overproduction of herbicide-sensitive target sites, (2) structural alteration of the target site, resulting in reduced herbicide affinity, and (3) detoxification by degradation of the herbicide before it reaches the target sites in the plant cell.

Resistance against glyphosate herbicide by genetic engineering has been achieved by (1) introducing a bacterial enzyme, 5-enolpyruvyl-shikimate-3-phosphate synthase (EPSP synthase) gene, under the control of T-DNA promoter into tobacco (Comai *et al.*, 1985) or tomato plants (Fillatti *et al.*, 1987) and (2) introducing a plant EPSP synthase gene under the control of CaMV 35S promoter into petunia (Shah *et al.*, 1986). In this case the inhibitory effect of glyphosate is encountered by overproduction of the plant enzyme. Phosphinotricine, a herbicide, inhibits the enzyme glutamine synthetase. This herbicide is detoxified by phospinothricin acetyl transferase, an enzyme from *Streptomyces hygroscopicus;* when cloned under CaMV 35S promoter and transferred to tobacco, tomato, and potato plants, a resistance against this herbicide is discernible (DeBlock *et al.*, 1987). The herbicide bromoxynil-resistant plants (tobacco) have been obtained by inserting the nitrilase gene (*bxn*) from *Klebsiella ozaenae,* which can convert bromoxynil to 3,5-dibromo-4 hydroxybenzoic acid. For expression in tobacco, this gene was placed under control of a light-regulated tissue-specific promoter, the ribulose bisphosphate carboxylase small subunit (Stalker *et al.*, 1988).

The herbicide atrazine inhibits photosynthesis by binding to the 32K protein of the chloroplast encoded by the psbA gene (Matto and Eldelman, 1987). A chimeric gene consisting of the coding sequence of a mutant psbA gene from *Amaranthus hybridus,* control sequences allowing its high-level expression as a nuclear gene and a sequence encoding a transit peptide for chloroplast targeting was constructed. The hybrid gene, when inserted into tobacco, yields an atrazine-tolerant plant (Cheung *et al.*, 1988). Interestingly, transgenic tobacco plants expressing a mutant acetolactate synthase (ALS) gene from tobacco or *Arabidopsis* have been found to be tolerant to sulfonylurea herbicides (Haughn *et al.*, 1988; Lee *et al.*, 1988). Sulfonylurea and imidazolinones inhibit ALS enzyme involved in branched amino acid biosynthesis such as leucine, isoleucine and valine (see Botterman and Leemans, 1988).

2.5. Engineering Stress Resistance

Plants are exposed to a wide variety of environmental stress. The stress could be mainly induced by heat, light, cold, heavy metals, anaerobic conditions, drought, and salinity. Like bacteria, many plants produce high concentrations of osmoprotectants, such as proline, proline-betaine, glycine-betaine, and alanine-betaine to maintain cell turgor pressure. Plant mutants that can overproduce appear to tolerate water stress also (Key *et al.*, 1985). Other than the production of osmoregulants, plants also synthesize new stress proteins. Tobacco cells adapted to grow in 1% NaCl produce increased level of two proteins and a unique protein not found in unadapted cells (Ericson and Alfinito, 1984). The heat shock proteins are also induced and reversed once the heat shock is removed (Schlesinger *et al.*, 1982; Lindquist, 1986). Drought condition or dehydration of plants can induce production of certain new proteins and some of them, which are also induced by the application of abscisic acid (ABA), are found to be rich in glycine (Gomez *et al.*, 1988; Mundy and Chua, 1988). When subject to heavy metals, plants can induce metallothionine-like protein (Rouser, 1984). In cadmium-resistant plants, a short cadmium binding peptide accumulates to accommodate a high amount of cadmium (Steffens *et al.*, 1986). Such cadmium binding peptides also confer metal resistance to yeasts, algae, and bacteria (Murasugi *et al.*, 1983). The two environmental factors affecting most of the crop plants are drought or water stress and salinity or salt stress. Waterlogging leads to anaerobiosis for plants, a condition under which alcohol dehydrogenase has been found to be induced to confer protection to the plants (Sachs and Freeling, 1978). Success in making the plant resistant to drought or salinity by inserting gene(s) is yet to be achieved.

2.6. Success in Producing Transgenic Plants from Cereals and Legumes

Several attempts have been made with wheat, barley, sorghum, maize, rice, and soybean for transformation, regeneration, and stabilization of inserted genes by direct or *Agrobacterium*-mediated techniques, but with only partial success (see Weising *et al.*, 1988). A protocol for easy, reproducible production of fertile transgenic cereals and legumes was not available until recently. Fertile transgenic rice plants (*Oriza sativa*) regenerated from transformed protoplasts have been obtained using vector pGL2, which carries bacterial hygromycin B resistance gene (*hph*) flanked by 35S promoter and polyA site from CaMV. The nonselectable gene encoding β-glucuronidase (GUS) was also transferred with the *hph* gene by electroporation and its expression was detected in the progeny of the stable transformants (Shimamoto *et al.*, 1989). Transformed embryo-derived cultures of two rice cultivars (*O. sativa*) using a different *Agrobacterium*-mediated gene transfer system have recently been obtained (Reineri *et al.*, 1990). However, regeneration of transformed cells to whole plants is still difficult to achieve. This supervirulent strain of *Agrobacterium* (A281) might be useful to transform *O. indica*, which is still very recalcitrant to transformation.

Transgenic soybean plants have been produced using an *Agrobacterium*-mediated gene transfer system. This procedure relied on a regeneration protocol in which shoot organogenesis was induced on cotyledons of soybean genotypes selected for susceptibility to *Agrobacterium* (Hinchee *et al.*, 1988). Stable transformation of soybean by particle acceleration has also been obtained (McCabe *et al.*, 1988). The vector pTV1100 includes both nopalin-neomycin phosphotransferase gene (*nos-nptII*) and CaMV 35S *GUS* gene in opposite orientation. Transgenic mung bean (*Vigna radiata*) plants have been obtained by *Agrobacterium*-mediated gene transfer systems (Biswas *et al.*, 1989). Both neomycin phosphotransferase and β-glucuronidase activities were expressed in the transformed plants. Stable transformation of mothbean (*Vigna aconitifolia*) by direct gene transfer through electroporation has been successfully worked out (Kohler *et al.*, 1987).

2.7.   Antisense RNA in Gene Regulation

Attempts to manipulate eukaryotic gene expression by the use of antisense RNA are now being made in order to inhibit certain gene functions deleterious to plants. The technique is based primarily on blocking the informational flow from DNA to RNA to protein, by introduction or generation of complementary RNA strand partly or fully to the sequence of the target mRNA. A short review on antisense genes in plants has already been published (Van der Krol *et al.*, 1988b). Antisense RNA is applied to the transcriptional products from two opposing DNA strands of the same region of segment of DNA or gene. This was first observed in phage lambda (Bovre and Szybalski, 1969). It was also reported that RNA complementary to ribosome binding sites including the Shine–Dalgarno sequence can function in nature as repressor for that gene (Simons and Kleckner, 1983). The technique is then based on blocking the mRNA translation by intro-duction of an RNA complementary to that particular mRNA. Thus an RNA duplex is formed between mRNA and antisense RNA. This duplex RNA is either rapidly degraded or the mRNA is impaired in the processing or it is blocked for translation (Inouye, 1988). The studies further indicate that the expression of both sense and antisense alpha amylase transcripts is developmentally controlled and both RNAs are present *in vivo* in barley aleurone tissue (Rogers, 1988). Plants were the first higher organism in which artificial antisense RNAs were tested (Ecker and Davis, 1986). The effective transient inhibition of chloram-phenicol acetyl transferase (CAT) activity in carrot cell cultures has been ob-served. Van der Krol *et al.* (1988a) have demonstrated the alteration of flower color pattern in transgenic petunia and tobacco plants on introduction of anti-sense RNA gene of chalcone synthase, which is involved in pigment formation. Various other reports thus far available with the plant systems in regulating the function of specific genes using antisense RNA technique are listed in Table II. It is apparent that this approach of engineering plants to reduce the undesirable characters or toxic substances and antimetabolites might have immense potential.

2.8.   Gene Tagging and Targeting in Plants

Improved methods for identification and isolation of new useful genes are of great importance for the betterment of crop plants. The cloning of transposon has allowed isolation of genes from several species by transposon-mediated gene tagging (Federoff *et al.*, 1984; Schmidt *et al.*, 1987). The demonstration that mobile elements isolated from maize are able to transpose when introduced to dicot plants indicates that this powerful technique is applicable to any plant species for which transformation and regeneration are possible (Baker *et al.*, 1987). It has also been shown that under appropriate transformation conditions, the T-DNA of the Ti plasmid of *Agrobacterium* can itself serve as an insertional

Table II
Antisense Genes Construct and Their Effectiveness

Construct	System	Results	Conclusion	Reference
1. Antisense Cat under promotor NOS/ PAL/CaMV-35S	Carrot cell	50–100-fold excess antisense RNA	95% inhibition of CAT activity	Ecker and Davis, 1986
2. Antisense NOS under CaMV 35S/NOS promotor	Regenerated tobacco plants	Not much in excess even under CaMV 35S promotor	Antisense NOS RNA degraded at faster rate	Rothstein et al., 1987
3. NOS template in reverse orientation under light activated promotor (Chl a/b binding protein)	Regenerated tobacco plants	3' region was effective in inhibition	Antisense RNA stable	Sandler et al., 1988
4. Antisense CAT gene under light activated promotor (Rubisco SS)	Tobacco	Lower level of antisense RNA than sense RNA under same construct	Antisense RNA is less stable	Delauney et al., 1988
5. Chimeric hygromycin resistance-antisense CAT gene and CaMV 35S promotor	Tobacco	High level of chimeric RNA containing antisense CAT RNA	Increased stability (due to active translation?)	Delauney et al., 1988
6. Antisense chsA cDNA under CaMV-35S promotor	Petunia, tobacco	High level of antisense RNA	90% inhibition in color formation in flowers	Van der Krol et al., 1988b
7. Antisense PG under CaMV-35S promotor	Tomato	Variation in antisense RNA	Inhibition in softening of tomato	Smith et al., 1988
8. Antisense Rubisco SS under CaMV-35S promotor	Tobacco	SS production inhibited parallel reduction in LS protein	Inhibition of Rubisco	Rodermel et al., 1988
9. Antisense CMV/PVX coat protein/CaMV 35S	Tobacco	LS mRNA steady level partially protected	Less effective than sense coat protein	Cuozzo et al.,1988

mutagen (Feldmann et al., 1989). The maize transposable element Ac has recently been introduced to the Arabidopsis genome by transformation and shown to transpose (Van Sluys et al., 1987). This opens the possibility of transposon tagging as a means of molecular cloning in Arabidopsis as well as in maize to identify a number of genes involved particularly in flower formation because several such mutants in Arabidopsis are already available.

Although production of transgenic plants is now possible, the insertion of foreign gene has so far been at random sites in the genome. Therefore, it is necessary that the gene be placed in the specific site. The protoplasts of transgenic plant carrying copies of a partial, nonfunctional drug resistance gene in the genome were used as recipients for DNA molecules containing the missing part of the gene. The integration of foreign DNA through homologous recombination within overlapping portions, resulting in the formation of an active gene in the host, can be achieved. This approach is referred to as gene targeting. It should be noted that the locus used as target has been artificially created by transformation. Success in producing transgenic plants with desired modifications to a specific nuclear gene has already been achieved in the case of tobacco (Paszkowski *et al.*, 1988). The gene-targeting frequency in this case has been noted as $0.5-4.2 \times 10^{-4}$. The frequency of gene targeting reported for mammalian systems varies from 10^{-5} to 10^{-2} depending on the cell type and the methods of DNA delivery (Lin *et al.*, 1985). In any case, gene targeting in the plant system is an important step for directed mutagenesis of the plant genome. This is useful in studying plant genome organization and alterations in a well-defined way in the plant breeding program.

2.9. RFLP Mapping in Plant Breeding

Restriction fragment length polymorphisms (RFLP) are the variations that are taken as changes in the length of defined DNA fragments produced by digestion of DNA with restriction endonucleases. RFLPs occur as a result of changes in base pair, particularly in the restriction enzyme recognition sites, and can be detected by hybridization of labeled DNA clones containing sequences that are homologous to a fragment of DNA in the chromosome. Hybridization with a unique cloned sequence allows identification of a specific chromosomal locus or region. Differences in fragment length as revealed in the gel electrophoregram are taken as alleles of that locus. Thus RFLPs serve as genetic markers in a manner analogous to conventional morphological or isozyme markers, but unlike most genetic markers, they are not correlated with the products of transcription and translation. The property of RFLPs to act as putative genetic markers is now being exploited to compare with chromosomal linkage maps already available for crop plants. For this purpose, (1) isolation of unique probes, (2) identification of polymorphic pattern with respect to specific probes, and (3) segregation analysis of RFLP pattern in F_1 and F_2 generations are essential. RFLP mapping on eukaryotic genetics was first applied in human genetics as early as 1980 (Botstein *et al.*, 1980). RFLP mapping in plant breeding has recently been discussed by Tanksley *et al.*, (1989). The utility of RFLP markers in plant breeding is based on finding tight linkages between these markers and genes of interest. Such linkage permits one to infer the presence of a desirable gene by assaying for RFLP markers. In tomato, RFLP markers have been identified that are tightly linked to genes for resistance to TMV, *Fusarium* wilt, bacterial speck, and root knot nematodes, as well as fruit-ripening properties (Paterson *et al.*, 1988). Similarly, linkages between RFLP markers and genes for resistance to maize dwarf mosaic virus and downy mildews have been worked out in maize and lettuce, respectively (Landry *et al.*, 1987; see Tanksley *et al.*, 1989). RFLP markers can be employed to detect genetic loci of complex traits. In that case a cross is made between two species that are genetically different for one or more characters, and segregating progeny are obtained from hybrid. A number of progeny are monitored for the character of interest and for their genotypes at RFLP marker loci at regular intervals through out the genome. A search is then made for associations between the segregating RFLP markers and the character of interest. If such associations are found, they should be due to linkage of the RFLP marker to a gene(s) affecting the character. Using a common set of clones for RFLP mapping, the degree to which the chromosome content and gene order have been conserved in tomato, potato, and pepper has already been determined (Tanksley *et al.*, 1988a; Bonierbale *et al.*, 1988). RFLP maps are now available for at least seven different plants (McCouch *et al.*, 1988; Chang *et al.*, 1988; Helentjaris, 1987; Landry *et al.*, 1987; Bonierbale *et al.*, 1988; Tanksley *et al.*, 1988b).

Because RFLP markers can be used so effectively to select for individuals with little unwanted donor DNA, they can be used not only to locate genes controlling a character from exotic germplasm, but also to expedite the transfer of small amounts of foreign chromosomal DNA containing the desired genes into commercial varieties. It is also likely that RFLP mapping will expedite acquisition of important genes from wild species, which may increase yield, resistance, or adaptibility to extreme environments. Genetic links may be established between related crop species through construction of RFLP maps based on common sets of clones. Finally, RFLP maps can be used increasingly to clone genes difficult to isolate by other means. Thus integration of RFLP techniques into plant improvement programs will (1) elucidate the pattern of movement of desirable genes among varieties (2) allow transfer of novel genes from related wild species, (3) make possible analysis of complex polygenic characters (quantitative trait loci) as Mendelian factors, (4) establish genetic relationship between sexually incompatible crop plants, and (5) allow identification of the RFLP pattern for specific disease resistance of plants.

2.10. Self-Incompatibility in Plants

In many plant families inbreeding is prevented by rejection of pollen tubes after their growth is arrested through the style. Rejection is controlled by the product of S genes, which have multiple alleles (Nasrallah and Nasrallah, 1986;

Cornish *et al.*, 1988). The problem of self-incompatibility is a fascinating subject, and studies now have delineated the nature and control of S genes. How the different alleles of S genes are generated and how the corresponding gene products influence self-incompatibility is the important question. S genes corresponding to the style glycoproteins of *Nicotiana alata* and *Brassica oleracea* have been cloned and sequenced. These genes illustrate examples of each of the two major self-incompatibility systems: gametophytic (*N. alata*) and sporophytic (*B. oleracea*). In sporophytic incompatibility, when an allele in the pollen parent (S1 S2) is matched with that of the pistil (S1 S2 or S1 S3), pollen germination is arrested at the stigma surface. In gametophytic incompatibility, when the pollen parent (S1 S2) is matched with either allele in the stylar tissue, growth of the pollen tube is arrested usually in the style. How the interaction of the pollen and style S gene products at the molecule level leads to arrest of pollen tube growth is under investigation. Sequence variability between alleles of the S gene in *N. alata* has been worked out (Ebert *et al.*, 1989). Comparison of the sequences shows that 56% of the predicted amino acids are conserved between the three alleles studied. The amino acid differences occur throughout the sequence but are concentrated in a series of hypervariable regions in the amino-terminal half of the molecule. The regions of homology include the amino-terminal sequence, most of the cysteine residues, and some glycosylation sites. This type of variable region interspersed with conserved sequences shows analogy with the structure of the immunoglobulin family. The cloned genes are not homologous between the gametophytic and sporophytic systems but display a similar pattern of extreme sequence variability. Despite the sequence variability between alleles within each of the systems, cysteine residues and potential glycosylation sites are relatively conserved (Nasrallah *et al.*, 1988). It is also interesting that a nuclear sequence associated with self-incompatibility in *N. alata* has been found to have homology with mitochondrial DNA (Bernatzky *et al.*, 1989).

Cytoplasmic male sterility (CMS) is the maternally inherited character observable in many higher plants that prevents the production of viable pollen without affecting female fertility. The CMS trait manifests itself in different ways among various plant species (Hanson and Conde, 1985). CMS is an important character in crop improvement through the exploitation of hybrid vigor. It has also been used in hybrid seed production of crops such as maize, sorghum, and bajra. The non-Mendelian inheritance pattern of CMS links it with mitochondrial dysfunction (Lonsdale *et al.*, 1984; Levings and Brown, 1989). Several studies have suggested a possible role for the low-molecular-weight DNAs in the mitochondria with CMS characters. The presence of two linear plasmids, S1 (6.4 kb) and S2 (5.4 kb), in maize is correlated with male sterility (Pring *et al.*, 1977). Similar DNA molecules, N_1 and N_2, have been found in the CMS line of sorghum (Dixon and Leaver, 1982). It is also evident that CMS–S pollen sterility is not strictly dependent on the presence of S elements (Escote *et al.*, 1985).

2.11. Pharmaceutical Products from Transgenic Plants

Plant genetic engineering has a potential for the production of valuable chemicals. In recent years attempts have been made to produce (1) antibodies, (2) enkephalins, and (3) albumin from transgenic plants. cDNAs obtained from mouse hybridoma mRNA were used to transform tobacco leaf discs through the expression vector pMON530 followed by regeneration of mature plants. Plants expressing single gamma or kappa immunoglobulin chains were crossed to yield progeny in which both chains were expressed. A functional antibody (IgG1) accumulated to 1.3% of total leaf protein in plants expressing full-length cDNAs containing leader sequences (Hiatl *et al.*, 1989). *Arabidopsis thaliana* 2S albumin gene was replaced partly by sequences encoding the neuropeptide leu-enkephalin (a pentapeptide) flanked by tryptic cleavage sites, and the chimeric gene was inserted into *Arabidopsis* and rape plants using vector pGSATE1. The chimeric 2S albumin isolated from the seeds of transgenic plants was digested with trypsin and after getting the peptide carboxypeptidase B was used to remove lysine for recovering pure leu-enkephalin (Vandekerckhove *et al.*, 1989). Using modified CaMV 35S promoter, the expression of chimeric genes encoding human serum albumin in transgenic potato and tobacco plants has been reported (Sijmons *et al.*, 1990).

3. PROBLEMS AND FUTURE OUTLOOK

A number of crop plants have already been developed by genetic engineering techniques yielding very promising results. Several problems are to be encountered, including (1) transformation of especially legumes and cereals, (2) regeneration of plants from these transformed cells, and (3) expression of inserted genes in a tissue-specific manner. With the model plants, such as *Arabidopsis*, tobacco, tomato, and potato, very interesting results have thus far been achieved indicating that more techniques will have to be developed to have minimal success with those recalcitrant crop plants that are still to be worked out in the laboratory. Although encouraging results have emerged in the area of making the plants herbicide, pest, and virus resistant, where a single gene transfer is involved, the genetics of resistance has not yet been understood properly. Apparent resistance or tolerance develops when a host plant carrying a resistance allele interacts with a pathogen race having a specific, complementary avirulence allele. If this is a workable hypothesis, then a mutation of a pathogen avirulence allele leading to an inactive gene product should result in a fully virulent type causing susceptibility. Similarly, a mutation in host resistance allele should also give rise to a susceptible phenotype. The preliminary experiments in this line do not support this hypothesis fully, suggesting that the resistance is genetically more complex. This presents a challenge to molecular biologists to apply recombinant DNA technology to make the important crop plants disease, pest, and nematode resistant. At present, no general algorithm can be prescribed to obtain the desired transgenic plant. Each plant will have to be worked out separately. The major crops that are currently being improved with the help of genetic engineering are soybean, cotton, rice, potato, and tomato.

The second aspect that is posing problems is the development of plants resistant to different environmental stresses, such as water, salt, high or low temperature, anaerobiosis, and light. It appears that some of the plant hormones might play a large role in inducing specific genes in order to exhibit the putative tolerance. Future studies are to be directed to isolate such genes that can confer resistance to different stresses. A similar problem is also involved in understanding the developmental process in the plant. The question arises as to how a specific gene is expressed only in a specific tissue, say leaf or root? So the search for specific promoters is worth investigating. Construction or isolation of promoters from the genomic clones to get tissue-specific expression is needed. Information about the sequences required for seed-specific expression of storage proteins is already available. Further, interesting results with CaMV 35S promoter are indicative that it contains two domains. One confers expression mainly in roots and the other confers expression in other tissues (Benfey et al., 1989). The alternate way to increase seed development is to increase the number of flowers. The number of genes required for flower formation and the gene products necessary to convert a vegetative bud to a flower bud are pertinent questions to be looked into. In fact, Arabidopsis is supplying a good deal of data on this aspect of flower formation, with recourse to targeting of genes as well as characterization of genes and their products. Self-incompatibility in higher plants not only provides a framework for understanding some aspects of sexual reproduction, but also supplements a general model for cell recognition/communication. However, demonstration that transformation with the cloned sequences can change the incompatibility specificity of a plant is still difficult. By careful selection of S alleles in donor and recipient, transformants modified in their self-incompatibility specificity should be identified. Molecular communication in plants, which is implicated in all aspects of transduction of signals, is at present amenable to genetic engineering methods to pinpoint component(s) responsible for it. The receptor concept at the membrane for signal response is not well worked out in plant systems. The results thus far generated suggest that the cell wall is a repository of signal molecules which, when released by the appropriate environmental or developmental stimulus, can lead to modulation of plant development (Varner and Lin, 1989).

Methods of transformation for insertion of genes in the mitochondria and chloroplasts are still rudimentary. Despite the reports of microprojectile (biolistic) and microinjection methods for transformation of chloroplasts and mitochondria, much work is needed to make these attempts perfect and successful (Boynton *et al.*, 1988; Cheung *et al.*, 1988). A major problem with direct gene transfer is the unpredictable pattern of foreign DNA integration. Rearrangement, truncation, and concatemerization of vector DNA during integration into plant genome are frequently observed. Despite these disadvantages, the technique of direct gene transfer to the organelles seems promising at present. The successful transformation of yeast mitochondria suggests that genetically polyploid and number of organelles may not limit transformation, provided that multiple copies of the donor DNA can be delivered and a strong selection for the inserted gene is applied (Johnston *et al.*, 1988).

Transgenesis to the improvement and enhancement of plant products as well as production of rare chemicals is the ultimate goal to be achieved. These efforts will definitely be complemented by the use of gene targeting to modify the host genome to reduce the unwanted functions. Many plant traits are found to be polygenic. It is necessary to develop techniques that combine the genes interspersed on separate linkage groups and to introduce the group of genes in plants in a single step to increase crop yields, to improve crop quality, and to reduce production costs. It is apparent that the problems are not insurmountable. The problems of enhancing efficiency of photosynthesis and of improving nitrogen fixation capability to a wide range of plants or even to bacteria are definitely complex and difficult to solve at present. The only hope is that, with increasing understanding of the molecular basis of these problems, new insights may emerge in the future. The risks of gene transfer into plants and the benefits of these engineered plants to growers, food processors, and consumers must be looked into carefully and cautiously.

4. CONCLUDING REMARKS

The ability to introduce foreign genes into a wide variety of plants has created a revolution in plant biology. Genetically engineered crop plants, including legumes and cereals resistant to herbicides, insects, and viruses, are appearing on the scene. Virus resistance alone could provide significant yield protection in important crop plants such as vegetables, corn, wheat, rice, and soybean. Genetic engineering has provided a large number of vector constructs and regulatory sequences that will allow for accurate targeting of gene expression to specific tissues of transgenic plants. However, a vast amount of work, particularly basic in nature, still needs to be done and should be given primary consideration.

ACKNOWLEDGMENTS. The author is grateful to the Council of Scientific and Industrial Research, Government of India, and Food and Agricultural Organization (U.N.) for financial support and to his colleagues for their help in preparation of this chapter.

5. REFERENCES

Abel, P. P., Nelson, R. S., De, B., Hoffman, N., Rogers, S. G., Fraley, R. T., and Beachy, R. N., 1986, Delay of disease development in transgenic plants that express the tobacco mosaic virus coat protein gene, *Science* **232**:738–743.

Akiyoshi, P., Klee, H., Amasino, R., Nester, E., and Gordon, M., 1984, T-DNA of *Agrobacterium tumefaciens* encodes an enzyme of cytokinin biosynthesis, *Proc. Natl. Acad. Sci. USA* **81**:5994–5998.

Baker, B., Couplaud, G., Fedoroff, N., Starlenger, P., and Schell, J., 1987, Phenotypic assay for excision of the maize controlling element Ac in tobacco, *EMBO J.* **6**:1547–1554.

Benfey, P. N., and Chua, N., 1989, Regulated genes in transgenic plants, *Science* **244**:174–181.

Benfey, P. N., Ren, L., and Chua, N. H., 1989, The CaMV 35S enhancer contains at least two domains which can confer different developmental tissue specific expression patterns, *EMBO J.* **8**:2195–2202.

Bernatzky, R., Anderson, M. A., and Clarke, A. E., 1989, Molecular genetics of self incompatibility in flowering plants, *Dev. Genet.* **9**:1–12.

Biswas, B. B., Pal, M., and Ghosh, U., 1989, Transgenic mung bean plants using agrobacterium mediated DNA transfer, *in EMBO Symposium on Molecular Communication in Higher Plants* p. 79, EMBL, Heidelberg.

Bleeker, A. B., Estelle, M. A., Somerville, C., and Kende, H., 1988, A dominant mutation confers insensitivity to ethylene in *Arabidopsis thaliana, Science* **241**:1086–1089.

Bohlmann, H., Clausen, S., Behnke, S., Giese, H., Hiller, C., Reimann-Philipp, U., Schrader, G., Barkholt, V., and Apel, K., 1988, Leaf specific thioneins of barley—A novel class of cell wall proteins toxic to plant pathogenic fungi and possibly involved in the defence mechanism of plants, *EMBO J.* **7**:1559–1565.

Boller, T., Gehri, A., Maunch, F., and Vogeli, U., 1983, Chitinase in bean leaf: Induction by ethylene, purification, properties and possible function, *Planta* **157**:22–31.

Bonierbale, M. W., Plaisted, R. L., and Tanksley, S. D., 1988, RFLP maps based on a common set of clones reveal modes of chromosomal evolution in potato and tomato, *Genetics* **120**:1095–1103.

Botstein, D., White, R. L., Skolnick, M., and Davis, R. W., 1980, Construction of genetic linkage map in man using restriction fragment length polymorphism, *Am. J. Hum. Genet.* **32**:314–331.

Botterman, J., and Leemans, J., 1988, Engineering herbicide resistance in plants, *Trends Genet.* **4**:219–222.

Bovre, K., and Szybalski, W., 1969, Patterns of convergent and overlapping transcription within b2 region of coliphage lambda, *Virology* **38**:614–626.

Boynton, J. E., Gillham, N. W., Harris, E. H., Hosler, J. P., Johnson, A. M., Jones, A. R., Randolph-Anderson, B. L., Robertson, D., Klein, T. M., Shark, K. B., and Sanford, J. C., 1988, Chloroplast transformation in *Chlamydomonas* with high velocity microprojectiles, *Science* **240**:1534–1538.

Braaksma, F. J., and Feenstra, W. J., 1982, Isolation and characterization of nitrate deficient mutants of *Arabidopsis thaliana, Theor. Appl. Genet.* **64**:83–90.

Brisson, N., and Hohn, T., 1986, Plant virus vectors: Cauliflower mosaic virus, *Methods Enzymol.* **118**:659–668.

Brisson, N., Paszkowski, J., Penswick, J. R., Gronenborn, B., Potrykus, I., and Hohn, T., 1984, Expression of a bacterial gene in plants by using a viral vector, *Nature* **310**:511–514.

Broglie, K. E., Gaynor, T. J., and Broglie, R. M., 1986, Ethylene regulated gene expression: Molecular cloning of the genes encoding an endochitinase from *Phaseolus vulgaris*, *Proc. Natl. Acad. Sci. USA* **83**:6820–6824.

Browse, J., Kunst, L., McCourt, P., and Somerville, C. R., 1987, *Arabidopsis, UCLA Symp. Mol. Cell. Biol.* **63**:437–447.

Chang, C., Bowman, J. C., Dejohn, A. W., Lander, E. S., and Meyerowitz, E. S., 1988, Restriction fragment length polymorphism linkage map for *Arabidopsis thaliana, Proc. Natl. Acad. Sci. USA* **85**:6856–6860.

Cheung, A. Y., Bogorad, L., Van Montagu, M., and Schell, J., 1988, Relocating a gene for herbicide tolerance: A chloroplast gene is coverted into a nuclear gene, *Proc. Natl. Acad. Sci. USA* **85**:391–395.

Chilton, M. D., Tepfer, D. A., Petit, D. C., Casse-Delbart, F., and Tempe, J., 1982, *Agrobacterium rhizogenes:* Inserts T-DNA into the genomes of the host plant cells, *Nature* **295**:432–434.

Collinge, D. B., and Slusarenko, A. J., 1987, Plant gene expression in response to pathogens, *Plant Mol. Biol.* **9**:389–410.

Comai, L., Facciotti, D., Hiatt, W. R., Thompson, G., Rose, R. E., and Stalker, D. M., 1985, Expression in plants of a mutant aro A gene from *Salmoella typhimurium* confer tolerance to glyphosate, *Nature* **317**:741–744.

Cornish, E. C., Anderson, M. A., and Clarks, A. E., 1988, Molecular aspects of fertilization in flowering plants, *Annu. Rev. Cell Biol.* **4**:209–228.

Cuozzo, M., O'Connell, K. M., Kaniewski, W., Fang, R-X., Chua, N-H., and Tumer, N. E., 1988, Viral protection in trangenic tobacco plants expressing the cucumber mosaic virus coat protein or its antisense RNA, *Biotechnology* **6**:549–557.

Darvill, A. G., and Albersheim, P., 1984, Phytoalexins and their elicitors—A defence against microbial infection in plants, *Annu. Rev. Plant Physiol.* **35**:243–275.

Davies, J. W., and Stanley, J., 1989, Geminivirus genes and vectors, *Trends Genet.* **5**:77–81.

DeBlock, M., Botterman, J., Vandewiele, M., Dockx, J., Thoen, C., Gossele, V., Mova, N. R., Thompson, C., Van Montagu, M., and Leemans, J., 1987, Engineering herbicide resistance in plants by expression of a detoxifying enzymes, *EMBO J.* **6**:2513–2518.

Delauney, A. J., Tabaeizadeh, Z., and Verma, D. P. S., 1988, A stable bifunctional antisense transcript inhibiting gene expression in transgenic plants, *Proc. Natl. Acad. Sci. USA* **85**:4300–4305.

Ditta, G., Stanfield, S., Corbin, D., and Helsinksi, D. R., 1980, Broad host range DNA cloning system for gram negative bacteria: Construction of a gene bank of *Rhizobium melioti, Proc. Natl. Acad. Sci. USA* **77**:7347–7351.

Dixon, R. A., 1986, The phytoalexin response: Elicitation, signalling and control of host gene expression, *Biol. Rev.* **61**:239–291.

Dixon, L. K., and Leaver, C. J., 1982, Mitochondrial gene expression and cytoplasmic male sterility in *sorghum, Plant Mol. Biol.* **1**:89–102.

Douglas, C. J., Staneloni, R. J., Rubin, R. A., and Nester, E. W., 1985, Identification and genetic analysis of an *Agrobacterium tumefaciens* chromosomal virulence region, *J. Bacteriol.* **161**:850–860.

Ebert, P. R., Anderson, M. A., Bernatzky, R., Altschuler, N., and Clarke, A. E., 1989, Genetic polymorphism of self-incompatibility in flowering plants, *Cell* **56**:255–262.

Ecker, J. R., and Davis, R. W., 1986, Inhibition of gene expression in plant cells by expression of antisense RNA, *Proc. Natl. Acad. Sci. USA* **83**:5372–5376.

Eckes, P., Donn, G., and Wengenmayer, F., 1987, Genetic engineering with plants, *Angew. Chem. Int. Ed. Engl.* **26**:382–402.

Ericson, M. C., and Alfinito, S. H., 1984, Protein produced during salt stress in tobacco cell culture, *Plant Physiol.* **74**:506–509.

Escote, L. J., Gabay-Laughnan, S. J., and Laughnan, J. R., 1985, Cytoplasmic reversion of fertility in cms-S maize need not involve loss of linear mitochondrial plasmids, *Plasmid* **14**:263–264.

Esquerre-Tugaye, M. T., and Lamport, D. T. A., 1979, Cell surfaces in plant microorganism interactions. I. A structural investigation of cell wall hydroxyproline rich glycoproteins which accumulate in fungus infected plants, *Plant Physiol.* **64**:314–319.

Esquerre-Tugaye, M. T., Mazan, D., Pelissier, B., Roby, D., and Toppan, A., 1984, Elicitors and ethylene trigger defence responses in plants, in *Ethylene: Biochemical, Physiological and Applied Aspects* (Y. Fuchs and Y. Chalutz, eds.), pp. 217–218, Martinus Nijhoff, The Hague.

Estelle, M. A., and Somerville, C., 1986, The mutants of *Arabidopsis, Trends Genet.* **2**:89–93.

Etessami, P., Callis, R., Ellwood, S., and Stanley, J., 1988, Delimitation of essential genes of cassava latent virus DNA2, *Nucleic Acids Res.* **16**:4811–4829.

Fedoroff, N., Furtek, D., and Nelson, O., Jr., 1984, Cloning of bronze locus in maize by a simple and generalizable procedure using the transposable controlling element activator (Ac), *Proc. Natl. Acad. Sci. USA* **81**:3825–3829.

Feldmann, K., Marks, M., Christianson, M., and Quatrano, R., 1989, A dwarf mutant of *Arabidopsis* generated by T-DNA insertion mutagenesis, *Science* **243**:1351–1354.

Fillatti, J. A., Sellmor, J., McCown, B., Heissig, B., and Comai, L., 1987, *Agrobacterium* mediated transformation and regeneration of *Populus, Mol. Gen. Genet.* **206**:192–199.

Fischhop, D. A., Bowdish, K. S., Perlak, F. J., Marrone, P. G., McCormick, S. M., Niedermeyer, J. G., Dean, D. A., Kusano-Kretzmer, K., Mayer, E. J., Rochester, D. E., Rogers, S. G., and Fraley, R. T., 1987, Insect tolerant transgenic tomato plants, *Biotechnology* **5**:807–813.

Fraley, R. T., Rogers, S. G., Horsh, R. B., Eichholtz, A. D., Flick, S. J., Fink, C. L., Hoffmann, N. L., and Sanders, P. R., 1985, The SEV system: A new disarmed Ti plasmid vector system for plant transformation, *Biotechnology* **3**:629–635.

French, R., Janda, M., and Ahlquist, P., 1986, Bacterial gene inserted in an engineered RNA virus: Efficient expression in monocotyledonous plant cells, *Science* **231**:1294–1297.

Friend, J., 1981, Plant phenolics, lignification and plant disease, in Progress in Phytochemistry, Vol. 7 (L. Rheinhold, J. B. Harborne, and T. Sewain, eds.), pp. 197–261, Pergamon Press, Oxford.

Friend, J., 1985, Phenolic substances and plant disease, *Annu. Proc. Phytochem. Soc.* **25**:367–392.

Garfinkel, D. J., and Nester, E. W., 1980, *Agrobacterium tumefaciens* mutants affected in crown gall tumerigenesis and octopine catabolism, *J. Bacteriol.* **144**:732–743.

Gasser, C. S., and Fraley, R. T., 1989, Genetically engineering plants for crop improvement, *Science* **244**:1293–1299.

Gomez, J., Sanchez-Martinez, D., Stiefel, V., Rigan, J., Puigdomenech, P., and Pages, M., 1988, A gene induced by the plant hormone abscisic acid in response to water stress encodes a glycine rich protein, *Nature* **334**:262–264.

Grand, C., and Rossignol, M., 1982, Changes in the lignification process induced by localized infection of muskmelons with *Colletotrichum lagenarium, Plant Sci. Lett.* **28**:103–110.

Grimsley, N., Hohn, B., and Walden, R., 1986, Agroinfection and alternative route for viral infection of plants by using Ti plasmid, *Proc. Natl. Acad. Sci. USA* **83**:3282–3286.

Guerche, P., Jouanin, L., Tepfer, D., Pelletier, G., 1987, Genetic transformation of oil seed rape (*Brassica napus*) by the Ri T-DNA of *Agrobacterium rhizogenes* and analysis of inheritance of the transformed phenotype, *Mol. Gen. Genet.* **206**:382–386.

Hahlbrock, K., Lamb, C. J., Purwin, C., Ebel, J., Fautz, E., and Schafer, E., 1981, Rapid response of suspension-cultured parsley cells to the elicitor from *Phytophthora megasperma* var. sojal. Induction of the enzymes of general phenyl propanoid metabolism, *Plant Physiol.* **67**:768–773.

Hanson, M. R., and Conde, M. F., 1985, Functioning and variation of cytoplasmic genomes: Lessons from cytoplasmic–nuclear interactions affecting male fertility in plants, *Int. Rev. Cytol.* **94**:213–267.

Haughn, G. W., Smith, J., Mazur, B., and Somerville, C., 1988, Transformation with a mutant *Arabidopsis* acetolactate synthase gene renders tobacco resistant to sulphonyl urea herbicides, *Mol. Gen. Genet.* **211**:266–271.

Helentjaris, T., 1987, A genetic linkage map for maize based on RFLPs, *Trends Genet.* **3**:217–221.

Hernalsteens, J. P., Van Vliet, F., de Beuckeleer, M., Depicker, A., and Engler, G., 1980, The *Agrobacterium tumefaciens* Ti plasmid as a host vector system for introducing foreign DNA in plant cells, *Nature* **287**:654–656.

Herrera-Estrella, L., Depicker, A., Van Montagu, M., and Schell, J., 1983a, Expression of chimeric genes transformed into plant cells using a Ti-plasmid derived vector, *Nature* **303**:209–213.

Herrera-Estrella, L., DeBlock, M., Messens, E., Harnalsteens, J. P., Van Montagu, M., and Schell, J., 1983b, Chimeric genes as dominant selectable markers in plant cells, *EMBO J.* **2**:987–995.

Hiatl, A., Cafferkey, R., and Bowdish, K., 1989, Production of antibodies in transgenic plants, *Nature* **342**:76–78.

Hilder, V. A., Gatehouse, A. M. R., Sheeraman, S. E., Barker, R. F., and Boulter, D., 1987, A novel mechanism of insect resistance engineered into tobacco, *Nature* **330**:160–163.

Hille, A., Purwin, C., and Ebel, J., 1982, Induction of enzymes of phytoalexin synthesis in cultured soybean cells by an elicitor from *phytophthora megasperma* f.sp. glycinea, *Plant Cell Rep.* **1**:123–127.

Hinchee, M. A. W., Connor-Ward, D. V., Newell, C. A., McDonnell, R. E., Sato, S. J., Gasser, C. S., Fischhoff, T. A., Re, B., Fraley, R. T., and Horsch, R. B., 1988, Production of transgenic soybean plants using agrobacterium mediated DNA transfer, *Biotechnology* **6**:915–922.

Howell, S. H., Walker, L. L., and Walden, R. M., 1981, Rescue of *in vitro* generated mutants of the cloned cauliflower mosaic virus genome in infected plants, *Nature* **293**:483–486.

Inouye, M., 1988, Antisense RNA: Its functions and applications in gene regulation—A review, *Gene* **72**:25–34.

John, M. E., and Long, C. M., 1990, Sequence analysis of arcelin 2, a lectin like plant protein, *Gene* **86**:171–176.

Johnston, S. A., Anziano, P. G., Shark, K., Sanford, J. C., and Butow, R. A., 1988, Mitochondrial transformation in yeast by bombardment with microprojectile, *Science* **240**:1538–1540.

Jones, J. D. G., Dean, C., Gidoni, D., Gilbert, D., Bond-Nutter, D., Lee, R., Bedbrook, J., and Dunsmuir, P., 1988, Expression of bacterial chitinase protein in tobacco leaves using two photosynthetic gene promoters, *Mol. Gen. Genet.* **212**:536–542.

Joos, H., Inze, D., Caplan, A., Sormann, M., Van Montagu, M., and Schell, J., 1983, Genetic analysis of T-DNA transcripts in nopaline crown galls, *Cell* **32**:1057–1067.

Karssen, C. M., Groot, S. P. C., and Koorneef, M., 1987, Hormone mutants and seed dormancy in *Arabidopsis* and tomato, in *Developmental Mutants of Higher Plants* (H. Thomas and D. Grierson, eds.), pp. 119–133, Cambridge University Press, Cambridge.

Key, J. L., Kimpel, J., Vierling, E., Lin, C. Y., and Nagao, R. T., 1985, in *Changes in Eukaryotic Gene Expression in Response to Environmental Stress* (B. G. Atkinson and D. B. Walden, eds.), pp. 327–348, Academic Press, New York.

Khurana, J. P., and Poff, K. L., 1988, Characterization of *Arabidopsis thaliana* mutants with altered phototropism, *Plant Physiol.* **86**:29–34.

Klee, H. J., Muskopf, Y. M., and Gasser, C. S., 1987, Cloning of an *Arabidopsis thaliana* gene encoding 5-enolpyruvylshikimate-3 phosphate synthase: Sequence analysis and manipulation to obtain glyphosate tolerant plants, *Mol. Gen. Genet.* **210**:437–442.

Klein, T. M., Wolf, E. D., Wu, R., and Sanford, J. C., 1987, High-velocity microprojectiles for delivering nucleic acids into living cells, *Nature* **327**:70–73.

Kohler, F., Golz, C., Eapen, S., Kohn, H., and Scheider, O., 1987, Stable transformation of moth bean *Vigna aconitifolia* via direct gene transfer, *Plant Cell Rep.* **6**:313–317.

Krens, F. A., Molendijk, L., Wullems, G. J., and Schilperoort, R. A., 1982, *In vitro* transformation of plant protoplasts with Ti plasmid DNA, *Nature* **296:**72–74.

Kuhn, D. N., Chappell, J., Boudet, A., and Hahlbrock, K., 1984, Induction of phenyl ammonia lyase and 4-coumarate: CoA ligase mRNAs in cultured plant cells by UV light or fungal elicitor, *Proc. Natl. Acad. Sci. USA* **81:**1102–1106.

Lamb, C. J., Lawton, M. D., and Dixon, R. A., 1989, Signals and transduction mechanisms for activation of plant defenses against microbial attack, *Cell* **56:**215–224.

Landry, B. S., Kesseli, R. V., Farrara, B., and Michelmore, R. W., 1987, A genetic map of lettuce (*Lactuca sativa* L) with restriction fragment length polymorphism, isozyme, disease resistance and morphological markers, *Genetics* **116:**331–337.

Lawton, M. A., Dixon, R. A., Hahlbrock, K., and Lamb, C. J., 1983, Elicitor induction of mRNA activity. Rapid effects of elicitor on phynylalanine ammonia-lyase and chalcone synthase mRNA activities in bean cells, *Eur. J. Biochem.* **130:**131–139.

Lee, S-C., and West, C. A., 1981, Polygalactouronase from *Rhizopus stolonifer*, an elicitor of casbene synthetase activity in castor bean (*Ricinus communis* L.) seedlings, *Plant Physiol.* **67:**633–639.

Lee, K., Townsend, J., Tepperman, J., Black, M., Chiu, C., Mazur, B., Dunsmuir, P., and Bedbrook, J., 1988, The molecular basis of sulphonylurea herbicide resistance in tobacco, *EMBO J.* **7:**1241–1248.

Legrand, M., 1983, Phenylpropanoid metabolism and its regulation in disease, in *Biochemical Plant Pathology* (J. Callow, ed.), pp. 367–384, Wiley, New York.

Levings, C. S., III, and Brown, G. G., 1989, Molecular biology of plant mitochondria, *Cell* **56:**171–179.

Lin, F. L., Sperle, K., and Stenberg, N., 1985, Recombination in mouse L-cells between DNA introduced into cells and homologous chromosomal sequences, *Proc. Natl. Acad. Sci. USA* **82:**1391–1395.

Lin, T. P., Casper, T., Somerville, C., and Preiss, J., 1988, Isolation and characterization of a starchless mutant of *Arabidopsis thaliana* (L) Heynh lacking ADP glucose pyrophosphorylase activity, *Plant Physiol.* **86:**1131–1135.

Lindquist, S., 1986, The heat–shock response, *Annu. Rev. Biochem.* **55:**1151–1191.

Lonsdale, D. M., Hodge, T. P., and Fauron, C. M. R., 1984, The physical map and organization of the mitochondrial genome from fertile cytoplasm of maize, *Nucleic Acids Res.* **12:**9249–9261.

Lorz, H., Baker, B., and Schell, J., 1985, Gene transfer to cereal cells mediated by protoplast transformation, *Mol. Gen. Genet.* **199:**178–182.

Matto, A. K., and Eldelman, M., 1987, Intramembrane translocation and posttranslation and palmitoylation of the chloroplast 32-KDa herbicide-binding protein, *Proc. Natl. Acad. Sci. USA* **84:**1497–1501.

Mauch, F., Hadwiger, L. A., and Boller, T., 1984, Ethylene: Symptom, not signal for the induction of chitinase and β-1,3-gluconase in pea pods by pathogens and elicitors, *Plant Physiol.* **76:**607–611.

Mazan, D., and Esquerre-Tugaye, M. T., 1986, Hydroxyproline rich glycoprotein accumulation in the cell walls of plants infected by various pathogens, *Physiol. Mol. Plant Pathol.* **29:**147–157.

McCabe, D. E., Swain, W. F., Martinell, B. J., and Christou, P., 1988, Stable transformation of soybean (*Glycine max*) by particle acceleration, *Biotechnology* **6:**923–926.

McCouch, S. R., Kochert, G., Yu, Z. H., Wang, Z. Y., Khush, G. S., Coffman, W. R., and Tanksley, S. T., 1988, Molecular mapping of rice chromosomes, *Theor. Appl. Genet.* **76:**815–829.

Meyerowitz, E. M., 1987, *Arabidopsis thaliana, Annu. Rev. Genet.* **21:**93–111.

Meyerowitz, E. M, 1989, *Arabidopsis,* a useful weed, *Cell* **56:**263–269.

Moffatt, B., and Somerville, C., 1988, Positive selection for male sterile mutants of *Arabidopsis* lacking adenine phosphoribosyl transferase activity, *Plant Physiol.* **86:**1150–1154.

Moreno, J., and Chrispeels, M. J., 1989, A lectin gene encodes the alpha amylase inhibitor of the common bean, *Proc. Natl. Acad. Sci. USA* **86:**7885–7889.

Mundy, J., and Chua, N-H., 1988, Abscisic acid and water stress induce the expression of a novel rice gene, *EMBO J.* **7:**2279–2286.

Murai, N., and Kemp, J., 1982, Octopine synthase mRNA isolated from sunflower crown gall callus is homologous to the Ti plasmid of *Agrobacterium tumefaciens, Proc. Natl. Acad. Sci. USA* **79:**86–90.

Murasugi, A., Wada, C., and Hayashi, Y., 1983, Occurrence of acid labile sulphide in cadmium binding peptide I from fission yeast, *J. Biochem.* **93:**661–664.

Nasrallah, M. E., and Nasrallah, J. B., 1986, Molecular biology of self incompatibility in plants, *Trends Genet.* **2:**239–243.

Nasrallah, J. B., Yu, S. M., and Nasrallah, M. E., 1988, Self incompatibility genes of *Brassica oleraceae:* Expression, isolation and structure, *Proc. Natl. Acad. Sci. USA* **85:**5551–5555.

Nichols, E. J., Beckman, J. M., and Hadwiger, L. A., 1980, Glycosidic enzyme activity in pea tissue and pea–*Fusarium solani* interactions, *Plant Physiol.* **66:**199–204.

Ohta, Y., 1986, High efficiency genetic transformation of maize by a mixture of pollen and exogenous DNA, *Proc. Natl. Acad. Sci. USA* **83:**715–719.

Osborne, T. C., Alexander, D. C., Sun, S. S. M., Cardona, C., and Bliss, F. A., 1988, Insecticidal activity and lectin homology of arcelin seed protein, *Science* **240:**207–210.

Paszkowski, J., Bam, M., Bogucki, A., and Potrykus, I., 1988, Gene targeting in plants, *EMBO J.* **7:**4021–4026.

Paterson, A. H., Lander, E. S., Hewitt, J. D., Paterson, S., Lincoln, S. E., and Tanksley, S. D., 1988, Resolution of quantitative traits into Mendelian factors by using a complete RFLP linkage map, *Nature* **335:**721–726.

Powell, P. A., Stark, D. M., Sanders, P. R., and Beachi, R. N., 1989, Protection against tobacco mosaci virus in transgenic plants that express tobacco mosaic virus antisense RNA, *Proc. Natl. Acad. Sci. USA* **86:**6949–6952.

Pring, D. R., Levings III C. S., Hu, W. W. L., and Timothy, D. H., 1977, Unique DNA associated with mitochondria in the "S"-type cytoplasm of male-sterile maize, *Proc. Natl. Acad. Sci. USA* **74:**2904–2908.

Reineri, D. M., Bottino, P., Gordon, M. P., and Nester, E. W., 1990, *Agrobacterium* mediated transformation of rice (*Oriza sativa* L.), *Biotechnology* **8:**33–38.

Rhodes, C. A., Pierce, D. A., Mettler, I. J., Mascarenhas, D., and Detmer, J. J., 1988, Genetically transformed maize plants from protoplasts, *Science* **240:**204–207.

Roby, D., Toppan, A., and Esquerre-Tugaye, M. T., 1985, Cell surfaces in plant microorganism interactions. V. Elicitors of fungal and of plant origin trigger the synthesis of ethylene and of cell wall hydroxyproline rich glycoprotein in plant, *Plant Physiol.* **77:**700–704.

Roby, D., Toppan, A., and Esquerre-Tugaye, M. T., 1986, Cell surfaces in plant-microorganism interactions. VI. Elicitors of ethylene from *Colletotrichum lagenarium* trigger chitinase activity in melon plants, *Plant Physiol.* **81:**228–233.

Rodermel, S. R., Merilyn, S. A., and Bogorad, L., 1988, Nuclear antisense gene—Inhibits RuBP carboxylase levels in chloroplasts of transformed tobacco plants, *Cell* **55:**673–681.

Rogers, J., 1988, RNA complementary to alpha amylase mRNA in barley, *Plant Mol. Biol.* **11:**125–138.

Rothstein, S. J., DiMaio, S., Strand, J. M., and Rice, D., 1987, Stable and heritable inhibition of the nopaline synthase in tobacco expressing antisense RNA, *Proc. Natl. Acad. Sci. USA* **85:**4300–4304.

Rouser, W., 1984, Isolation and partial purification of Cadmium-binding protein from roots of the grass *Agrostis gigantia, Plant Physiol.* **74:**1025–1029.

Sachs, M. M., and Freeling, M., 1978, Selective synthesis of alcohol dehydrogenase during anaerobic treatment of maize, *Mol. Gen. Genet.* **161:**111–115.

Sanchez-Serrano, J., Schmidt, R., Sohell, J., and Willmitzer L., 1986, Nucleotide sequence of proteinase inhibitor II encoding cDNA of potato (*Solanum tuberosum*) and its mode of expression, *Mol. Gen. Genet.* **203:**15–20.

Sandler, S. J., Stayton, M., Townsend, J. A., Ralston, M. L., Bredbrook, J. R., and Dunsmuir, P., 1988, Inhibition of gene expression in transformed plants by antisense RNA, *Plant Mol. Biol.* **11:**301–311.

Schell, J., 1987, Transgenic plants as tools to study the molecular organization of plant genes, *Science* **237:**1176–1182.

Schlesinger, N. J., Ashburner, M., and Tessieres, A. (eds.), 1982, *Heat Shock from Bacteria to Man,* Cold Spring Harbor Laboratories, Cold Spring Harbor, NY.

Schlumbaum, A., Mauch, F., Vogeli, U., and Boller, T., 1986, Plant chitinases are potent inhibitors of fungal growth, *Nature* **324:**665–667.

Schmidt, R., Burr, F., and Burr, B., 1987, Transposon tagging and molecular analysis of the maize regulatory locus opaque-2, *Science* **238:**960–963.

Shah, D., Harsh, R. B., Lee, H., Kishore, G., Winter, J., Tumer, N., Hironka, C., Sanders, P., Gasser, C., Alkent, S., Siegal, N., Rogers, S., and Fraley, R., 1986, Engineering herbicide tolerance in transgenic plants, *Science* **233:**478–481.

Shimamoto, K., Terada, R., Izawa, T., and Fujimoto, H., 1989, Fertile transgenic rice plants regenerated from transformed protoplasts, *Nature* **338:**273–276.

Sijmons, P. C., Dekker, B. M. M., Schrammeijer, B., Verwoerd, T. C., Van der Elzen, P. J. M., and Hoekema, A., 1990, Production of correctly processed human serum albumbumin in transgenic plants, *Biotechnology* **8:**217–221.

Simons, R. W., and Kleckner, N., 1983, Translational control of IS-10 transposition, *Cell* **34:**683–691.

Smith, D. A., and Banks, S. W., 1986, Biosynthesis, elicitation and biological activity of isoflavonoid phytoalexins, *Phytochemistry* **25:**979–995.

Smith, C. J. S., Watson, C. F., Ray, J., Bird, C. R., Morris, P. C., Schuch, W., and Grierson, D., 1988, Antisense RNA inhibition of polygalacturonase gene expression in transgenic tomatoes, *Nature* **334:**724–726.

Somerville, C., and Grill, E., 1989, Genetic methods of gene isolation in *Arabidopsis,* in *EMBO Symposium on Molecular* Communication in Higher Plants, pp. 57–58, EMBL, Heidelberg.

Stalker, D. M., McBride, K., and Malyj, L. D., 1988, Herbicide resistance in transgenic plants expressing a bacterial detoxification gene, *Science* **242:**419–422.

Stanley, J., and Townsend, R., 1986, Infectious mutants of cassava latent virus generated *in vivo* from intact recombinant clones containing single copies of the genome, *Nucleic Acids Res.* **14:**5981–5998.

Stanley, J., Markham, P. G., Callis, R. J., and Pinner, M. S., 1986, The nucleotide sequence of an infectious clone of the gemini virus beet curly top virus, *EMBO J.* **5:**1761–1767.

Steffens, J., Hunt, D., and Williams, B., 1986, Accumulation of non protein metal binding polypeptides (gamma-glutamyl-cysteinyl) *n*-glycine in selected cadmium resistant tomato cells, *J. Biol. Chem.* **261:**13879–13882.

Tanksley, S., Miller, J., Paterson, A., and Bernatzky, R., 1988a, Molecular mapping of plant chromosomes, in *Chromosome Structure and Function* (J. F. Gustafson and R. Appels, eds.), pp. 157–173, Plenum Press, New York.

Tanksley, S. D., Bernatz, R., Lapitan, N. L., and Prince, J. P., 1988b, Conservation of gene repertoire but not gene order in pepper and tomato, *Proc. Natl. Acad. Sci. USA* **85:**6419–6423.

Tanksley, S. D., Young, N. D., Paterson, A. H., and Bonierbales, M. W., 1989, RFLP mapping in plant breeding: New tools for an old science, *Biotechnology* **7:**257–262.

Tepfer, D., 1984, Transformation of several species of higher plants by *Agrobacterium rhizogenes:* Sexual transmission of the transformed genotype and phenotype, *Cell* **37:**959–967.

Thomashow, M., Hugly, S., Buchholz, W., and Thomashow, C., 1986, Molecular basis for the auxin-independent phenotype of crown gall tumor tissues, *Science* **231:**616–618.

Topfer, R., Gronenborn, B., Schell, J., and Steinbiss, H., 1989, Uptake and transient expression of chemeiric genes in seed derived emryos, *Plant Cell* **1:**133–139.

Vaeck, M., Reynaerts, A., Hofte, H., Jansens, S., Benckelean, M. D., Dean, C., Zabeau, M., Van Montagu, M., and Leemans, J., 1987, Transgenic plants protected from insect attack, *Nature* **328:**33–37.

Van Haute, E., Joos, H., Maes, M., Warren, G., Van Montagu, M., and Schell, J., 1983, Intergeneric transfer and exchange recombination of restriction fragments cloned in pBR322: A novel strategy for reversed genetics of the Ti plasmids of *Agrobacterium tumefaciens, EMBO J.* **2:**411–417.

Van der Krol, A. R., Lemting, E. E., Veenstra, J., Van der Meer, I. B., Koes, R. E., Gerats, A. G. M., Mol, J. N. M., and Stuitje, A. R., 1988a, An antisense chalcone synthase gene in transgenic plants inhibits flower pigmentation, *Nature* **333:**866–869.

Van der Krol, A. R., Mol, J. N. M., and Stuitje, A. R., 1988b, Antisense gene in plants: An overview, *Gene* **72:**45–50.

Van Larebeke, N., Engler, G., Holsters, M., Van den Elsaker, S., Zaeneu, I., Schilperoort, R. A., and Schell, J., 1974, Large plasmid in agrobacterium tumefaciens essential for crown gall–inducing ability, *Nature* **252:**169–170.

Van Sluys, M. A., Temple, J., and Fedoroff, N., 1987, Studies on the introduction and mobility of the maize activator element in *Arabidopsis thaliana* and *Daucus carota, EMBO J.* **6:**3881–3889.

Vance, C. E., Kirk, T. K., and Sherwood, R. T., 1980, Lignification as a mechanism of disease resistance, *Annu. Rev. Phytopathol.* **18:**259–288.

Vandekerchkhove, J., Van Damme, J., Van Lijsebettens, M., Botterman, J., DeBlock, M., Vandewiele, M., De Clercq, A., Leemans, J., Van Montagu, M., and Krebbers, E., 1989, Enkephalins produced in transgenic plants using modified 2S seed storage proteins, *Biotechnology* **7:**929–932.

Van Loon, L. C., 1985, Pathogenesis related proteins, *Plant Mol. Biol.* **4:**111–116.

Varner, J. E., and Lin, L-S., 1989, Plant cell wall architecture, *Cell* **56:**231–239.

Vilaine, F., and Casse-Delbart, F., 1987, A new vector derived from *Agrobacterium rhizogenes* plasmids: A micro-Ri plasmid and its use to construct a mini Ri plasmid, *Gene* **55:**105–114.

Ward, A., Etessami, P., and Stanley, J., 1988, Expression of a bacterial gene in plants mediated by infectious gemini virus DNA, *EMBO J.* **7:**1583–1587.

Watson, B., Currier, T. C., Gordon, M. P., Chilton, M. D., and Nester, E. W., 1975, Plasmid required for virulence of *Agrobacterium tumefaciens, J. Bacteriol.* **123:**255–264.

Weising, K., Schell, J., and Kahl, G., 1988, Foreign genes in plants: Transfer, structure, expression, and applications, *Annu. Rev. Genet.* **22:**421–477.

White, R. F., 1979, Acetylsalicylic acid (aspirin) induces resistance to tobacco mosaic virus in tobacco, *Virology* **99:**410–412.

Wilson, L. G., Fry, J. C., 1986, Extensin—A major cell wall glycoprotein, *Plant Cell Env.* **9:**239–260.

Yadav, N. S., Van der Leyden, J., Bennet, D., Barnes, W. H., and Chilton, M. D., 1982, Short direct repeats flank the T-DNA on a nopaline Ti plasmid, *Proc. Natl. Acad. Sci. USA* **79:**6322–6326.

Yanofsky, M. F., Porter, S. G., Young, C., Albright, L. M., Gordon, M. P., and Nester, E. W.,

1986, The *vir* D operon of *Agrobacterium tumefaciens* encodes a site specific endonuclease, *Cell* **47**:471–477.

Zaenen, I., Van Larebeke, N., Teuchy, H., Van Montagu, M., and Schell, J., 1974, Supercoiled circular DNA in crown-gall inducing *Agrobacterium* strains, *J. Mol. Biol.* **86**:109–127.

Zambryski, P., Holsters, M., Kruger, K., Depicker, A., Schell, J., Van Montagu, M., and Goodman, H., 1980, Tumor DNA structure in plant cells transformed by *A. tumefaciens*, *Science* **209**:1385–1391.

Zambryski, P., Joos, H., Genetello, C., Leemans, J., Van Montagu, M., and Schell, J., 1983, Ti plasmid vector for the introduction of DNA into plant cells without altering their regeneration capacity, *EMBO J.* **2**:2143–2150.

Zambryski, P., Tempe, J., and Schell, J., 1989, Transfer and function of T-DNA genes from *Agrobacterium* Ti and Ri plasmids in plants, *Cell* **56**:193–201.

Zhang, W., and Wu, R., 1988, Efficient regeneration of transgenic plants from rice protoplasts and correctly regulated expression of the foreign gene in the plants, *Theor. Appl. Genet.* **76**:835–840.

Chapter 2

Molecular Biology of Legume Vicilin-Type Seed Storage Protein Genes

Paula P. Chee and Jerry L. Slightom

1. INTRODUCTION

The major seed storage proteins of legumes are globulins (packaged into cotyledonary protein storage bodies), which are represented in most legumes by two different types of polypeptides; the nonglycosylated 11S fraction (called legumins) and the glycosylated 7S proteins (called vicilins) (Derbyshire *et al.*, 1976). This chapter focuses on the investigations of the vicilin-type storage protein genes, which encode vicilin of pea (*Pisum sativum*), the α'-subunit of β-conglycinin of soybean (*Glycine max*), and phaseolin of common bean (*Phaseolus vulgaris*).

In the seed, the 7S proteins consist of multisubunit combinations of approximately 150–220 kilodaltons (kDa) (Derbyshire *et al.*, 1976; Blagrove *et al.*, 1984); however, the individual protein subunits vary in size between and within legume species. The principal vicilin-type polypeptide subunits found are: the α-(51–53 kDa), β-(47–48 kDa), and γ-(43–46 kDa) subunits of phaseolin in *P. vulgaris* (Brown *et al.*, 1981); the α'-(76 kDa), α-(72 kDa), and β-(53 kDa)

Paula P. Chee and Jerry L. Slightom   Molecular Biology, The Upjohn Company, Kalamazoo, Michigan 49007.

Subcellular Biochemistry, Volume 17: Plant Genetic Engineering, edited by B. B. Biswas and J. R. Harris. Plenum Press, New York, 1991.

subunits of β-conglycinin in *G. max* (Derbyshire *et al.*, 1976; Meinke *et al.*, 1981); and the convicilin (70 kDa) and vicilin subunits (50 kDa) in *P. sativum* (Higgins, 1984). The phaseolin and β-conglycinin vicilin-type proteins are not substantially altered by post-translational processing, while vicilins of *P. sativum* are extensively cleaved into a series of small polypeptides (Higgins, 1984). Vicilin polypeptides (50 kDa) of *P. sativum* are cleaved into polypeptides of 34, 27, 25, and 12 kDa in size (Higgins, 1984).

The three phaseolin polypeptide types make up about 50% of the protein in the *P. vulgaris* seed (Ma and Bliss, 1978). However, molecular analysis of individual members of the phaseolin gene family shows that these polypeptides are encoded by only two unique gene types, the α- and β-type genes (Sun *et al.*, 1981; Slightom *et al.*, 1983, 1985). Each phaseolin gene type is encoded by at least three or four gene copies per haploid genome (Talbot *et al.*, 1984). Heterogeneity among the phaseolin subunits appears to be the result of incomplete addition, or partial degradation, of *N*-linked oligosaccharide side chains on the α- and β-type polypeptides (Bollini *et al.*, 1983; Paaren *et al.*, 1987; Sturm *et al.*, 1987). In *G. max*, the α-subunits of β-conglycinin appear to be encoded by a small number of gene, one to two genes each per haploid genome (Goldberg *et al.*, 1981), while the β-subunits of β-conglycinin are encoded by a larger number of genes (Goldberg *et al.*, 1981; Tierney *et al.*, 1987; Harada *et al.*, 1989). The pea vicilin storage proteins are encoded by at least four genetic loci of which three are known to encode the vicilin-type proteins (Lycett *et al.*, 1983; Higgins, 1984); the related convicilin-type polypeptides appear to be encoded by two genes (Bown *et al.*, 1988).

Amino acid determinations and deduced amino acid sequences of vicilin-type gene sequences shows that they all lack sulfur-containing amino acids—containing less than 1%. For these proteins to be nutritionally balanced they should have between 3 and 6% sulfur-containing amino acids. Thus a goal of molecular biology research, other than to obtain an understanding of gene structure and function, is to incorporate more sulfur amino acids into these genes by using either DNA mutations or/and insertions. Alternatively, nutritional improvement could be obtained by expressing high sulfur storage protein (HSSP) genes isolated from a nonlegumin plant source, such as using the 2S albumin storage protein gene of Brazil nut (*Bertholletia excelsa*) (Altenbach *et al.*, 1987). However, before such genetic engineering approaches can be used successfully, a large amount of basic molecular biology information concerning the regulation of seed storage protein genes needs to be determined. The types of amino acid changes (or types of foreign gene–derived HSSP) that can be tolerated by the seed expression and storage systems also need to be determined. In addition, efficient plant transformation and regeneration procedures for important cultivars of soybean, pea, and common bean species need to be developed.

2. VICILIN GENE STRUCTURAL AND FUNCTIONAL SEQUENCES

2.1. Comparative Analyses of Vicilin-Type Genes

Phaseolin, pea vicilin, and the α'-subunit of β-conglycinin polypeptides differ considerably in both size and antigenic identity (Doyle et al., 1985, 1986). However, a comparison of their nucleotide sequences shows that they do share both structural and sequence identities (Doyle et al., 1986; Higgins et al., 1988). Doyle et al. (1986) reported the complete nucleotide sequence for the α'-subunit of β-conglycinin (Gma-α'), a total of 3636 base pairs (bp), and compared its nucleotide and amino acid sequences with that obtained from the β-phaseolin gene (Pvu-β), about 2800 bp. Similarly, Higgins et al. (1988) reported the complete nucleotide sequence of a pea vicilin gene (2766 bp) and its comparison with Pvu-β. In this chapter we will refer to this pea vicilin gene as "Psa-V." These nucleotide sequence alignments showed that all three genes have the same basic structural organization, with six exons being separated by five introns. This same exon and intron organization is also found for the β-subunit of β-conglycinin of soybean (Harada et al., 1989) and convicilin of pea (Bown et al., 1988).

Table I summarizes the shared identities found between comparison of the Pvu-β and Gma-α' and Pvu-β and Psa-V genes using the previously published

Table I
Nucleotide Sequence Identity between Pvu-β and Gma-α' and Pvu-β and Psa-V

Structural region	Pvu-β[a] versus Gma-α'		Pvu[b]-β versus Psa-V	
	bp[c] match/ bp compared	% identity	bp match/ bp compared	% identity
5'-Untranslated	39/70	56	18/43	42
Exon-1	237/314	75	181/302	60
Intron-1	52/82	63	33/75	44
Exon-2	123/195	63	120/177	68
Intron-2	60/87	69	26/89	29
Exon-3	67/81	83	56/81	69
Intron-3	61/97	63	39/103	38
Exon-4	143/233	61	130/250	52
Intron-4	91/123	74	49/128	38
Exon-5	198/261	76	170/263	65
Intron-5	74/113	65	44/102	43
Exon-6	115/180	64	53/177	30
3'-Untranslated	92/135	68	52/91	57

[a]Calculated from the Pvu-β and Gma-α' comparison presented by Doyle et al. (1986).
[b]Calculated from the Pvu-β and Psa-V comparison presented by Higgins et al. (1988).
[c]Gaps are included in these calculations and they are treated as one position regardless of length.

alignments (Doyle *et al.*, 1986; Higgins *et al.*, 1988). The *Pvu*-β and *Psa*-V genes share an overall identity of about 52%, in sequences that extend between the 5'- and 3'-untranslated region (Table I), with the exons and introns showing considerably different degrees of shared sequence identities, 57 and 38%, respectively. The overall identity between the *Gam*-α' and *Pvu*-β genes is significantly higher, about 69%, with exons and introns showing similar degrees of identity, 70 and 68%, respectively. The degree of identity shared among the exons regions of these genes ranges between 30% (*Pvu*-β versus *Psa*-V exon 6) and 83% (*Pvu*-β versus *Gma*-α' exon 3); exon 3 shares the highest degree of identity for both gene comparisons (Table I). The most noticeable difference is found in exon 1, where *Gma*-α' contains a 522-bp insertion (encoding 174 amino acids) that is not present in exon 1 of either *Pvu*-β or *Psa*-V genes. This insertion results in the addition of 20 kDa to the *Gma*-α' polypeptide and thus accounts for most of the difference in size between this polypeptide and *Pvu*-β and *Psa*-V polypeptides. A clearer description of the evolutionary relationships among these seed storage protein genes was obtained by Gibbs *et al.* (1989). This analysis shows that soybean β-conglycinin subunit genes (both α and β) are more closely related to phaseolin than is the pea *Psa*-V gene.

2.2. Identification of Potential *Cis*-Acting Regulatory Elements

The *cis*-acting regulatory elements common to most eukaryotic genes, CCAAT- and TATA-elements (Efstratiadis *et al.*, 1980), are also present in similar locations (-77 and -31 bp 5' of the cap site, respectively) in these plant genes. Multiple CCAAT- and TATA-elements are found in the *Pvu*-β sequence, but not in the *Psa*-V and *Gma*-α' genes; the multiple TATAA-elements in phaseolin are believed to be responsible for the numerous cap sites found in phaseolin mRNAs. At least 12 different capped mRNA species have been identified by S1 nuclease analysis (Slightom *et al.*, 1985). The nucleotide sequences surrounding the TATA-elements of these genes match the "plant consensus," while their CCAAT-elements are more similar to the mammalian consensus sequence (Messing *et al.*, 1983). In addition to these regulatory elements, these 5'-flanking DNAs contain sequence regions that share identity with another mammalian (SV-40 type) *cis*-acting regulatory element which has a consensus sequence of GTGGAAAG (Gruss, 1984). However, because these seed storage protein genes are subject to similar tissue-specific and developmental regulation, it is conceivable that they could share other *cis*-acting regulatory signals that are unique to control their own tissue and developmental specific expression (see below).

One additional well-known *cis*-acting regulatory element (AATAAA) is located in the 3'-untranslated DNA region, and it is used to signal mRNA splicing and polyadenylation. The 3'-untranslated regions of *Gma*-α' and *Pvu*-β gene transcripts are both 135 bp in length and appear to share the same degree of identity (68%) as the coding regions (Table I). This region of the pea vicilin gene transcript is 101 bp in length and shares 57% identity with the Pvu-β 3'-untranslated region (Table I). Nucleotide sequence identity near the poly(A) signal is not well conserved; the Gma-α' gene contains overlapping poly(A) signals, which distort any alignment in this area (Doyle et al., 1986).

2.3. Searching for Seed-Specific *Cis*-Acting Regulatory Elements

Experiments involving the transfer of these genes into the genome of tobacco or petunia, via *Agrobacterium tumefaciens* vector systems, suggest that DNA elements that control tissue-specific and developmental expression of these genes are located within about 800 bp of the cap sites (Beachy et al., 1985; Sengupta-Gopalan et al., 1985; Chen et al., 1986; Higgins et al., 1988). The most extensively studied of these three genes for sequences involved in gene regulation has been the soybean Gma-α' gene. Beachy and co-workers examined the 5'-flanking DNA region of the Gma-α' gene for DNA sequence elements that interact with petunia and tobacco seed developmental regulatory factors. Transcriptional control of most eukaryotic genes requires the presence of a TATA-element for proper initiation of transcripts and, to a lesser extent, a CCAAT-element, which appears to modulate the level of expression (Wasylyk et al., 1983; Myers et al., 1986). In some cases, enhancer elements have been found to greatly affect the level of gene expression in a tissue-specific manner (Gruss, 1984). Pvu-β, Gma-α', and Psa-V storage protein genes contain sequences that match these identified regulatory elements (Doyle et al., 1986; Higgins et al., 1988); a comparison of the 5'-flanking regions (about 400 bp) shows an overall shared identity of 66% for Pvu-β versus Gma-α' and 60% for Pvu-β and Psa-V (data not shown).

Chen et al. (1986) used the nucleotide sequence comparison presented by Doyle et al. (1986) as a guide to construct a series of deletion mutants of the Gma-α' gene to test the effect of these known regulatory elements and to identify other potential regulatory elements. Each mutated Gma-α' gene was transferred and integrated into the petunia genome. Figure 1 shows a multiple alignment comparison of the Gma-α', Pvu-β, and Psa-V 5'-flanking regions along with the location of several of the deletions used in the Gma-α' study (Chen et al., 1986). Chen et al. (1986) measured the levels of Gma-α' protein derived from each deletion and characterized this level as a percentage of Gma-α' protein produced by the native construction in transgenic petunia seeds (Beachy et al., 1985).

Deletion of the SV-40 mammalian enhancer type sequence at position −560 (Doyle et al., 1986) shows little effect on the level of expression, indicating that this sequence element does not play a role in regulating the expression of this gene in petunia seeds. Little effect on the level of expression was also observed

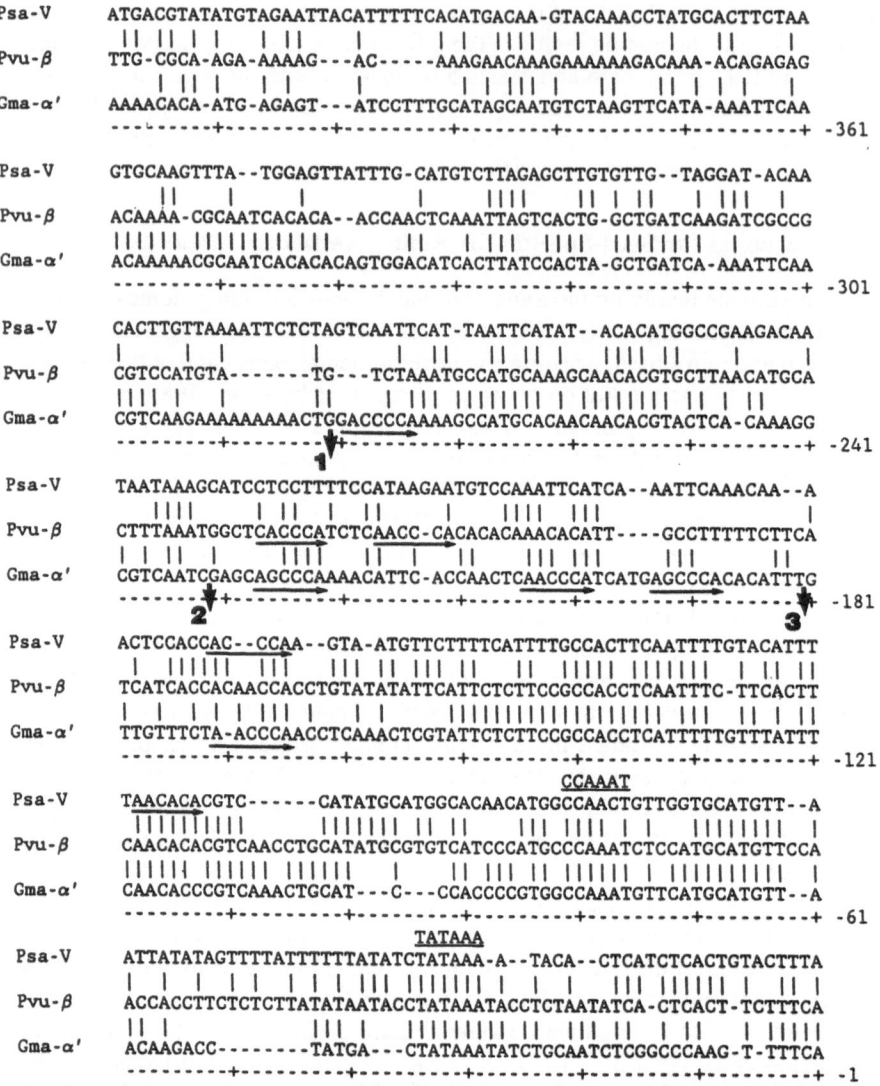

FIGURE 1. Alignment of promoter regions from *Psa*-V, *Pvu*-β, and *Gma*-α' seed storage protein genes. Nucleotide sequences were aligned using the multiple alignment program CLUSTAL (Higgins and Sharp, 1988). The nucleotide numbering systems starts at the transcriptional initiation site found for each gene (Slightom *et al.*, 1983; Doyle *et al.*, 1986; Higgins *et al.*, 1988). The location of gene regulatory elements is indicated by the consensus sequences written above the alignment, and the locations of the developmental enhancer hexanucleotide elements (Chen *et al.*, 1986) are indicated by horizontal arrows. The vertical arrows numbered 1, 2, and 3 show the end-points of deletion of the *Gma*-α' gene promoter which Chen *et al.* (1986) used for searching for functional *cis*-acting regulatory sequences (see text).

for the deletions at positions −457 and −257 (Doyle *et al.*, 1986; Chen *et al.*, 1986). The deletion of sequences 5′ of arrow 1 in Figure 1 resulted in little loss (about 20%) in the level of *Gma*-α′ gene expression, and the expression level was reduced a bit more when sequences 5′ of the second arrow in Figure 1 were deleted (35% loss of activity). However, when sequences 5′ of arrow 3 in Figure 1 were deleted, the expression of the *Gma*-α′ gene was severely limited (96% loss of activity). These results suggested to Chen *et al.* (1986) that DNA elements important for controlling the expression of the *Gma*-α′ gene must be located within this region (see region between arrows 2 and 3 in Figure 1). Similar analysis of deletions located near the CCAAT- and TATA-elements showed no detectable protein expression (Chen *et al.*, 1986). This series of experiments clearly showed that the genetic information located between positions −232 and −181 (Figure 1) of the *Gma*-α′ gene are required for recognition of petunia embryo-specific developmental regulatory factors. However, additional support for the importance of this region, by testing similar deletions in either the *Pvu*-β or *Psa*-V gene, has not been reported.

A search for potential nucleotide sequence enhancer-type elements located between these positions revealed the presence of an imperfect direct repeat of 28 nucleotides and five smaller (G + C)-rich repeats (AGCCCA or AACCCA) four of which are located within the larger repeat (Doyle *et al.*, 1986; Chen *et al.*, 1986). Chen *et al.* (1986) suggested that the smaller direct repeat elements may provide the genetic information responsible for regulating the level of expression of the *Gma*-α′ gene in developing seeds of transgenic petunia plants, and if this is true, these sequences may also be important in regulating the expression of this gene in developing soybean seeds. This hypothesis is supported by finding sequences that match these short repeats in similar locations of the *Pvu*-β gene, (CACCCA) and (AACCCA) and the sequence (CACCCA) in *Psa*-V; the locations of these sequence elements are indicated by arrows in Figure 1. This hexanucleotide is also found in the soybean gene encoding the β-subunit of β-conglycinin (Harada *et al.*, 1989) and in the pea convicilin gene (Bown *et al.*, 1988).

The most convincing evidence that supports the importance of these hexanucleotides in regulating the expression of the *Gma*-α′ comes from the finding that they are capable of binding proteins present in crude nuclear extracts isolated from immature soybean seeds (Allen *et al.*, 1989). The hexanucleotide AACCCA has been defined as the likely protein recognition site for binding the potential *trans*-acting protein factors using S1 mapping (Allen *et al.*, 1989). As Figure 1 shows, this hexanucleotide sequence is found in about the same location of *Pvu*-β (two repeats) and in *Psa*-V (only one element is found). Test for similar interaction between the *Pvu*-β 5′ flanking DNA and proteins present in immature *P. vulgaris* seeds did not reveal any proteins that bind specifically to this hexanucleotide, but another protein binding factor was found. This protein factor, however, binds to a

DNA region located about 660 bp 5' of the *Pvu*-β gene mRNA translation initiation site (Bustos *et al.*, 1989). Thus, it is not clear whether the hexanucleotide sequences present in the *Pvu*-β and *Psa*-V genes actually function in the manner that they appear to function in the *Gma*-α' gene. Although, Allen *et al.* (1989) have suggested that two AACCCA repeats are necessary for protein binding, neither the *Pvu*-β nor the *Psa*-V gene contains two closely spaced repeats of this hexanucleotide.

2.4. Importance of 5'-Untranslated Region in Aiding Translation

It is well known that gene promoter strength is important in obtaining high levels of mRNAs; however, few direct comparison of gene promoter strengths have been reported. In an interesting study recently reported by Riggs *et al.* (1989), the luciferase gene was used as a reporter to test the regulation of the bean seed storage proteins phytohemagglutinin (PHA-L) and *Pvu*-β promoters (including 5'-untranslated sequences) in transgenic tobacco seeds. Although the amount of protein represented in the bean seeds differs considerably for PHA-L and *Pvu*-β, the study by Riggs *et al.* (1989) found that both promoters were equally active in the expression of luciferase mRNA in tobacco seeds. However, transgenic tobacco plants containing the *Pvu*-β–luciferase gene fusion produced eightfold more luciferase protein than that found in plants containing the PHA-L–luciferase gene fusion. These results suggest that promoter strength is not the only major factor in regulating the level of gene expression; the 5'-untranslated sequences also play an important role, by increasing either mRNA stability and/or translatability. The 5'-untranslated regions of the PHA-L and *Pvu*-β genes differ considerably, with their length being the most dramatic difference: 11 versus 76 nucleotides (Slightom *et al.*, 1983; Hoffman and Donaldson, 1985). The importance of 5'-untranslated sequences has also been noted in studies where the 5'-untranslated regions of plant viruses (tobacco mosaic and alfalfa mosaic viruses) were used to enhance the expression of reporter genes such as chloramphenicol acetyltransferase and neomycin phosphotransferase (Sleat *et al.*, 1988). Experiments are needed to test additional 5'-untranslated sequences to determine their range of translational enhancement. The addition of an optimal 5'-untranslated sequence would provide a convenient way to increase translational efficiencies.

3. EXPRESSION OF BEAN SEED STORAGE PROTEIN GENES IN FOREIGN PLANT TISSUES

3.1. Expression of the Phaseolin Gene in Foreign Plant Tissues

The transfer and integration of foreign DNAs into the genome of various dicotyledonous plant species via *Agrobacterium tumefaciens* is quite routine.

The first plant-derived gene to be transferred into a foreign plant tissue was the phaseolin storage protein gene, *Pvu*-β; it was transferred into the sunflower genome and expressed in undifferentiated sunflower tissue (Murai *et al.*, 1983). Analysis of this transformed sunflower callus for phaseolin mRNA showed the presence of a 1700-nucleotide mRNA that is identical in size to that found in the developing bean cotyledon (Murai *et al.*, 1983). Finding this correct-size phaseolin mRNA indicates that the five intron sequences of this *Pvu*-β gene (a total of 515 nucleotides) have been removed to obtain a mature phaseolin mRNA. This result suggests that plant mRNA splicing mechanisms are conserved, at least between widely divergent dicot plant species.

In undifferentiated callus tissue (sunflower and tobacco), full-length polypeptides were difficult to detect (Murai *et al.*, 1983; Chee *et al.*, 1986) as the protein is rapidly degraded. The expected size for the β-type phaseolin polypeptide is 48 kDa (Brown *et al.*, 1981; Slightom *et al.*, 1983), while that found in the undifferentiated tissues is nearer 46 kDa. The difference between the observed and expected sizes suggested that posttranslational processing of the phaseolin polypeptides was not complete. This could be due to either incorrect signal peptide cleavage or incomplete glycosylation of the polypeptides. Correct or nearly correct posttranslational processing appeared to occur when the β-type phaseolin gene was expressed in the developing tobacco seeds (Sengupta-Gopalan *et al.*, 1985). However, even in the seeds, some of the phaseolin polypeptides were degraded by cleavage at a set of distinct proteolytic cleavage sites. This degradation appears to involve the same cleavage sites found in germinating bean seeds (Boylan and Sussex, 1987) because the protein degradation patterns appear to be quite similar. Why some of the phaseolin polypeptides are unstable in tobacco seeds is not presently clear; it could possibly be due to the accumulation of excess protein, more than can be stored in the tobacco seed protein bodies, or due to the protein being deposited in locations where it is exposed to proteolytic enzymes. Analysis of seed storage protein bodies shows that a major portion of the phaseolin polypeptides are correctly targeted to the amorphous matrix of the tobacco seed protein bodies (Greenwood and Chrispeels, 1985). Similar results have been observed with the expression of other vicilin-type storage protein genes in tobacco seeds: the *Gma*-α' gene (Beachy *et al.*, 1985; Chen *et al.*, 1986), the β-subunit of β-conglycinin (Bray *et al.*, 1987), and the pea vicilin gene, *Psa*-V (Higgins *et al.*, 1988).

3.2. Engineering of an Intronless Phaseolin Gene into an Avirulent Ti Vector

The transfer and expression of a *Pvu*-β gene that lacked all five of the phaseolin gene intron sequences (a phaseolin minigene, or *miniPvu*-β) in tobacco tissues was reported by Chee *et al.* (1986). The level of phaseolin peptides arising from expression of the *miniPvu*-β gene was similar to that found for the

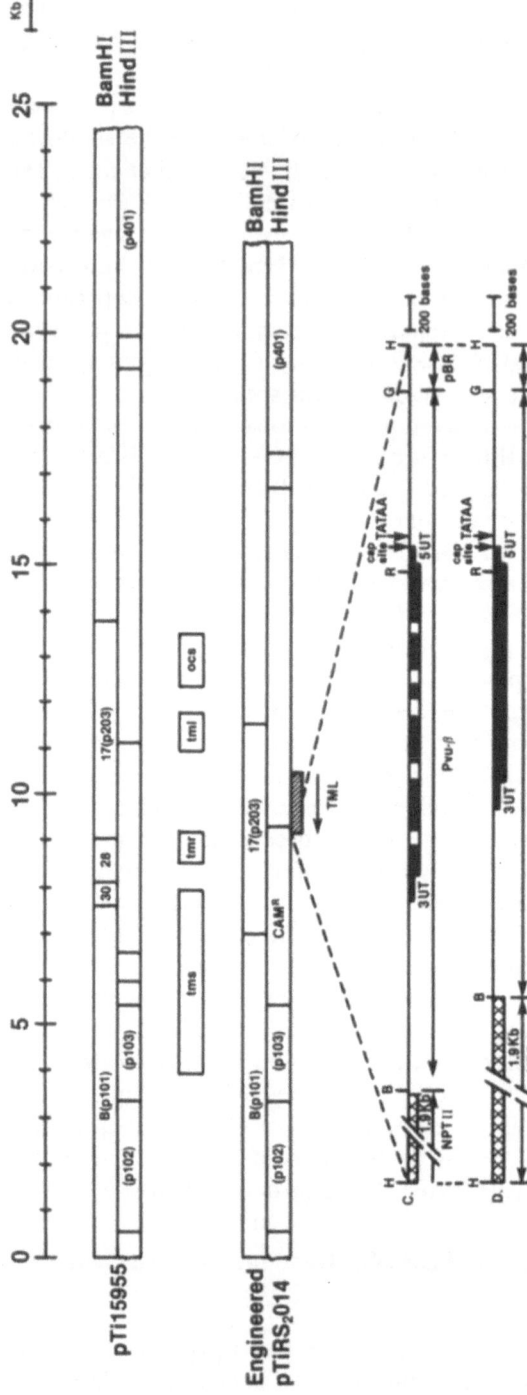

FIGURE 2. Construction of pTiRS$_2$014 and derivatives containing the *Pvu*-β or *miniPvu*-β gene. The map of the parent Ti plasmid, pTi15955, is shown along with the position of *BamHI* and *HindIII* sites. Clones obtained from this region, p101, p203, p102, p103, and p401, have been described by Barker *et al.* (1983). The location of the oncogenic loci (tumor morphology shooting, *tms*; tumor morphology rooting, *tmr*; and tumor morphology large, *tml*) and the octopine gene are shown below the pTi15955 map. The map of the engineered pTiRS$_2$014Cm^r plasmid is shown below the map of pTi15955. *Pvu*-β and *miniPvu*-β gene constructions (Chee *et al.*, 1986) were cloned into the shuttle vector p111 (Murai *et al.*, 1983) to obtain plasmids p111-Rpβ NPTII (Murai *et al.*, 1983) and p111-Rpβ/cDNA (Chee *et al.*, 1986), which contain the *Pvu*-β or *miniPvu*-β gene, respectively, located within the *tml* locus of the T-DNA *BamHI* fragment 17. Restriction enzyme sites are: B, *BamHI*; R, *EcoRI*, S, *SmaI*; H, *HindIII*. The letter G denotes the location of the ligated *BamHI* site (from pBR322) and the *BglII* site upstream from the phaseolin gene (Slightom *et al.*, 1983). The hatched region shown below the map of pTiRS$_2$014 shows the location of the *tml* locus, which was repositioned after the deletion of the *tms* and *tmr* loci; this locus was also inactivated due to the positioning of the phaseolin genes. Cross-hatched regions show the location of the NPT II gene and the solid black regions indicated phaseolin-coding DNAs.

40

native *Pvu*-β gene, which indicted that, at least in undifferentiated tissues, intron splicing was not necessary for biogenesis of stable mRNA molecules. However, whether the absence of these five introns would effect developmental expression of the *Pvu*-β in seeds has only recently been determined (Chee *et al.*, 1991). In this and the next section we briefly describe some of the experiments needed for engineering the *miniPvu*-β gene into an avirulent *Agrobacterium* strain and its transfer into tobacco tissues and analysis for phaseolin peptides in the seeds of R_1- and R_2-generation tobacco plants.

The experiment to transfer the *miniPvu*-β gene first involved the construction of a Ti plasmid that does not contain the oncogenic loci (*tms2* and *tmr*), as outlined in Figure 2: an avirulent *Agrobacterium* ti plasmid referred to as pTiRS$_2$2014Cm^r. A comparison of the *BamHI* and *HindIII* maps obtained from the T-DNA region of pTiRS$_2$014Cm^r with that of the parent Ti plasmid, pTi15955, is shown in Figure 2. The procedure used to transfer the pRK290-based shuttle plasmid p111-Rp3.8B, which contains the *Pvu*-β gene, and p1112p3.8B/cDNA, which contains the *miniPvu*-β, into the T-DNA region of pTiRS$_2$014Ch^r involved the use of the triparental marker transfer procedure (Ruvkin and Ausubel, 1981). DNA hybridization of *HindIII*-digested Ti-plasmid DNAs, isolated from engineered *Agrobacterium* strains TiRS$_2$014phas and TiRS$_2$014phasmin, with ^{32}P-labeled phaseolin DNA showed that the phaseolin genes are present within the *HindIII* fragments, measuring about 6.1 and 5.6 kb, respectively (Figure 3). These results indicate that the *Pvu*-β or *miniPvu*-β gene regions of these shuttle vectors (p111-Rp3.8β and p111Rp3.8β/cDNA) had been correctly transferred and integrated into the avirulent T-DNA plasmid (pTiRS$_2$014Cm^r) of these *Agrobacterium* strains. The *BamHI* digestions and subsequent hybridization results show *BamHI* fragments of about 6.4 and 5.9 kb, respectively (Figure 3). These results show, as expected, that the phaseolin genes contained within these avirulent *Agrobacterium* Ti plasmids are a fusion of the 2.2-kb (3′-end) of the T-DNA *BamHI*-17 fragment (see Figure 2) with the 4.2- or 3.7-kb *Pvu*-β or *miniPvu*-β gene-containing fragments, respectively. Thus the two engineered *Agrobacterium* plasmids, pTiRS$_2$014phas and pTiRS$_2$014-phasmin, contain the expected phaseolin gene construction.

3.3.   Transfer of *MiniPvu*-β Gene into the Tobacco Genome

These mutant *Agrobacterium* strains, TiRS$_2$014phas and TiRS$_2$014phasmin, were used to transfer the *Pvu*-β and *miniPvu*-β genes into tobacco tissues using the procedures described by Chee *et al.* (1986). Tobacco shoots were grown without selection, and shoots containing the T-DNA region were identified by assaying for the production of octopine (Petit *et al.*, 1983). Gene transfers were confirmed by analysis of genomic DNAs isolated from leaf materials followed by digestion with either *HindIII* or *BamHI*. Blot hybridization results for

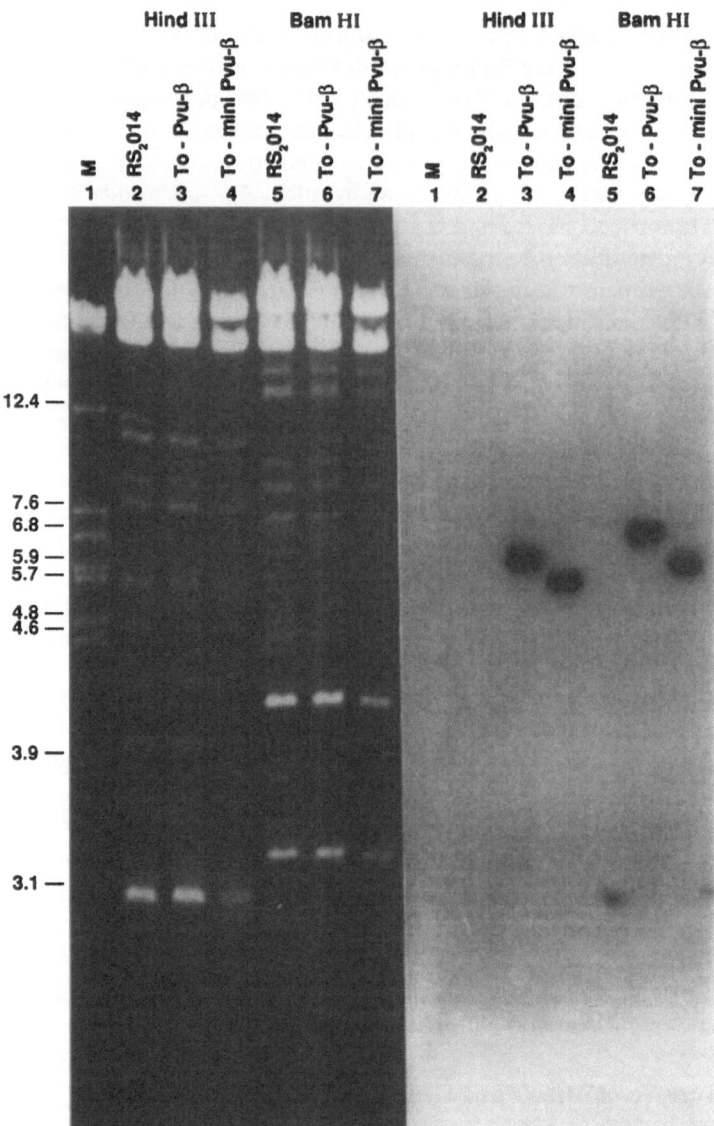

FIGURE 3. Determination of orientation of *Pvu*-β and *miniPvu*-β gene construction derivatives of pTiRS₂014. Ti DNAs were digested with either *HindIII* or *BamHI*, fractionated in a 0.7% agarose gel, and blotted onto nitrocellulose. Insertion of the phaseolin gene from the T-DNA *BamHI* fragment 17 is indicated by the presence of the hybridizing band located at about 6.1 kb (*Pvu*-β) and 5.6 kb (*miniPvu*-β) for the *HindIII* digestion and 6.4 kb (*Pvu*-β) and 5.9 kb (*miniPvu*-β) for *BamHI* digestion. Intensely staining DNA bands at approximately 12, 3.7, and 2.2 kb are from the plasmid pPH1J1 used in the triparental mating procedure (Ruvkin and Ausubel, 1981). The lane labeled M contains DNA size standards described previously by Chee *et al.* (1986).

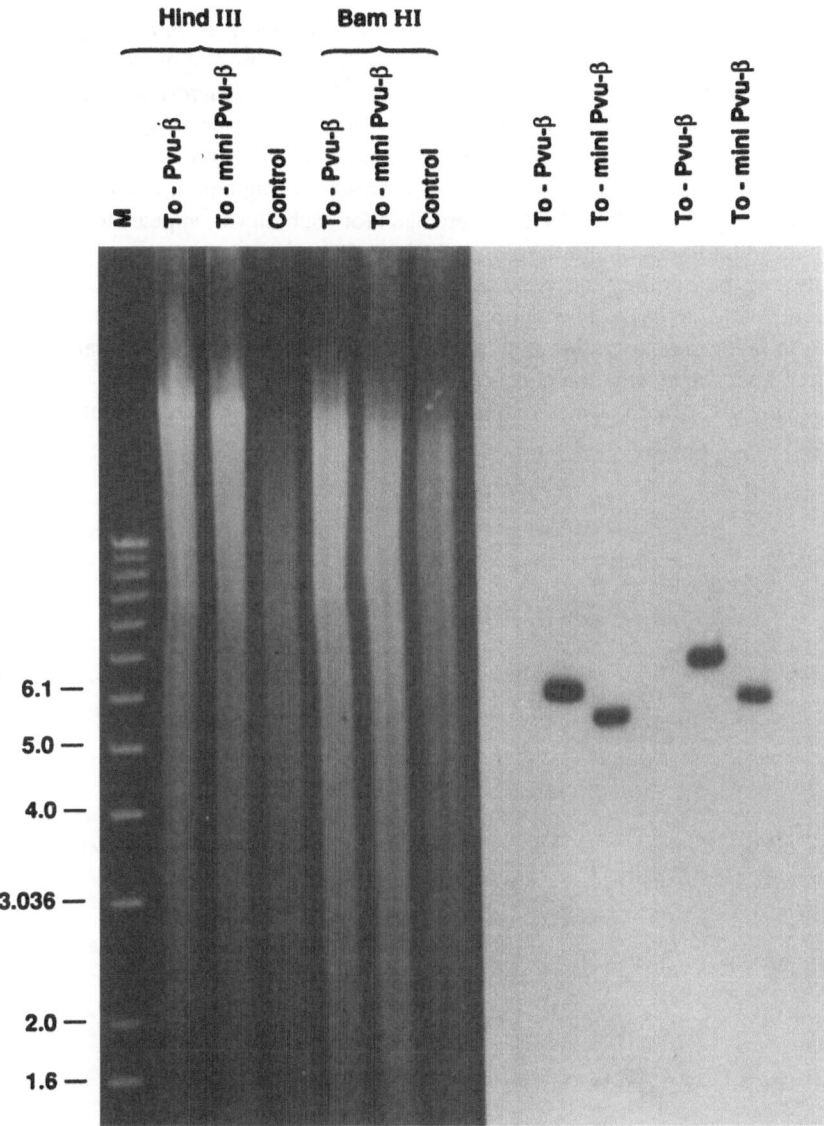

FIGURE 4. Analysis of *Pvu*-β and *miniPvu*-β genes transferred and integrated within the genome of two tobacco lines. Total tobacco DNAs were isolated from tobacco leaf tissues transformed with pTiRS$_2$014 (control lane), pTiRS$_2$014Phas (To-*Pvu*-β lanes), and pTiRS$_2$014Phasmin (To-*miniPvu*-β lanes). DNA samples (6 μg) were digested with either *HindIII* or *BamHI* followed by electrophoresis in a 0.7% agarose gel and blotted onto nitrocellulose paper. Nitrocellulose blots were hybridized against the phaseolin 2.5-kb *EcoRI* to *BamHI* fragment (Slightom *et al.*, 1983). The ethidium bromide–stained gel is shown at the left, and following hybridization against ^{32}P-labeled phaseolin probe, its radiographic results are shown at the right. The DNA size standard lane is indicated by the letter M.

two selected R_1 plant lines, which contain either the *Pvu*-β or *miniPvu*-β gene, are shown in Figure 4. Because these phaseolin hybridizing fragments (shown in Figure 4) were identical to those in the corresponding *Agrobacterium* (see Figure 3), the phaseolin gene transfers were complete and not subject to any noticeable rearrangements. The finding of one uniform hybridizing fragment also supports this interpretation. The intensity of these hybridizing fragments, combined with the fact that equal amounts of DNA were used for each digest, appears to indicate that these plant lines contain the same number of phaseolin gene copies. Approximately one phaseolin gene copy is contained within each plant line. T-DNA transfers using *cis*-type Ti vectors derived from *Agrobacterium* strain 15955 appear to favor the integration of a low number of T-DNA copies (generally only one copy) per transformation event (Murai *et al.*, 1983; Sengupta-Gopalan *et al.*, 1985). This is especially true when transformed tobacco plants are obtained using nonselective conditions.

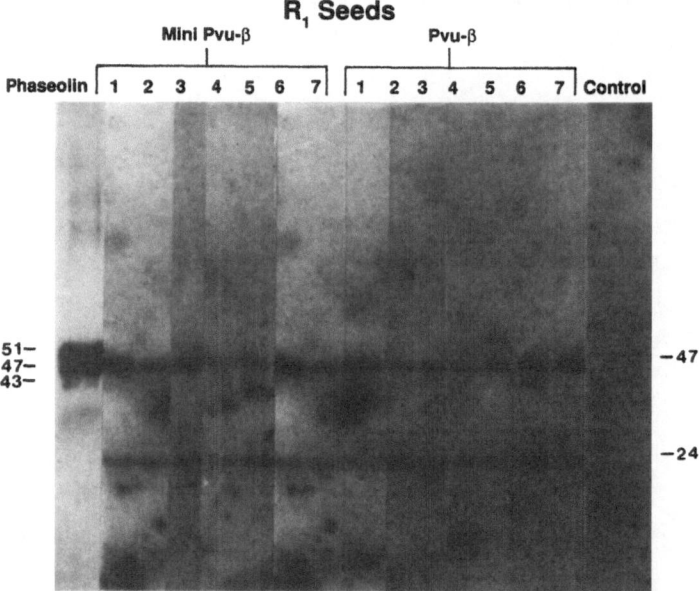

FIGURE 5. Immunological detection of phaseolin polypeptides in tobacco seeds derived from several transformed R_1 tobacco lines. Total soluble proteins were isolated from a single seed obtained from seven independent R_0 tobacco plant lines transformed with either the *Pvu*-β or *miniPvu*-β gene construction of phaseolin. The lane labeled "phaseolin" contains 25 ng of phaseolin purified from bean seeds. Lanes labeled control contain proteins extracted from seeds transformed with only the vector, pTiRS$_2$014, which did not contain either phaseolin gene construction. Treatment of filters and the identification of phaseolin polypeptides are identical to those described by Chee *et al.* (1986). Protein band sizes are in kDa.

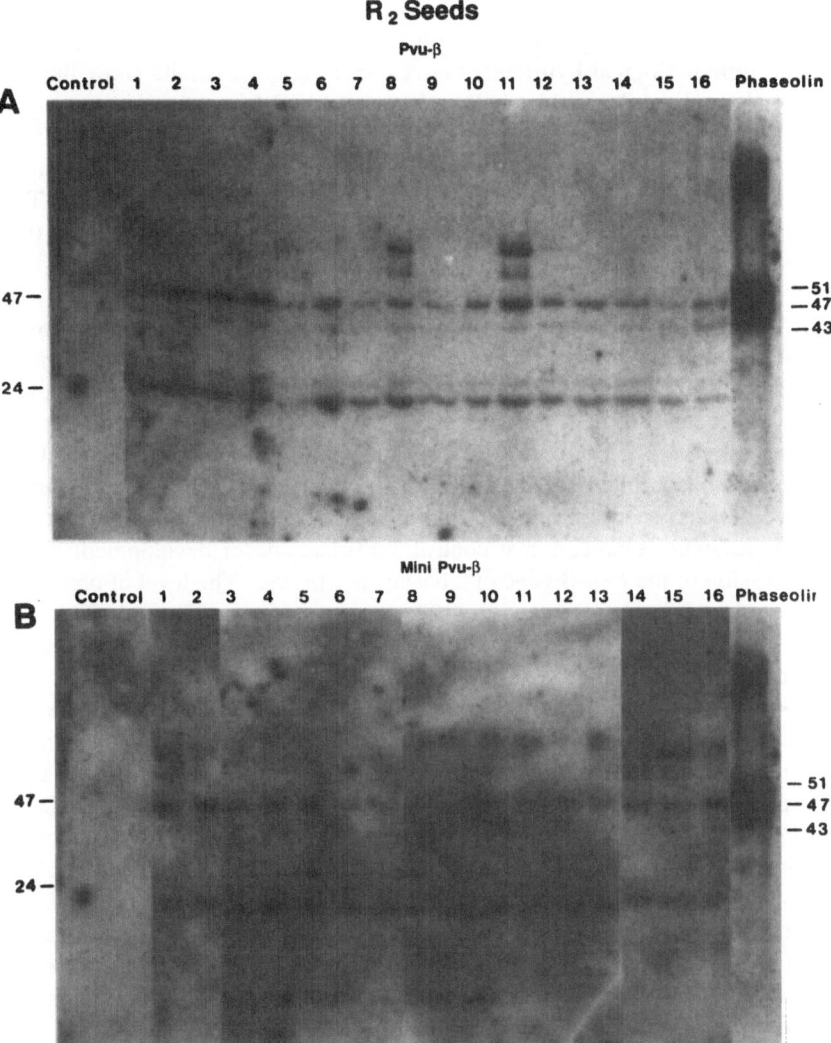

FIGURE 6. Immunological detection of phaseolin polypeptides in R_2 tobacco seeds obtained from trnasformed tobacco plants. Total soluble proteins were isolated from R_2 seeds derived from 16 independent R_1 plants. (A) Results obtained from seeds expressing the *Pvu*-β gene; (B) results from seeds expressing the *miniPvu*-β gene. The lane labeled "phaseolin" contains 25 ng of purified bean phaseolin protein and the lane labeled control contains proteins extracted from plants transformed with the vector, pTiRS$_2$014, as described in Figure 5. Protein band sizes are in kDa.

Protein extracts from transformed seeds were subjected to one-dimensional SDS–acrylamide gel electrophoresis followed by blotting onto nitrocellulose filters and immunological hybridization. Autoradiographic results show the detection of phaseolin polypeptides in R_1 tobacco seeds obtained from seven plants transformed with either the Pvu-β or $miniPvu$-β gene (Figure 5). This analysis was confirmed by a similar assay using R_2 seeds obtained from 16 R_2 plants (Figure 6), which were derived from R_1 seed lots described in Figure 5. The presence of phaseolin immunoreactive polypeptides smaller than 48 kDa (the size expected for the fully glycosylated β-type phaseolin peptides) indicates that the full-length and possibly aggregate phaseolin polypeptides are being degraded. These degradation products are similar in size (24 kDa) to the phaseolin immunoreactive band observed in sunflower and tobacco callus tissues (Murai *et al.*, 1983; Chee *et al.*, 1986), in seeds (Sengupta-gopalan *et al.*, 1985), and even in the native bean (Boylan and Sussex, 1987). These results indicate that as found with the Pvu-β gene, the $miniPvu$-β gene is also expressed in the seeds of transformed tobacco plants. Thus, because the $miniPvu$-β gene is expressed in transgenic tobacco seeds, it appears that these intron sequences do not contain nucleotide sequence elements that control the tissue and/or developmental specific expression of the Pvu-β gene in tobacco seed tissues. The level of phaseolin protein produced in the R_2 seeds of tobacco containing either the Pvu-β or $miniPvu$-β gene appears to be similar, and in both cases protein products are localized mainly in the matrix of the protein bodies (Chee *et al.*, 1991).

Based on the information derived from these experiments, the level of expression of a particular class of genes, the vicilin-type seed storage protein genes, may be independent of the presence of its native introns. In contrast, *Adh1* gene expression has shown a great dependence on the presence of introns (Callis *et al.*, 1987). The absence of the five phaseolin introns appeared not to affect the developmental regulation or the level of phaseolin protein found in mature tobacco seeds. This is an encouraging result, because the expression levels of modified and even foreign genes in the seeds of plants may not be subject to variations due to the presence or absence of specific seed storage protein gene intron sequences. Thus the design of genes to produce proteins to alter the nutritional quality of bean seeds may not need to be engineered to contain seed storage protein gene intron sequences.

4. EXPRESSION OF OTHER FOREIGN SEED STORAGE PROTEIN GENES IN TOBACCO

Changing the amino acid composition of bean seeds can be approached by making modifications directly to one of the major seed storage proteins described above; however, many amino acid codons would need to be modified to substantially alter the protein composition obtained from bean seeds. An alternative strategy is to obtain genes from other sources that encode proteins that are high in the amino acid trait(s) to be altered. Several types of genes have been suggested for increasing the content of sulfur amino acids in beans; the high sulfur-containing zein protein genes (the high sulfur-containing genes are referred to as the γ- and δ-types; Thompson and Larkins, 1989), the Brazil nut 2S albumin gene (Altenbach *et al.*, 1987), and synthetic genes (Hoffman *et al.*, 1988; Yang *et al.*, 1989). The γ- and δ-type zein genes encode polypeptides that contain about 16% and 22% sulfur, respectively (Pedersen *et al.*, 1986; Kirihara *et al.*, 1988) and the Brazil nut gene encodes a polypeptide that contains 25% sulfur (Altenbach *et al.*, 1987).

4.1. Expression of High Sulfur-Containing Seed Storage Protein Genes in Tobacco

Hoffman *et al.* (1987) used the *Pvu*-β promoter to express the zein γ-gene in tobacco seeds and found as much as 1.6% of the total seed protein was zein that was deposited in the crystalloid component of vacuolar protein bodies. This experiment was important in that it showed that a storage protein of a completely different protein class, known as the monocot water-insoluble prolamine-type protein, could be expressed and deposited in the protein bodies of dicot plants. Improvement in the targeting and transport of a monocot protein may be accomplished by using translational fusion of the *Pvu*-β amino acid signal peptide (Slightom *et al.*, 1983) connected to the zein coding region. Using a similar strategy, Altenbach *et al.* (1989) expressed the 2S Brazil nut gene in tobacco seeds using the Brazil nut signal peptide to target the protein to the tobacco seed storage protein bodies. The total seed protein isolated from transgenic tobacco seeds was found to contain about 8% Brazil nut protein, which resulted in increasing the tobacco seed methionine content by about 30%.

The expression of both the zein and Brazil nut genes in tobacco shows the feasibility of using foreign seed storage protein genes to alter seed protein amino acid compositions. The fact that these experiments found much more 2S Brazil nut protein in the tobacco seeds than the zein protein (even though the same *Pvu*-β gene promoter was used) is probably due to fact that the Brazil nut 2S protein is in the same class of water-soluble seed storage proteins normally found in the seeds of dicot plant species. Most dicot plants contain a 2S albumin fraction, but the water-insoluble prolamine protein fraction is limited to monocot species (Thompson and Larkins, 1989). Thus, even though zein protein is found in tobacco seeds, the feasibility of using it to alter the amino acid composition of seed proteins is questionable because of its insolubility and low level of expression. These problems could possibly be solved by using the phaseolin signal peptide for better targeting to seed storage protein bodies, or by the addition of amino acid residues, which would increase the water solubility of the zein polypeptides.

4.2. Expression of Synthetic Seed Storage Protein Genes

The use of synthetic genes for altering amino acid composition is an interesting concept; however, stability of the proteins in the seed environment may be a problem. Hoffman *et al.* (1988) reported on the expression of a phaseolin gene that had been modified by the insertion of 45 bp that encoded several methionine codons. This modified phaseolin gene did express in transgenic tobacco seeds, producing both mRNA and protein. However, a major portion of the protein produced was not deposited correctly and was degraded, which was the major reason why little of this modified phaseolin protein was found in the tobacco seeds. Instability of this modified phaseolin protein may have been the result of selecting a poor position for placement of the methionine-rich insert. Using X-ray crystallography data, Lawrence *et al.* (1990) suggested that the insert was positioned in an internal helix region that forms a major structural component of the phaseolin trimer. The inclusion of 15 amino acids into this structural region could have distorted the protein and resulted in its not being correctly deposited in the phaseolin trimer structure. Without the correct storage of this protein, its degradation was inevitable.

Yang *et al.* (1989) synthesized a gene that contained 80% of the essential amino acids and expressed it at a low level in potato tubers using the nopaline synthase promoter. The targeting of this type of a synthetic gene for expression in plant seeds has not yet been reported. The experiments described by Hoffman *et al.* (1988) and Yang *et al.* (1989) suggest that the use of synthetic genes is interesting, but until more information is known concerning the type of protein structures required for proper seed storage, such experiments should be approached with caution.

5. CONCLUDING REMARKS

The advent of molecular biology methods of cloning and modifying genes and gene transfer technologies has greatly enhanced our knowledge of how genes function. This ability has led to a revolution in the plant sciences, with many new genes being isolated and studied. Among these plant gene systems, the seed storage protein genes have been at the forefront of these investigations. The identification of the first seed storage protein gene structure was accomplished less than 10 years ago, and today much is known about the location of the genetic elements that control tissue-specific and developmental expression. However, much remains to be learned about how these genetic elements interact with *trans*-acting protein factors.

In addition to learning more about the regulation of seed storage proteins, more information is needed concerning their structures within seeds. A knowledge of which protein structures are important for proper targeting to seed storage protein bodies and for stable storage within the protein body is also needed. Such knowledge would greatly aid the selection of the type of foreign proteins (such as that derived from the Brazil nut gene) that could be tolerated if expressed and deposited within the seed storage protein bodies. The combination of protein structural information and gene transfer technologies for soybean, pea, and common bean cultivars will utlimately lead to development of improved cultivars of these plant species designed to satisfy the specific nutritional needs of different animal groups (poultry, fish, hogs, cattle, etc.) and humans.

ACKNOWLEDGMENTS. We thank Roger Drong for his helpful comments and help in proofreading the manuscript.

6. REFERENCES

Allen, R. D., Bernier, F., Lessard, P. A., and Beachy, R. N., 1989, Nuclear factors interact with a soybean β-conglycinin enhancer, *Plant Cell* **1:**623–631.

Altenbach, S. B., Pearson, K. W., Leung, F. W., and Sun, S. S. M., 1987, Cloning and sequence analysis of a cDNA encoding a Brazil nut protein exceptionally rich in methionine, *Plant Mol. Biol.* **8:**239–250.

Altenbach, S. B., Pearson, K. W., Meeker, G., Saraci, L. C., and Sun, S. S. M., 1989, Enhancement of the methionine content of seed proteins by the expression of a chimeric gene encoding a methionine-rich protein in transgenic plants, *Plant Mol. Biol.* **13:**513–522.

Barker, R. F., Idler, K. B., Thompson, D. V., and Kemp, J. D., 1983, Nucleotide sequence of the T-DNA region from the *Agrobacterium tumefaciens* octopine Ti plasmid pTi15955, *Plant Mol. Biol.* **2:**335–350.

Beachy, R. N., Chen, Z.-L., Horsch, R. B., Rogers, S. G., Hoffmann, N. J., and Fraley, R. T., 1985, Accumulation and assembly of soybean β-conglycinin in seeds of transformed petunia plants, *EMBO J.* **4:**3047–3053.

Blagrove, R. J., Lilley, G. G., Van Donkelaar, A., Sun, S. M., and Hall, T. C., 1984, Structural studies of a French bean storage protein: Phaseolin, *Int. J. Biol. Macromol.* **6:**137–141.

Bollini, R., Vitale, A., and Chrispeels, M. J., 1983, *In vivo* and *in vitro* processing of seed reserve protein in the endoplasmic reticulum: Evidence for two glycosylation steps, *J. Cell Biol.* **96:**999–1007.

Bown, D., Ellis, T. H. N., and Gatehouse, J. A., 1988, The sequence of a gene encoding convicilin from pea (*Pisum sativum* L.) shows that convicilin differs from vicilin by an insertion near the N-terminus, *Biochem. J.* **251:**717–726.

Boylan, M. T., and Sussex, I. M., 1987, Purification of an endopeptidase involved with storage-protein degradation in *Phaseolus vulgaris* L. cotyledons, *Planta* **170:**343–352.

Bray, E. A., Naito, S., Pan, N-S., Anderson, E., Dube, P., and Beachy, R. N., 1987, Expression of the β-subunit of β-conglycinin in seeds of transgenic plants, *Planta* **172:**364–370.

Brown, J. W. S., Bliss, F. A., and Hall, T. C., 1981, Linkage relationships between genes controlling seed proteins in French bean, *Theor. Appl. Genet.* **60:**251–258.

Bustos, M. M., Guiltinan, M. J., Jordano, J., Begum, D., Kalkan, F. A., and Hall, T. C., 1989, Regulation of β-glucuronidase expression in transgenic tobacco plants by an A/T-rich, *cis-*acting sequence found upstream of a French bean β-phaseolin gene, *Plant Cell* **1:**839–853.

Callis, J., Fromm, M., and Walbot, V., 1987, Introns increase gene expression in cultured maize cells, *Genes Dev.* **1:**1183–1200.

Chee, P. P., Klassy, R. C., and Slightom, J. L., 1986, Expression of a bean storage protein "phaseolin minigene" in foreign plant tissues, *Gene* **41:**47–57.

Chee, P. P., Jones, J. M., and Slightom, J. L., 1991, Expression of bean storage protein minigene in tobacco seeds: Introns are not required for seed specific expression, *J. Plant Physiol.* **137:**402–408.

Chen, Z-L., Schuler, M. A., and Beachy, R. N., 1986, Functional analysis of regulatory elements in a plant embryo-specific gene, *Proc. Natl. Acad. Sci. USA* **83:**8560–8564.

Derbyshire, E., Wright, D. J., and Boulter, D., 1976, Legumin and vicilin, storage proteins of legume seeds, *Phytochemistry* **15:**3–24.

Doyle, J. J., Ladin, B. F., and Beachy, R. N., 1985, Antigenic relationship of legume seed proteins to the 7S seed storage protein of soybean, *Biochem. Syst. Ecol.* **13:**123–132.

Doyle, J. J., Schuler, M. A., Godette, W. D., Zenger, V., Beachy, R. N., and Slightom, J. L., 1986, The glycosylated seed storage proteins of *Glycine max* and *Phaseolus vulgaris:* Structural homologies of genes and proteins, *J. Biol. Chem.* **261:**9228–9238.

Efstratiadis, A., Posakony, J. W., Maniatis, T., Lawn, R. M., O'Connell, C., Spritz, R. A., DeRiel, J. K., Forget, B. G., Weissman, S. M., Slightom, J. L., Blechl, A. E., Smithies, O., Baralle, F. E., Shoulders, C. C., and Proudfoot, N. J., 1980, The structure and evolution of the human β-globin gene family, *Cell* **21:**653–668.

Gibbs, P. E. M., Strongin, K. B., and McPherson, A., 1989, Evolution of legume seed storage proteins—A domain common to legumins and vicilins is duplicated in vicilins, *Mol. Biol. Evol.* **6:**614–623.

Goldberg, R. B., Hoschek, G., Ditta, G. S., and Breidenbach, R. W., 1981, Developmental regulation of cloned superabundant embryo mRNAs in soybean, *Dev. Biol.* **83:**218–231.

Greenwood, J. S., and Chrispeels, M. J., 1985, Correct targeting of the bean storage protein phaseolin in the seeds of transformed tobacco, *Plant Physiol.* **79:**65–71.

Gruss, P., 1984, Magic enhancers? *DNA* **3:**1–5.

Harada, J. J., Barker, S. J., and Goldberg, R. B., 1989, Soybean β-conglycinin genes are clustered in several DNA regions and are regulated by transcriptional and posttranscriptional processes, *Plant Cell* **1:**415–425.

Higgins, D. G., and Sharp, P. M., 1988, CLUSTAL: A package for performing multiple sequence alignment on a microcomputer, *Gene* **73:**237–244.

Higgins, T. J. V., 1984, Synthesis and regulation of the major proteins in seeds, *Annu. Rev. Plant Physiol.* **35:**191–221.

Higgins, T. J. V., Newbigin, E. J., Spencer, D., Llewellyn, D. J., and Craig, S., 1988, The sequence of a pea vicilin gene and its expression in transgenic tobacco plants, *Plant Mol. Biol.* **11:**683–695.

Hoffman, L. M., and Donaldson, D. D., 1985, Characterization of two *Phaseolus vulgaris* phytohemagglutin genes closely linked on the chromosomes, *EMBO J.* **4:**883–889.

Hoffman, L. M., Donaldson, D. D., Bookland, R., Rashka, K., and Herman, E. M., 1987, Synthesis and protein body deposition of maize 15-kd zein in transgenic tobacco seeds, *EMBO J.* **6:**3213–3221.

Hoffman, L. M., Donaldson, D. D., and Herman, E. M., 1988, A modified storage protein is synthesized, processed, and degraded in the seeds of transgenic plants, *Plant Mol. Biol.* **11:**717–729.

Kirihara, J. A., Petri, J. B., and Messing, 1988, Isolation and sequence of a gene encoding a methionine-rich 10-kDa zein protein from maize, *Gene* **71:**359–370.

Lawrence, M. C., Suzuki, E., Varghese, J. N., Davis, P. C., Van Donkelaar, A., Tulloch, P. A., and Colman, P. M., 1990, The three-dimensional structure of the seed storage protein phaseolin at 3 angstrom resolution, *EMBO J.* **9:**9–15.

Lycett, G. W., Delanuney, A. J., Gatehouse, J. A., Gilroy, J., Cory, R. R. D., and Boulter, D., 1983, The vicilin gene family of pea (*Pisum sativum* L.): A complete cDNA coding sequence for preprovicilin, *Nucleic Acids Res.* **11**:2367–2380.

Ma, Y., and Bliss, F. A., 1978, Seed proteins of common bean, *Crop Sci.* **17**:431–437.

Meinke, D. W., Chen, J., and Beachy, R. N., 1981, Expression of storage-protein genes during soybean seed development, *Planta* **153**:130–139.

Murai, N., Sutton, D. W., Murray, M. G., Slightom, J. L., Merlo, D. J., Reichert, N. A., Sengupta-Gopalan, C., Stock, C. A., Barker, R. F., Kemp, J. D., and Hall, T. C., 1983, Phaseolin gene from bean is expressed after transfer to sunflower via tumor-inducing plasmid vectors, *Science* **222**:476–482.

Messing, J., Geraghty, D., Heidecker, G., Hu, N.-T., Kridl, J., and Rubenstein, I., 1983, Plant gene structure, in *Genetic Engineering of Plants* (T. Kosugl, C. P. Meredith, and A. Hollaender, eds.), pp. 211–227, Plenum Press, New York.

Myers, R. M., Tilly, K., and Maniatis, T., 1986, Fine structure genetic analysis of a β-globin promoter, *Science* **232**:613–618.

Paaren, H. E., Slightom, J. L., Hall, T. C., Inglis, A. S., and Blagrove, R. J., 1987, Purification of a seed glycoprotein: N-terminal and deglycosylation analysis of phaseolin, *Phytochemistry* **26**:335–343.

Pedersen, K., Agros, P., Naravana, S. V. L., and Larkins, B. A., 1986, Sequence analysis and characterization of a maize gene encoding a high-sulfur zein protein of M_r 15,500, *J. Biol. Chem.* **261**:6279–6284.

Petit, A., David, C., Dahl, G. A., Ellis, J. G., Guyon, P., Casse-Delbart, F., and Tempe, J., 1983, Further extension of the opine concept: Plasmids in *Agrobacterium rhizogenes* cooperate for opine degradation, *Mol. Gen. Genet.* **190**:204–214.

Riggs, C. D., Hunt, D. C., Lin, J., and Chispeels, M. J., 1989, Utilization of luciferase fusion genes to monitor differential regulation of phytohemagglutinin and phaseolin promoters in transgenic tobacco, *Plant Sci.* **63**:47–57.

Ruvkin, G. B., and Ausubel, F. M., 1981, A general method for site-directed mutagenesis in prokaryotes, *Nature* **289**:85–88.

Sengupta-Gopalan, C., Reichert, N. A., Barker, R. F., Hall, T. C., and Kemp, J. D., 1985, Developmentally regulated expression of the bean β-phaseolin gene in tobacco seed, *Proc. Natl. Acad. Sci. USA* **82**:3320–3324.

Sleat, D. E., Hull, R., Turner, P. C., and Wilson, T. M. A., 1988, Studies on the mechanism of translational enhancement by the 5′-leader sequence of tobacco mosaic virus RNA, *Eur. J. Biochem.* **175**:75–86.

Slightom, J. L., Sun, S. M., and Hall, T. C., 1983, Complete nucleotide sequence of a French bean storage protein gene: Phaseolin, *Proc. Natl. Acad. Sci. USA* **80**:1897–1901.

Slightom, J. L., Drong, R. F., Klassy, R. C., and Hoffman, L. M., 1985, Nucleotide sequences from phaseolin cDNA clones: The major storage proteins from *Phaseolus vulgaris* are encoded by two unique gene families, *Nucleic Acids Res.* **13**:6483–6498.

Sturm, A., Van Kuik, J. A., Vliegenthart, J. F. G., and Chrispeels, M. J., 1987, Structure, position, and biosynthesis of the high mannose and complex oligosaccharide side chains of the bean storage protein phaseolin, *J. Biol. Chem.* **262**:13392–13403.

Sun, S. M., Slightom, J. L., and Hall, T. C., 1981, Intervening sequences in a plant gene-comparison of the partial sequence of cDNA and genomic DNA of French bean phaseolin, *Nature* **289**:37–41.

Talbot, D. R., Adang, M. J., Slightom, J. L., and Hall, T. C., 1984, Size and organization of a multigene family encoding phaseolin, the major seed storage protein of *Phaseolus vulgaris* L., *Mol. Gen. Genet.* **198**:42–49.

Thompson, G. A., and Larkins, B., 1989, Structural elements regulating zein gene expression, *BioEssays* **10**:108–113.

Tierney, M. L., Bray, E. A., Allen, R. D., Ma, Y., Drong, R. F., Slightom, J., and Beachy, R. N., 1987, Isolation and characterization of a genomic clone encoding the β-subunit of β-conglycinin, *Planta* **172**:356–363.

Wasylyk, B., Waslylyk, C., Augereau, P., and Chambon, P., 1983, The SV40 72 bp repeat preferentially potentiates transcription starting from proximal natural or substitute promoter elements, *Cell* **32**:503–514.

Yang, M. S., Espinoza, N. O., Nagpala, P. G., Dodds, J. H., White, F. F., Schnorr, K. L., and Jaynes, J. M., 1989, Expression of a synthetic gene for improved protein quality in transformed potato plants, *Plant Sci.* **64**:99–111.

Chapter 3

Haploid Plants from Tissue Culture
Application in Crop Improvement

Robert A. Morrison, David A. Evans, and Zhegong Fan

1. INTRODUCTION

Haploid plants are of great interest to geneticists and plant breeders. Haploids offer geneticists the opportunity to examine genes in the hemizygous condition and facilitate identification of new mutations. Plant breeders value haploids as a source of homozygosity following chromosome doubling from which efficient selection of both quantitative and qualitative traits is accomplished (Griffing, 1975). As a tool in crop improvement strategies, it is imperative that haploids are produced from individual heterozygous genotypes in sufficiently large numbers to compensate for undesirable gene combinations that result from linkage and random assortment via meiosis. Just as plant breeders advance large populations through several generations (to maximize variability in later generations), application of haploidy in plant breeding is dependent on the ability to produce a haploid population of sufficient size to accommodate selection of desired gene combinations.

Recent advances in plant tissue culture methodologies have enabled the production of haploid plants in sufficient numbers for crop improvement applica-

Robert A. Morrison, David A. Evans, and Zhegong Fan   DNA Plant Technology Corporation, Cinnaminson, New Jersey 08077.

Subcellular Biochemistry, Volume 17: Plant Genetic Engineering, edited by B. B. Biswas and J. R. Harris. Plenum Press, New York, 1991.

tions (Keller *et al.*, 1987; Morrison and Evans, 1988). Subsequent to tissue culture production, factors including chromosome doubling and tissue culture–induced variation play an integral role in the development of new crop varieties. The interaction between plant breeding requirements and plant tissue culture methods will be the focus of this chapter. A summary of recent work aimed at increasing efficiencies of haploid plant tissue cultures will precede a discussion of secondary factors, including chromosome doubling and gametoclonal variation. This will be followed by specific examples of crop improvement based on haploid plants.

2. SUMMARY OF TISSUE CULTURE METHODOLOGIES

Haploids may be produced from several approaches, including culture of anthers, microspores, ovaries, and ovules. In terms of efficiency of haploid production, anther and microspore culture have been very successful. The disadvantage of ovary/ovule culture is that the potential number of haploids generated from an ovule would be much less than the number recovered from anther/microspore culture even through the frequencies of responding male and female gametophytic cells are the same (Keller *et al.*, 1987).

Progress has been made in isolated microspore culture in rapeseed (Swanson *et al.*, 1987), corn (Pescitelli *et al.*, 1990), and rice (Datta *et al.*, 1990). In comparison to anther culture, it has been demonstrated that microspore culture can be more efficient in embryo production (Siebel and Pauls, 1989a), in regeneration of a random gametic array from the microspore population (Datta *et al.*, 1990), in *in vitro* selection of mutants (Swanson *et al.*, 1988), and in genetic transformation (Neuhaus *et al.*, 1987).

It was previously noted that the application of *in vitro* haploids in plant breeding programs was highly dependent on the capability of producing haploid plants in sufficiently high numbers. This capability is, in turn, dependent on not only the available tissue culture methodologies, but also the genotype of the plant from which gametes will be obtained for *in vitro* culture (Lazar *et al.*, 1990; Hayes and Chen, 1989). Since many factors influence *in vitro* haploid production in similar ways, we will concentrate our discussion on the current knowledge of these factors related to the efficiency of haploid production.

2.1. Anther/Microspore Culture

Depending on conditions, microspores can undergo two different developmental pathways. Under the normal condition, a microspore, formed after meiosis, undertakes the gametophytic route, in which the male gametophyte initiates a cell division to produce the vegetative and the generative cells. The generative cell undergoes the second division to produce two sperm cells that will participate in fertilization to form a new zygote. No further cell divisions occur once sperm are produced. When a microspore is induced to undertake the sporophytic pathway, multiple cell divisions precede the development of embryos or callus. Once a microspore is triggered to undertake the sporophytic pathway, it will not revert back to gametophytic development. Apparently, the fundamental step of anther/microspore culture is to induce microspores to undergo sporophytic development. The process of anther/microspore culture can be divided into three phases, induction, embryo/callus formation, and plant regeneration. Induction involves the diversion of microspores from the gametophytic pathway to sporophytic development. This is followed by either embryo or callus formation. Generally, it is difficult to separate these phases on the tissue culture level. The last phase involves regeneration of haploid plants from embryos or calli. Usually, culture medium composition and growth hormone (type and concentration) play an essential role in this aspect.

For a microspore to respond to the induction, it must be viable under *in vitro* conditions and be of a specific developmental stage during the gametophytic pathway. Studies suggest that microspores become responsive under *in vitro* conditions from later uninucleate to early binucleate stage (Morrison and Evans, 1988; Keller *et al.*, 1987). Microspores within anthers are not developmentally synchronized and may be at different cytological phases. Thus, treatments in anther culture systems may extend the viability of microspores, permitting them to develop to the specific stage along the gametophytic pathway resulting in increased efficiency.

Several factors may influence the viability and development of microspores in culture, including donor plant growth, pretreatment, and culture medium composition (Phippen and OcKendon, 1990). The importance of donor plant growth on anther/microspore culture has been well documented (Dunwell and Perry, 1973; Keller and Stringham, 1978; Foroughi-Wehr and Friedt, 1984; Simmonds, 1989). In general, vigorous growing plants provide a high proportion of responding microspores *in vitro*. We recently found that microspores isolated from plants of *Brassica napus* grown at 10°/5°C (day/night) had higher viability in the first 2 days of culture than those grown at 20°/15°C (unpublished results). Even though evidence has not been accumulated in other species, it cannot be ruled out that the physiological conditions of donor plants is highly correlated to the viability of microspores *in vitro*.

Pretreatment of flower buds and inflorescences at reduced temperature, especially in cereals, has a favorable effect on anther/microspore culture. Foroughi-Wehr and Wenzel (1989) speculated that one possible role of this treatment was the elimination of a weak or nonviable microspore such that vigorous material was enriched. This treatment also increased the number of microspores arrested at a specific developmental stage.

The importance of culture medium composition in the induction process has been demonstrated in some species. In barley and tobacco, isolated microspores require nutrient starvation for several days to induce androgenesis (Knudsen *et al.*, 1989; Sunderland and Dunwell, 1977). In *B. napus*, high concentration of sucrose in the medium is essential for success of anther/microspore culture (Keller *et al.*, 1987). Dunwell and Thurling (1985) observed that the role of high sucrose concentration in *Brassica* anther culture was to increase the viability of microspores. A similar observation was made by Finnie *et al.* (1989) among cultured anthers of barley.

The culture condition that allows tobacco (*Nicotiana tabacum*) microspores to complete the gametophytic pathway *in vitro* has been defined (Benito *et al.*, 1988). It appears the system provides a convenient way to manipulate microspore development and may lead to the induction of haploids in crop species that have been recalcitrant. In *B. napus*, microspores of the variety "topas" cultured at 25°C displayed a series of cytological events that paralleled that of gametophytic development (data unpublished). Since the mature microspores obtained *in vitro* had not been used in pollinations to test seed set, it was not possible to conclude that the pathway observed *in vitro* was identical to the *in vivo* gametophytic development. It should be noted that prolonged development of microspores has been considered to be a primary factor responsible for high frequency of embryo production via isolated microspore culture for these two species.

Little is known of the mechanisms that transform microspores from gametophytic to sporophytic development. In *B. napus*, microspores need to be placed under 32°C for at least 8 hr to induce embryo formation. The longer the high temperature treatment, the higher the numbers of embryos produced. However, longer than 4 days of the high temperature treatment reduced embryo formation. Cytologically, microspores cultured under 25°C follow the sequence of gametophytic pathway resulting in a cell with one defused vegetative nucleus defused and two condensed generative nuclei. A structure similar to a pollen tube is eventually formed, and the cell content is discharged from the tube, resulting in degeneration of microspores. Under 32°C culture condition, uninucleate microspores divide to form two equal-size nuclei, which give rise to subsequent divisions leading to embryo formation. For binucleate microspores, the vegetative cell serves as the progenitor of embryos (data unpublished). It appears that elevated temperature is the inducing factor that switches the development of microspores. A similar beneficial effect of high temperature has been reported in *Datura* (Sopory and Maheswari, 1976), wheat (Ouyang *et al.*, 1983), pepper (Dumas de Vaulx *et al.*, 1982), and eggplant (Dumas de Vaulx and Chambonnet, 1982), although cytological studies have not been conducted. Other culture treatments, such as cold shock, centrifugation, and irradiation, may also serve the same induction function.

Genotypic difference in response to anther/microspore culture have been widely observed in many species. In wheat anther culture, genetic factors controlling embryo production, plant regeneration, and albino/green ratio have been identified using Chinese spring monosomics (Agache *et al.*, 1989). Genes on 1D chromosome and 5BL chromosome arm increased the frequency of embryo formation. Lines possessing the 1BL/1RS wheat-rye translocation chromosome have shown the best androgenic response. A gene that increased albino frequency was located on the Chinese spring, wheat 5B chromosome. One way to minimize the genotypic effect may be to transfer these genes into different agronomic backgrounds to broaden the genetic basis for anther/microspore culture. It should also be understood that each genotype may have specific culture requirements, including growth condition, pretreatments, culture medium, and culture environment. Identifying the optimal conditions for each genotype will minimize genotypic effects as well. Gland *et al.* (1988) have shown that with manipulation of medium components, pH, and incubation temperature, five spring and one winter rapeseed lines produced the same amount of embryos through isolated microspore culture.

Although many studies have been conducted to increase efficiency, it appears that techniques of anther/microspore culture are still rather empirical. Treatments that ensure microspore viability and development *in vitro* should contribute to the increased efficiencies. More precise understanding of the induction and embryo/callus formation may facilitate the design of new techniques that would widen the application of haploid technology to those crop species in which anther/microspore culture is not efficient for practical breeding.

2.2. Ovary/Ovule Culture

The regeneration of haploid plants from female tissues (megaspores) has been reported for several crops, including barley, wheat, rice, tobacco, and sugar beet (Bossoutrot and Hosemans, 1985; Yang and Zhou, 1982). These reports featured the culture of unpollinated ovaries or ovules that formed haploid plants at frequencies too low for application in plant breeding programs. The disparity between efficiencies of megaspore- and microspore-based haploid systems appears to be related to induction of genes for cell division and morphogenesis. While microspores of several crops have been induced with relative ease, megaspores have remained incalcitrant.

Interspecific and intergeneric hybridization is one induction method that has proven valuable for producing megaspore-derived haploids. Crops that have benefitted from this method include wheat (Suenaga and Nakajima, 1989) and barley (Hayes and Chen, 1989; Pickering, 1983). Stimulation of the megaspore toward sporophytic morphogenesis is effected by pollination of wheat or barley plants with pollen of *Hordeum bulbosum* and of wheat plants with pollen of *Zea mays*. After fertilization, the chromosomes of the pollen parent are selectively eliminated, resulting in haploid embryos that must be rescued utilizing *in vitro* culture.

Success in producing barley haploids via interspecific hybridization with *H. bulbosum* is dependent on both barley and *bulbosum* genotypes. Pickering (1983) observed significant effects of both barley and *bulbosum* genotypes for initiation and embryo development, while only the barley genotypes affected the regeneration of plants from the initiated embryos. A similar study by Hayes and Chen (1989) demonstrated significant genotypic variation for haploid production efficiency (total of haploid plants) among 10 barley cultivars. This study also noted a significant genotype × culture technique interaction when comparing *in vitro* floret and tiller cultures. In a separate report, Chen and Hayes (1989) described the efficiency of the floret method over the culture of individual tillers. A mean of 41.6 haploid plants per 100 pollinated florets was achieved using floret culture compared to only 13.5 haploid plants per 100 pollinated florets using individual tiller culture. Chen and Hayes attributed the higher frequency to the formation of larger, differentiated embryos from the floret cultures.

It is apparent from review of recent reports that haploid plant production from both male and female gametes is highly dependent on both the genotypes of the plants and an interaction of the genotype with other factors. Since a breeding program cannot realistically concentrate on a narrow set of breeding lines, overcoming the barriers that prevent certain genotypes from responding is of utmost importance. Also, incremental increases in efficiencies among responding genotypes is of equal importance in full application of these techniques.

3. GAMETOCLONAL VARIATION

Genetic variation among plants regenerated from cultured tissues is a common observance. This was termed somaclonal variation because it was first reported from studies of plants regenerated from somatic tissues; subsequent reports have documented genetic changes among plants regenerated from cultured gametes (Morrison and Evans, 1988). The genetic basis of the variation is comprised of several sources, including elimination of residual heterozygosity and mutation. A distinct feature of gametoclonal variation (compared to somaclonal) is the direct expression of recessive mutations induced by the culture process in the regenerated plants. In plants derived from cultured microspores, the gametophytic cell type from which the plants regenerate appears to be involved in the occurrence of variation. For example, in tobacco, it is known that haploids originate from the vegetative cell (Sunderland and Dunwell, 1977), which undergoes substantial metabolic activity preceding anthesis, including limited DNA synthesis and extensive degradation and resynthesis of cytoplasmic components. DNA analysis of androgenetic plants of tobacco demonstrated that these plants had a higher DNA content than control plants (Dhillon *et al.*, 1983). Scanning microdensitometry and DNA denaturation studies further revealed that the androgenetic tobacco plants, as a population, had a higher mean percentage of condensed or heterochromatic DNA and that changes in base sequence had occurred during androgenesis and/or subsequent chromosome doubling. Because these changes occurred at a time before androgenetic mitosis was initiated, they were inherited in successive cell generations and ultimately manifested in regenerated plants.

Not all haploid plants regenerated from microspores originate from the vegetative cell. Many are derived from the generative cell, which may not undergo extensive nuclear DNA changes. For instance, Rode *et al.* (1987) detected only minor genetic variation at the DNA level among DH lines of a spring wheat cultivar. Using DNA–DNA molecular hybridization analysis of the starting cultivar and first- and second-cycle DH lines, they identified a single change in the nontranscribed spacer region of ribosomal DNA of a first-cycle DH, which remained stable after a second cycle of *in vitro* androgenesis. They did not correlate this molecular change with a recognizable phenotype.

Organelle genome modification via androgenesis is another aspect of gametoclonal variation that can affect the application of tissue culture haploids in crop improvement. While Charmet *et al.* (1985) failed to detect changes in mt DNA restriction fragment patterns of variant triticale DH lines, plastome variation in a tobacco DH (Bhaskaran *et al.*, 1983) indicated that the chloroplast genome is susceptible to genetic change via the tissue culture process. Further evidence for this premise is the preponderance of albino plants regenerated from microspore-derived callus in anther cultures of Gramineae crops (Bullock *et al.*, 1982; Ouyang *et al.*, 1983). Working with albino anther-derived plants of rice, Sun *et al.* (1979) demonstrated that plastids of these plants lacked ribosomes because of an absence of 23S and 16S ribosomal RNA. Because of this deficiency, plastome-encoded proteins were not synthesized. Using molecular techniques, Day and Ellis (1984) revealed that these albino haploids resulted from extensive deletions in their plastomes. The heterogeneity of deletions among the albino haploids (only a single common restriction fragment) explains why not all haploids exhibited the albino phenotype.

The regeneration of haploid and doubled-haploid plants from microspore-derived callus as opposed to embryos is another facet of gametoclonal variation that should be considered when employing tissue culture haploids in a breeding program. Somaclonal variation associated with the callus phase may be observed among the regenerated plants, as described by Schaeffer *et al.* (1984) and Schaeffer (1982). Segregation for yield components among third- and fourth-generation plants derived from rice anther cultures indicated the occurrence of genetic changes. The plants, spontaneous DHs regenerated from microspore-derived callus, were found to be different from colchicine-induced DHs among which no segregation was observed. Thus it appeared that variation for yield components occurred after the chromosome number had doubled spontaneously. This concept is exemplified by the work of Chen *et al.* (1983), who described independent assortment of unlinked genes in anther-derived rice plants. A single spontaneous DH was homozygous for all marker loci except the waxy locus, indicating that a mutation occurred after the chromosome number had doubled.

Two reports involving wheat DHs exemplify the occurrence of gametoclonal variation among androgenetic plants regenerated from microspore-derived callus. Marburger and Jauhar (1989) evaluated 22 DH lines obtained from anther-derived callus of the wheat cultivar "Chris." Androgenetic plants were colchicine-induced or spontaneously chromosome-doubled, resulting in a DH population that was two generations removed from the possible carryover effects of the tissue culture procedure. The majority of DHs were shorter, had fewer kernels per spike, and lower 1000 Kernel weight and yield per tiller compared to the cultivar "Chris," although only a few differences were found to be significant. Identified differences in isozyme patterns and meiotic abnormalities among DH lines that were significantly different may be the basis of the observed gametoclonal variation. In a related study, Baenziger *et al.* (1989) determined the level of gametoclonal variation among DHs of the wheat cultivars "Kitt" and "Chris." By comparing DHs, as populations, to lines derived from single seed descent and the respective starting cultivars, significant yield reduction was observed in the "Kitt" DHs, but not in the "Chris" DH population. Since the genetic component of variance for yield of the DH populations was significantly larger than the respective DH populations, they attributed the differences to gametoclonal variation induced by the tissue culture process. They stated that the lower average grain yield of the DH populations was due to a larger group of low-yielding DHs compared to SSD lines. The observed gametoclonal variation did not appear to be a hindrance to wheat breeding since expected population means, based on expected gains from selection, were similar for SSD and DH populations.

In contrast to the rice and wheat reports was work by Orton and Browers (1985) involving plants regenerated from anther cultures of broccoli. Anther-derived plants in this study were regenerated from microspore-derived embryos and not from an intervening callus. Isozyme analysis of 762 plants regenerated from anther cultures of four different hybrid cultivars resulted in only a single plant harboring each of the heterozygous codominant isozyme markers of the hybrid cultivars. Orton and Browers attributed this result to sporophytic origin of the heterozygous plant, although an alternative possibility could have been an unreduced gamete.

While the report of Orton and Browers (1985) demonstrated that in this system tissue culture–induced variation was not the cause of undesired heterozygotes, genetic variation was observed among the broccoli DHs. Observed variation was in the form of nonrandom segregation of alleles within loci resulting in a skewing in favor of a fast migrating allele at two marker loci. Because of this result, they concluded that plant populations derived from microspores via broccoli anther culture did not represent a random gametic array. This suggests that differences existed in the androgenetic capacity of one parent of the hybrid cultivars used as anther source resulting in a preponderance of alleles from the better responding genotype (parent). This androgenetic selection is analogous to gametophytic selection that occurs during pollen formation. A similar observation was made among a limited number of pepper DHs derived from cultured anthers of an interspecific hybrid (Morrison *et al.*, 1986). All but a single DH plant harbored alleles (isozyme loci) of the cultivated parent that had a higher androgenetic capacity. In work with disease-resistance genes among tobacco DHs derived from a multiple disease-resistant hybrid, Burk *et al.* (1979) found that haploids with the allele for resistance to tobacco mosaic virus occurred more frequently than expected and haploids with the allele for resistance to root knot nematodes occurred less frequently than expected.

Androgenetic selection, described in several reports, may be of benefit if the preponderant alleles are of the desired genotype. However, if alleles of the desired genotype occur at a frequency lower than expected from random assortment, then a greater number of haploid plants will be required to compensate for disparate genotypic ratios among DH populations. In addition to disparate genetic ratios, the occurrence of other random variations (nuclear and plastome mutations, deletions, amplifications), most of which will prove deleterious, warrants an efficient haploid production system capable of producing DHs in sufficient numbers to compensate for any deleterious changes that may occur.

4. CHROMOSOME DOUBLING METHODS

In order to utilize haploid plants in breeding programs, the chromosome number must be doubled to restore fertility for subsequent seed collection. For reasons stated in the previous section, it is imperative that the method by which the chromosome number is doubled be one that results in little or no variation among the resulting DHs. In some anther culture systems (Morrison *et al.*, 1986; Schaeffer *et al.*, 1984) DHs occur spontaneously and thus it is not necessary to treat haploid plants for ploidy elevation. It is necessary to minimize or eliminate the possibility of variation occurring in DH cells before they regenerate into whole plants. This can be accomplished by optimizing regeneration protocols for callus- and embryo-mediated regeneration. For systems that require a chromosome-doubling method, two techniques have been used separately or in combination to produce DHs. Both techniques have been implicated in inducing genetic variation and thus must be used with great care.

The most common method for chromosome doubling is treatment of haploid meristems with colchicine. Devaux (1989) reported efficiency in producing DHs derived from a number of barley hybrids via colchicine treatment of haploids plants. However, while the combined efficiency for all hybrids was over 70%, significant genotypic differences were observed with some hybrids producing colchicine-induced DHs at only 40%; other genotypes had an efficiency of over 90%. This substance, a mitotic poison, acts by inhibiting formation of tubulin, the major component of spindle fibers, such that chromosomes are not able to move to separate daughter cells. DNA replication occurs in line with the cell cycle resulting in increased ploidy. It does appear to have effects similar to those of other known mutagens resulting in simple and complex mutations in several crops (Francis and Jones, 1989; Downes and Marshall, 1983). This mutagen action appears to be related to timing and duration of application along with colchicine concentration. While colchicine treatment of haploid plant meristems results in haplo–diplo mosaics, DHs are easily recovered by identifying sectors with viable seed.

Regeneration of plants from haploid explants *in vitro* is another method by which doubling of chromosome number can be achieved (Hoffmann *et al.*, 1982; Burk and Matzinger, 1976; Kasperbauer and Collins, 1972). This method relies on the regeneration of plants from leaf midrid tissue which is comprised of some cells that undergo endopolyploidy related to their function. These cells, however, are susceptible to accumulating somaclonal variation which may be manifested in resulting DH plants. In order to minimize somaclonal variation, it is necessary to optimize regeneration protocols such that plants regenerate efficiently and rapidly.

A combination of the colchicine and *in vitro* methods of chromosome doubling was reported by Wan *et al.* (1989) using anther-derived maize callus. While no difference in chromosome doubling was observed between 0.025 and 0.05% colchicine levels, a significant difference was observed among three treatment durations (24, 48, 72 hr). Only 50% of plants regenerated from callus treated with colchicine for 24 hr were DH, while all regenerated plants were DH after 72 hr of treatment. It was indicated that no ploidy chimeras were regenerated since most of the DH plants produced seeds after self-pollinations. Abnormal plants among the regenerates were observed. These were not attributed to the colchicine treatment, however, since identical abnormalities (stunted growth, terminal ear, abnormal ear) existed among the plants regenerated from untreated control calli. The variation was thought to be due to the tissue culture conditions.

The conclusion from this study was that combining colchicine treatment with *in vitro* regeneration resulted in an efficient chromosome-doubling method. Concerning the total population of DH plants, it was suggested that only vigorous plantlets should be identified for introduction in a plant breeding program. This further supports the premise of the development of an efficient haploid tissue culture system capable of producing an adequate number of haploid plants. In addition, the method by which the chromosome number is doubled should be equally efficient such that the DH population is of sufficient size (after elimination of deleterious variants) to select desired gene combinations.

5. COMPARISON OF DOUBLED-HAPLOID AND CONVENTIONAL INBREEDING METHODS

From a theoretical standpoint, it seems apparent that the reduced time frame associated with the DH method would result in a more efficient breeding process compared to conventional inbreeding techniques. Indeed, Griffing (1975) postulated that when total plant numbers were restricted, as is the case in most breeding programs, the DH method provided considerable advantage over standard inbreeding/selection methods. Yonezawa et al. (1987) concurred with this theory only when (1) a small number of loci were involved, (2) desirable alleles were recessive to undesirable ones, and (3) the loci were not closely linked. In the case of quantitative traits comprising loci for which linkage disequilibrium seemed apparent, Yonezawa et al. (1987) suggested that the DH method offered less advantage and recommended production of haploids from selected F_2 plants as opposed to F_1 hybrids.

In a study comprising DH and single-seed-descent (SSD) lines derived from two barley hybrids, Choo et al. (1982) observed identical distribution between the DH and SSD lines with respect to grain yield, heading date, and plant height. They mentioned that although the SSD method provided increased opportunity for recombination (based on multiple rounds of gametogenesis compared to a single event for DHs), it did not produce a sample of recombinants that differed significantly from the DH sample. Contrary to this result were the findings of Powell et al. (1986) working with spring barley lines produced by SSD, pedigree inbreeding, and doubled-haploidy. While they found each method capable of producing recombinant inbred populations that could outperform the higher scoring parent, significant differences were observed between the SSD and DH lines for thousand-grain weight. In agreement with the results of Yonezawa et al. (1987), they attributed this finding to coupling linkages (disequilibrium) involving epistatic genes.

Subsequent to the report by Powell et al. (1986) was a similar study by Caligari et al. (1987) comprising inbred lines derived from five spring barley hybrids utilizing DH and SSD methods. Similar to Powell et al. (1986), they observed inbreds in both samples that exceeded the higher scoring parent for characters of agronomic importance. In fact, transgressive segregation among genes for agronomic performance was evident since several inbreds from each sample surpassed F_1 hybrid performance even in cases of heterosis. They also

observed significant differences between the DH and SSD sample means for
thousand-grain weight and plant height again indicating a greater proportion of
recombinants among the SSD sample. However, they observed that among several characteristics, the proportion of lines exceeding the higher-scoring parent in
the DH sample was significantly greater than in the SSD sample. Thus, although
the SSD method produced a greater number of recombinants, the DH method
produced a greater number of "desirable" recombinants, which they attributed to
coupling linkages among preferred traits.

In a study that compared wheat DHs to inbreds selected from the pedigree
system, Winzeler *et al.* (1987) found no significant differences between the two
samples and suggested that DHs be produced from F_1 hybrids. In fact, they
found the DH sample yielded a more precise disease assessment and had higher
protein levels compared to the lines derived from inbreeding.

6. TRANSFER OF SPECIFIC TRAITS VIA HYBRID SORTING

The goal of genetic engineering is to manipulate or transfer specific traits
within a variety without compromising desirable traits of the cultivar. Although
hybrid sorting is less precise compared to recombinant methods, hybrid sorting
affords the capability of gene transfer similar to conventional inbreeding but in a
shorter time frame. Examples of gene transfer via hybrid sorting are listed in
Table I.

Anther-derived barley plants were produced from cultured anthers of
hybrids between a barley yellow mosaic virus (BaYMV)-resistant six-rowed
cultivar and several susceptible two-rowed varieties (Foroughi-Wehr and Friedt,

Table I
Examples of Gene Transfer via Hybrid Sorting

Crop	DH method	Genes/trait transferred	Reference
Barley	Anther culture	Virus resistance	Foroughi-Wehr and Friedt, 1984
Barley	Interspecific hybridization	Mildew resistance	Powell *et al.*, 1984
Wheat	Anther culture	Multiple disease resistance	Parisi and Picard, 1986
Tobacco	Anther culture	Alkaloid concentration	Ostrem *et al.*, 1986
Canola	Microspore culture	Glucosinolate levels	Siebel and Pauls, 1989b
Rice	Anther culture	Isozyme markers	Guiderdoni *et al.*, 1989
Perennial ryegrass	Anther culture	Isozyme markers	Hayward *et al.*, 1990

1984). A total of 132 androgenetic plants were produced and subsequently screened for resistance to BaYMV resistance. Sixty percent of the DHs were resistant while 23% combined the resistance with the desired two-rowed charac-teristics. Using the interspecific hybridization method, Powell *et al.* (1984) produced a set of haploid barley plants from a hybrid between mildew-resistant and mildew-sensitive cultivars. By screening for resistance at the haploid stage, they were able to identify recombinants harboring the resistance gene and thus discard susceptible genotypes prior to chromosome doubling.

Success in transferring genetic traits via hybrid sorting was not limited to barley. Ostrem *et al.* (1986) successfully manipulated genes for alkaloid con-centration by producing plants from cultured anthers of hybrids between low- and high-alkaloid breeding lines. Chemical analysis of haploid plants revealed a distinct segregation between low-, intermediate-, and high-alkaloid genotypes. Subsequent chromosome doubling of plants with the desired alkaloid level re-sulted in no discrepancies between ploidy levels for alkaloid concentration.

The improvement of rapeseed and oilseed rape cultivars has been hindered by the relatively high levels of toxic glucosinolates in many varieties. Since breeding for low-glucosinolate rapeseed varieties is a major priority, Siebel and Pauls (1989b) examined the prospects of hybrid sorting via microspore culture to decrease glucosinolate levels in some lines or maintain low glucosinolate levels in other varieties. Hybrids from low-by-high and low-by-low parent crosses were used as microspore sources. Since the seed was the desired product for analysis of glucosinolate levels, resulting microspore-derived plants were screened for spontaneous diploids prior to chemical analysis. The results of glucosinolate analysis contradicted other studies which indicated that gametoclonal variation hindered identification of low-glucosinolate DHs. Although additive effects of the dominant high-glucosinolate alleles resulted in a skewing in favor of high glucosinolate genotypes among the DHs from the high-by-low glucosinolate hybrids, sufficient numbers of low-glucosinolate DHs were detected for selection of improved lines. In the sample derived from low-by-low glucosinolate hybrids, only low-glucosinolate genotypes were observed.

7. DEVELOPMENT OF CROP VARIETIES VIA *IN VITRO* HAPLOID METHODS

Despite the problems associated with the *in vitro* methods and resulting plants, a number of crop varieties have been developed through the production of haploid plants (Table II). The most comprehensive effort in application of *in vitro* haploids in varietal development programs occurred in China (Hu and Zeng, 1984). Using anther culture methods, several varieties of rice were produced in an abbreviated frame of time compared to conventional inbreeding methods.

Table II
Crop Varieties Developed via *In vitro* Haploids

Variety name	Crop	Method	Reference
Hua Yu No. 1,2	Rice	Anther culture	Hu and Zeng, 1984
Xin Xiu	Rice	Anther culture	Hu and Zeng, 1984
Tang Huo No. 2	Rice	Anther culture	Hu and Zeng, 1984
Jingdan 2288	Wheat	Anther culture	Hu and Zeng, 1984
Danyu No. 1	Tobacco	Anther culture	Hu and Zeng, 1984
Qun Hua	Maize	Pollen	Wu *et al.*, 1980
Shan Hua 7706	Rice	Anther culture	Hu and Zeng, 1984
Hua Han Zao 77001	Rice	Anther culture	Zhang, 1980
Rodeo	Barley	Bulbosum	Campbell *et al.*, 1984
NC744	Tobacco	Anther culture	Chaplin *et al.*, 1980
Mingo	Barley	Bulbosum	Ho and Jones, 1980
Jinhua No. 1	Wheat	Anther culture	Hu *et al.*, 1983
F211	Tobacco	Anther culture	Nakamura *et al.*, 1975
Florin	Wheat	Anther culture	de Buyser *et al.*, 1987
Gwuylan	Barley	Anther culture	Baenziger *et al.*, 1984
Bell Sweet	Pepper	Anther culture	Anonymous, 1988

Rice varieties Hua Yu No. 1 and 2 were released from a breeding program that combined methods of bulk, pedigree selection, and anther culture. The resulting varieties combined high yield, resistance to bacterial blight, and wide adaptability. It was projected that cultivation of these varieties will reach several thousand hectares within a single region in China. By far the most popular anther-derived rice variety, Xin Xiu, was selected for high yield; over 100,000 hectares of the variety have been cultivated in Eastern China (Hu and Zeng, 1984).

Hu and Zeng (1984) also reported the development of a wheat variety, Jingdan 2288, using anther culture. The variety combined characteristics derived from two different parents, including large spikes, high kernel number, vigorous tillering, and resistance to short stem, lodging, stripe rust, and powdery mildew. A detailed account of the development of the anther-derived wheat variety "Florin" was provided by de Buyser *et al.* (1987). A total of 41 DH lines were obtained from cultured anthers of the hybrid cross "Wizard" × "Iena." From this population 18 DH lines were selected on the basis of morphological characteristics, reaction to various diseases, as well as resistance to lodging. Two DH lines were advanced for upscale testing, one a colchicine-induced DH ("Florin") the other a spontaneously doubled DH. In 2 successive years the DH "Florin" was superior to four of the most widely grown wheat varieties in France.

The maize breeding line "Qun Hua" developed by Wu *et al.* (1980) is an example of the utilization of *in vitro* haploids in generating improved hybrid varieties. Superior combining ability of this anther-derived DH line was apparent during evaluation of 62 crosses using "Qun Hua" as a common parent. Ninety percent of the crosses had yields higher than those of the parents, and several crosses combined the high-yield characteristic with disease resistance.

The production of tobacco haploid plants from anthers cultured *in vitro* resulted in the release of an elite breeding line. Chaplin *et al.* (1980) released the line NC744, which combined the multiple disease-resistance traits of "Coker 86" with Potato Virus Y (PVY) resistance from "VY32" (Coker 86 lacked resistance to PVY). Using a hybrid-sorting strategy, haploid plants were produced from the hybrid between "Coker 86" and "VY32." The resulting DH line, released as NC744, harbored resistance to PVY (from VY32), blankshank, and bacterial wilt (from Coker 86) along with other attributes necessary in a flue-cured variety, including high yield and acceptable chemical characteristics. In addition to PVY resistance, the line also exhibited resistance to tobacco etch and tobacco vein mottle viruses.

8. FUTURE PROSPECTS

Previous reviews touted the "promise" of *in vitro*–derived haploids in crop improvement strategies. The present review clearly demonstrates that *in vitro* haploid methods are here and now. This statement is exemplified by the list of varieties and breeding lines derived from DHs that have been released thus far (Table II). In addition, Devaux (1989) indicated that DH production is widely used in many barley breeding programs. Thus it can be inferred that haploid breeding will continue to become commonplace in an ever-increasing number of plant breeding programs.

In addition to its application in producing inbreds from hybrid combinations, *in vitro* haploids offer promise in novel crop improvement strategies. One such strategy is haploid (microspore) mutagenesis aimed at rapid generation of inbred mutant lines (Swanson *et al.*, 1989). They utilized an *in vitro* canola microspore system coupled with chemical mutagenesis (ethyl nitrosourea) followed by *in vitro* selection with an imidazolinone derivative. They succeeded in producing five fertile DH imidazolinone-tolerant canola lines of the variety Topas which resisted field-recommended levels of the commercial herbicides Assert and Pursuit. Two of the mutants were classified as semidominant and unlinked with each other. When the two mutations were combined in the form of an F_1 hybrid, the hybrid proved superior in imidazolinone tolerance to either of the homozygous mutants alone.

The transfer of novel genes via haploid tissue culture systems is another crop improvement strategy for future application. An advantage of haploid over diploid tissue for transformation is immediate fixation of the transferred gene in resulting transgenic plants. Also, by potential reduction in copy number within transformed haploid tissue, more efficient selection for transformed traits may be attained. This concept was put forth by Swanson and Erickson (1989) in their transformation work with haploid Canola tissue. Microspore-derived embryos were recovered after transformation with a disarmed octopine-producing strain of Agrobacterium tumefaciens harboring a binary Ti-plasmid vector. The plasmid conferred resistance to kanamycin and hygromycin. Several embryos exhibited green sectors in the presence of 50 mg/liter kanamycin. These embryos were subsequently regenerated into plants and their chromosome numbers doubled. Southern hybridization analysis of DH progeny revealed two plants as having two insertion sites each while a third plant had only a single insertion site. Bioassays using microspores of the transgenic plants grown in the presence of kanamycin demonstrated that the DH with a single copy of the foreign gene tolerated kanamycin at 20 times the lethal level for transgenic DHs with two copies of the gene.

9. REFERENCES

Agache, S., Bachelier, B., deBuyser, J., Henry, Y., and Snape, J., 1989, Genetic analysis of anther culture response in wheat using aneuploid, chromosome substitution and translocation lines, *Theor. Appl. Genet.* **77**:7–11.

Anonymous, 1988, Bellsweet pepper, *Seed Industry*, May:28.

Baenziger, P. S., Kudirka, D. T., Schaeffer, G. W., and Lazar, M. D., 1985, The significance of doubled-haploid variation, in *Gene Manipulation in Plant Improvement* (J. P. Gustafson, ed.), pp. 385–414, Plenum Press, New York.

Baenziger, P. S., Wesenberg, S. D., Smail., V. M., Alexander, W. L., and Schaeffer, G. W., 1989, Agronomic performance of wheat doubled-haploid lines derived cultivars by anther culture, *Plant Breeding* **103**:101–109.

Benito, M. R. M., Macke, F., Alwen, A., and Heberle-Bors, E., 1988, *In-situ* seed production after pollination with *in vitro*–matured, isolated pollen, *Planta* **176**:145–148.

Bhaskaran, S., Smith, R. H., and Finer, J. J., 1983, Ribulose bisphosphate carboxylase activity in anther-derived plants of *Saintpaulia ionatha* Wendl. Shag. *Plant Physiol.* **73**:639–642.

Bossoutrot, D., and Hosemans, D., 1985, Gynogenesis in *Beta vulgaris* L.: From *in vitro* culture of unpollinated ovules to the production of doubled-haploid plants in soil, *Plant Cell Rep.* **4**:300–303.

Bullock, W. P., Baenziger, P. S., Schaeffer, G. W., and Bottino, P. J., 1982, Anther culture of wheat (*Triticum aestivum*) F₁s and their reciprocal crosses, *Theor. Appl. Genet* **62**:155–159.

Burk, L. G., and Matzinger, 1976, Variation among anther-derived double-haploids from an inbred line of tobacco, *J. Hered* **67**:381–384.

Burk, L. G., Chaplin, J. F., Gooding, G. V., and Powell; N. T., 1979, Quantity production of anther-derived haploids from a multiple disease resistant tobacco hybrid. I. Frequency of plants with resistance or susceptibility to tobacco mosaic virus (TMV), potato virus Y (PVY), and root knot (RK), *Euphytica* **28**:201–208.

Caligari, P. D. S., Powell, W., and Jinks, J. L., 1987, A comparison of inbred lines derived by doubled haploidy and single seed descent in spring barley (*Hordeum vulgare*), *Ann. Appl. Biol.* **111**:667–675.

Campbell, K. W., Blawn, R. I., and Ho, K. M., 1984, Rodeo barley, *Canadian Journal of Plant Science* **64**:203–205.

Chaplin, J. F., and Burk, L. G., 1984, Registration of LMAFC 34 tobacco germplasm, *Crop Sci.* **24**:1220.

Chaplin, J. F., Burk, L. G., Gooding, G. V., and Powell, N. T., 1980, Registration of NC744 tobacco germplasm, *Crop Sci.* **20**:677.

Charmet, G., Vede., F., Bernard, M., Bernard, S., and Mathieu, C., 1985, Cytoplasmic variability in androgenetic doubled-haploid lines of triticale, *Agronomie* **5**:709–717.

Chen, C. C., Chu, W. L., Yu, L. J., Ren, S. S., and Yu, W. J., 1983, Genetic analysis of anther-derived plants of rice: Independent assortment of unlinked genes, *Can. J. Genet. Cytol.* **25**:324–328.

Chen, F. Q., and Hayes, P. M., 1989, A comparison of *Hordeum bulbosum*–mediated haploid production efficiency in barley using *in vitro* floret and tiller culture, *Theor. Appl. Genet.* **77**:701–704.

Choo, T. M., Reinbergs, E., and Park, S. J., 1982, Comparison of frequency distributions of doubled-haploid and single seed descent lines in barley, *Theor. Appl. Genet.* **61**:215–218.

Datta, S. K., Datta, K., and Potrykus, I., 1990, Embryogenesis and plant regeneration from microspores of both "Indica" and "Japonica" rice (*Oryza sativa*), *Plant Sci.* **67**:83–88.

Day, A., and Ellis, T. H. N., 1984, Chloroplast DNA deletions associated with wheat plants regenerated from pollen: Possible basis for maternal inheritance of chloroplasts, *Cell* **39**:359–368.

de Buyser, J., Henry, Y., Lonnet, P., Hertzog, R., and Hespel, A., 1987, "Florin": A doubled-haploid wheat variety developed by the anther culture method, *Plant Breeding* **98**:53–56.

Devaux, P., 1989, Variations in the proportions of fertile colchicine-treated haploid plants derived from winter barley plants, *Plant Breeding* **103**:247–250.

Dhillon, S. S., Wernsman, E. A., and Miksche, J. P., 1983, Evaluation of nuclear DNA content and heterochromatin changes in anther-derived dihaploids in tobacco (*Nicotiana tabacum* cv Coker 139), *Can. J. Genet. Cytol.* **25**:169–173.

Downes, R. W., and Marshall, D. R., 1983, Colchicine-induced variants in sunflower, *Euphytica* **32**:757–766.

Dumas de Vaulx, R., and Chambonnet, D., 1982, Culture in vitro d'antheres d'aubergine (*Solanum melongena* L.): Stimulation de la production de plantes au moyen detraitements a +35 C associes a de faibles teneurs ensubstances de croissance, *Agronomie* **2**:983–988.

Dumas de Vaulx, R., Chambonnett, D., and Sibi, M., 1982, Stimulation of *in vitro* androgenesis in pepper (Capsicum annuum) by elevated temperature treatments, in *Variability in Plants Regenerated from Tissue Culture* (E. Earle, and Y. Demarly, eds.), pp. 92–98, Praeger, New York.

Dunwell, J. M., and Perry, E., 1973, The influence of *in vivo* growth conditions of *N. tabacum* plants on the *in vitro* embryogenetic potential of their anthers, *John Innes Inst. Rep.* **64**:69–70.

Dunwell, J. M., and Thurling, N., 1985, Role of sucrose in microspore embryo production in *Brassica napus* ssp. *oleifera*, *J. Exp. Bot.* **36**:1478–1491.

Finnie, S. J., Powell, W., and Dyer, A. F., 1989, The effect of carbohydrate composition and concentration on anther culture response in barley (*Hordeum vulgare* L.), *Plant Breeding* **103**:110–118.

Foroughi-Wehr, B., and Friedt, W., 1984, Rapid production of recombinant barley yellow-mosaic virus resistant *Hordeum vulgare* lines by anther culture, *Theor. Appl. Genet.* **67**:377–382.

Foroughi-Wehr, B., and Wenzel, G., 1989, Androgenetic haploid production, *IAPTC Newslett.* **58**:11–18.

Francis, A., and Jones, R. N., 1989, Heritable nature of colchicine induced variation in diploid *Lolium perenne*, *Heredity* **62**:407–410.

Gland, A., Lichter, R., and Schweiger, H. G., 1988, Genetic and exogenous factors affection embryogenesis in isolated microspore cultures of *Brassica napus* L., *J. Plant Physiol.* **132**:613–617.

Griffing, B., 1975, Efficiency changes due to use of doubled-haploids in recurrent selection methods, *Theor. Appl. Genet.* **46**:367–386.

Guiderdoni, E., Glaszmann, J. C., and Courtois, B., 1989, Segregation of 12 isozyme genes among double haploid lines derived from a *japonica* × *indica* cross of rice (*Oryza sativa* L.), *Euphytica* **42**:45–53.

Hayes, P. M., and Chen, F. Q., 1989, Genotypic variation for *Hordeum bulbosum* mediated haploid production in winter and facultative barley, *Crop Sci.* **29**:1184–1188.

Hayward, M. D., Olesen, A., Due, I. K., Jenkins, R., and Morris, P., 1990, Segregation of Isozyme marker loci amongst androgenetic plants of *Lolium perenne* L., *Plant Breeding* **104**:68–71.

Ho, K. M., and Jones, G. E., 1980, Mingo barley, *Canadian Journal of Plant Science* **60**:279–280.

Hoffmann, F., Thomas, E., and Wenzel, G., 1982, Anther culture as a breeding tool in rape. II. Progeny analyses of androgenetic lines and induced mutants from haploid cultures, *Theor. Appl. Genet.* **61**:225–232.

Hu, H., and Zeng, J. Z., 1984, development of new varieties via anther culture, in *Handbook of Plant Cell Culture*, Vol. 3 (P. V. Ammirato, D. A. Evans, W. R. Sharp, and Y. Yamada, eds.), pp. 65–90, Macmillan, New York.

Hu, D., Tang, Y., Yuan, Z., and Wong, J., 1983, The induction of pollen sporophyte of winter wheat and the development of the new variety Jinghaul, *Scientia Agr. Sinica* **1**:29–35.

Kasperbauer, M. J., and Collins, G. B., 1972, Reconstitution of diploids from leaf tissue of anther-derived haploids in tobacco, *Crop Sci.* **12**:98–101.

Keller, W. A., and Stringham, G. R., 1978, Production and utilization of microspore-derived haploid plants, in *Frontiers of Plant Tissue Culture* (T. Thorpe, ed.), pp. 113–122, University of Calgary Press, Calgary, Alta., Canada.

Keller, W. A., Arnison, P. G., and Cardy, B. J., 1987, Haploids from gametophytic cells-recent developments and future prospects, in *Plant Tissue and Cell Culture* (C. E. Green, D. A. Somers, W. P. Hackett, and D. D. Biesboer, eds.), pp. 223–241, Alan R. Liss, New York.

Knudsen, S., Due, I. K., and Andersen, S. B., 1989, Components of response in barley anther culture, *Plant Breeding* **103**:241–246.

Lazar, M. D., Schaeffer, G. W., and Baenziger, P. S., 1990, The effects of interactions of culture environment with genotype on wheat (*Triticum aestivum*) anther culture response, *Plant Cell Rep.* **8**:525–529.

Marburger, J. E., and Jauhar, P. P., 1989, Agronomic, isozyme, and cytogenetic characteristics of "Chris" wheat doubled-haploids, *Plant Breeding* **103**:73–80.

Morrison, R. A., and Evans, D. A., 1988, Haploid plants from tissue culture: New plant varieties in a shortened time frame, *Biotechnology* **6**:684–690.

Morrison, R. A., Koning, R. E., and Evans, D. A., 1986, Anther culture of an interspecific hybrid of *Capsicum*, *J. Plant Physiol.* **126**:1–9.

Nakamura, A., Yamada, T., Oka, M., Tatemichi, Y., Egushi, K., Ayabe, T., and Kobayashi, K., 1975, Studies on the haploid method of breeding by anther culture in Tobacco. V. Breeding of mild flue-cured variety F211 by haploid method, *Bulletin Iwata Tobacco Experiment Station* **7**:29–39.

Neuhaus, G., Spangenberg, G., Scheid, O. M., and Schweiger, H. G., 1987, Transgenic rapeseed plants obtained by the microinjection of DNA into microspore-derived embryoids, *Theor. Appl. Genet.* **75**:30–36.

Orton, T. J., and Browers, M. A., 1985, Segregation of genetic markers among plants regenerated from cultured anthers of broccoli (*Brassica oleracea* var. "*italica*"), *Theor. Appl. Genet.* **69**:637–643.

Ostrem, J. A., Litton, C. C., and Collins, G. B., 1986, Isolation of dark tobacco breeding lines differing in alkaloid concentration via anther-derived haploids, *Plant Breeding* **96**:224–231.

Ouyang, J. W., Zhou, S. M., and Jia, S. E., 1983, The response of anther culture to culture temperature in *Triticum aestivum, Theor. Appl. Genet.* **66:**101–109.

Parisi, L., and Picard, E., 1986, Disease response of doubled-haploid lines and their original cultivars in wheat (*Triticum aestivum* L.), *Zeitschrift Pflanzenzüchtg* **96:**63–78.

Pescitelli, S. M., Johnson, C. D., and Petolino, J. F., 1990, Isolated microspore culture of maize: Effects of isolation technique, reduced temperature, and sucrose level, *Plant Cell Rep.* **8:**628–631.

Phippen, C., and Ockendon, D. J., 1990, Genotype, plant, bud size and media factor affecting anther culture of cauliflowers (*Brassica oleracea* var. *botrytis*), *Theor. Appl. Genet.* **79:**33–38.

Pickering, R. A., 1983, The influence of genotype on doubled-haploid barley production, *Euphytica* **32:**863–876.

Powell, W., Asher, M. J. C., Wood, W., and Hayter, A. M., 1984, The manipulation of mildew resistance genes in a barley breeding programme by the use of doubled haploids, *Plant Breeding* **93:**43–48.

Powell, W., Caligari, P. D. S., and Thomas, W. T. B., 1986, Comparison of spring barley lines produced by single seed descent, pedigree inbreeding, and doubled-haploidy, *Plant Breeding* **97:**138–146.

Rode, A., Hartman, C., Benslimane, A., Picard, E., and Quetier, F., 1987, Gametoclonal variation detected in the nuclear ribosomal DNA from doubled-haploid lines of a spring wheat (*Triticum aestivum* L., cv. "Cesar"), *Theor. Appl. Genet.* **74:**31–37.

Schaeffer, G. W., 1982, Recovery of heritable variability in anther-derived doubled-haploid rice, *Crop Sci.* **22:**1160–1164.

Schaeffer, G. W., Sharpe, R. T., and Cregan, P. B., 1984, Variation for improved protein and yield from rice anther culture, *Theor. Appl. Genet.* **67:**383–389.

Siebel, J., and Pauls, K. P., 1989a, A comparison of anther and microspore culture as a breeding tool in *Brassica napus, Theor. Appl. Genet.* **78:**473–479.

Siebel, J., and Pauls, K. P., 1989b, Alkenyl glucosinolate levels in androgenic populations of *Brassica napus, Plant Breeding* **103:**124–132.

Simmonds, J., 1989, Improved androgenesis of winter cultivars of *Triticum aestivum* L. in response to low temperature treatment of donor plants, *Plant Sci.* **65:**225–231.

Sopory, S. K., and Maheswari, S. C., 1976, Development of pollen embryoids in anther cultures of *Datura innoxia*. I. General observation and effects of physical factors, *J. Exp. Bot.* **27:**49–57.

Suenaga, K., and Nakajima, K., 1989, Efficient production of haploid wheat (*Triticum aestivum*) through crosses between Japanese wheat and maize (*Zea mays*), *Plant Cell Rep.* **8:**263–266.

Sun, C. S., Wu, S. C., Wang, C. C., and Chu, C. C., 1979, The deficiency of soluble proteins and plastid ribosomal RNA in the albino pollen plantlets of rice, *Theor. Appl. Genet.* **55:**193–197.

Sunderland, N., and Dunwell, J. M., 1977, Anther and pollen culture, in *Plant Tissue and Cell Culture* (H. E. Street, ed.), pp. 223–265, Blackwell, Oxford.

Swanson, E. B., and Erickson, L. R., 1989, Haploid transformation in *Brassica napus* using an octopine-producing strain of *Agrobacterium tumefaciens, Theor. Appl. Genet.* **78:**831–835.

Swanson, E. B., Coumans, M. P., Wu, C. H., Barsby, T. L., and Beversdorf, W. D., 1987, Efficient isolation of microspores and production of microspore-derived embryos from *Brassica napus, Plant Cell Rep.* **6:**94–97.

Swanson, E. B., Coumans, M. P., Brown, G. L., Patel, J. D , and Beversdorf, W. D., 1988, The characterization of herbicide tolerant plants in *Brassica napus* L. after *in vitro* selection of microspores and protoplasts, *Plant Cell Rep.* **7:**83–87.

Swanson, E. B., Herrgsell, M. J., Arnoldo, M., Sippell, D. W., and Wong, R. S. C., 1989, Microspore mutagenesis and selection: Canola plants with field tolerance of the imidazolinones, *Theor. Appl. Genet.* **78:**525–530.

Wan, T., Petolino, J. F., and Widholm, J. M., 1989, Efficient production of doubled-haploid plants through colchicine treatment of anther-derived maize callus, *Theor. Appl. Genet.* **77:**889–892.

Winzeler, H., Schmid, J., and Fried, P. M., 1987, Field performance of androgenetic doubled-haploid spring wheat lines in comparison with lines selected by the pedigree system, *Plant Breeding* **99**:41–48.

Wu, J. L., Zhong, Q. L., Nong, F. H., and Chang, T. M., 1980, Embryogenesis in corn culture, *Acta Physiology* **6**:221–224.

Yang, H. Y., and Zhou, C., 1982, *In vitro* induction of haploid plants from unpollinated ovaries and ovules, *Theor. Appl. Genet.* **63**:97–104.

Yonezawa, K., Nomura, T., and Sasaki, Y., 1987, Conditions favoring doubled-haploid breeding over conventional breeding of self-fertilizing crops, *Euphytica* **36**:441–453.

Zhang, Z. H., 1980, Application of anther culture technique to rice breeding, Symposium Celebrating the 20th Anniversery of the International Rice Research Institute, Los Baños, Philippines.

Chapter 4

Plant Transposable Elements

Nihal K. Notani

1. INTRODUCTION

Transposable elements were first detected in maize by Barbara McClintock and reported in the 1950s. These were revealed by experiments that were designed for a cytogenetic study involving the short arm of chromosome 9. It was in the progeny of plants undergoing the chromosomal type of breakage–fusion–bridge cycle that a burst of somatic instability and mutations appeared. Initially, Mc-Clintock focused on the breakage events occuring at a specific locus between *waxy (wx)* gene and the centromere on the short arm of chromosome 9. She termed this "the standard locus of *Ds (Dissociation)*." However, these breakage events required the presence of another autonomous element, which she termed the *activator (Ac)*. She noted cases in which *Ds* activity disapeared from the *standard* locus and reappeared elsewhere in the genome. In addition, somatic instability was observed at several gene loci. Such mutator activity was also under the control of the *Ac*. McClintock made an operational distinction between genes and transposable elements. Genes occupy fixed loci and transposable elements can move from one location to another. Because the frequency and temporal occurrence of somatic instability were under the control of the *Ds* state and *Ac* dose, respectively, she called these "controlling elements." McClintock (1951) also characterized another family of transposable elements which she

Nihal K. Notani   Biomedical Group, Bhabha Atomic Research Centre, Bombay 400085, India.

Subcellular Biochemistry, Volume 17: Plant Genetic Engineering, edited by B. B. Biswas and J. R. Harris. Plenum Press, New York, 1991.

termed *Supprssor-mutator (Spm)*, which was independently isolated and characterized by Peterson (1961), but termed *Enhancer (En)*.

Brink and Nilan (1952) also showed that *P-vv* (variegated pericarp and cob) has an *Ac*-like movable element conjoined to *P-RR* (red pericarp and cob), responsible for the variegated phenotype. Moreover, *M-P* (modulator of *P* locus) acted autonomously; i.e., it affected the *P* locus directly. Modulator could also activate *Ds* at the standard locus. The phenotype of light-variegated pericarp was shown to be due to the presence of an additional copy of *M-P* present in the genome. Similarly, twin sectors of light variegated and red in variegated background were explained as arising from transposition of *M-P* from a duplicated segment of chromosome (including *P* locus) to a nearby unreplicated region, where it is duplicated again (Greenblatt, 1984).

Another transposable element, *Mu (Mutator)*, discovered by Robertson (1978) raised the frequency of germinal mutations by a factor of about 50. However, its rules of inheritance are somewhat complex, apparently because of its inactivation.

Three transposable elements, *Tam* 1, *Tam* 2, and *Tam* 3, have been studied in *Antirrhinum majus* (see, for example, Carpenter *et al.*, 1988). These have characteristics similar to the elements in maize. Recently, wrinkled-seed character *(rugosus)*, originally studied also by Mendel, has been shown to be due to the insertion of a transposable element (Bhattacharya *et al.*, 1990).

2. MOLECULAR CHARACTERIZATION OF TRANSPOSABLE ELEMENTS

2.1. *Ac* Family

Ac DNA from *wx-m7* and *wx-m9* alleles has been sequenced. Both have given identical sequences of 4563 bp. *Ac* has 11 bp imperfect inverse repeat (IR) ends. It has two open-reading frames (ORFs) (Pohlman *et al.*, 1984; Muller-Neumann *et al.*, 1984). It makes a target duplication of 8 bp.

There are three classes of *Ds*, all of which can be activated by *Ac*. The first class has *Ds* elements that are internal deletions of *Ac*. The second class of *Ds* has examples in which several hundred base pairs are homologous to *Ac* at each end but internal sequences are nonhomologous. A third class of *Ds* has similarities to *Ac* mainly in their ends and short adjacent sequences (Baker *et al.*, 1987). *Ds* elements have perfect IRs. Following excision, a footprint, a slightly altered sequence, is left behind.

Ac has at least two *trans*-acting functions: (1) it "catalyzes" its own and *Ds* transposition, and (2) increased doses of *Ac* inhibit transposition. Deletions in

ORF1 and ORF2 of *Ac* do not complement each other, suggesting that both work as a single unit (Dooner *et al.*, 1986). A 3.5-kb mRNA with abundance of 1-3 × 10^7 of poly-A RNA is found only in *Ac* lines. This corresponds to the long ORF encoding 807 amino acids. Four introns are removed (Kunze *et al.*, 1987).

In certain cases, insertion of elements into exons does not abolish total activity. Inserts of 4.3 or 1.5 (kb) *Ds* in *wx* gene yield wild-type RNA. cDNA analysis revealed that *Ds* is spliced out of RNA (Wessler *et al.*, 1987).

In pea, Mendel's recessive allele *wrinkled* apparently contains a 0.8-kb insertion that is very similar to the *Ac/Ds* family of transposable elements (Bhattacharya *et al.*, 1990). *Tam* 3 of antirrhinum is more akin to *Ac/Ds*. *Tam* 3 has a length of 3.5 kb at Nivea (Chalcone synthase) locus, 12 bp perfect IRs, and makes a target duplication of 8 bp (Sommer *et al.*, 1985).

2.2. *Spm/En* Family

Spm has two highly structured termini of about 200 bp in length. It has 13 bp IR ends. Generically, *Spm* is similar to *Tam* 1 and *Tam* 2 of antirrhinum and *Tgm of Glycine max* (Gierl *et al.*, 1985).

En-1 sequenced from *wx*-844 is 8287 bp long. It has 11 exons and the first intron is 4434 nucleotides long. It has two ORFs of 2714 and 761 bp (Pereira *et al.*, 1986). RNA splicing of a transposable element in *bz*-m 13 CS9 permits gene expression. This allele contains a 902-bp defective *Spm* in the second exon. Transcription proceeds through *dSpm* but is then spliced out. About 40–50% of the enzymatic activity is recovered in the absence of *Spm* (Kim *et al.*, 1987). A new allele of *C-I*-Bombay has been isolated that is apparently made somatically unstable by an *Spm*-like transposable element (Allagikar *et al.*, 1990). The *C-I* allele is known to give two types of somatic instability: (1) color dots in colorless background or (2) colorless areas in colored background. The *C-I* Bombay mutable allele belongs to the former class.

2.3. *Mu*-1 Family

Two sizes of *Mu* (Mutator) have been isolated and characterized: one is 1.7 kb in size and the other 1.4 kb. The only difference between the two seems to be that an additional 385-bp block is present in the 1.7-kb element. Thus, it has been suggested that the 1.4-kb version arises from the 1.7-kb element by a deletion (Taylor and Walbot, 1987).

Mutations induced by *Mu* appear to be of the deletion kind. Robertson and Stinard (1987) observed that at least 12 of the *yg2* mutants were due to deletions involving also the linked locus *wd*. Robertson *et al.* (1988) found no relationship between somatic and germinal *Mu* activity.

2.4.   *Cin* 4 Elements

Cin 4 is considered a nonviral retroposon. It was detected as an insert in *A*1 gene of maize. *Cin* 4 is 1.1 kb long. It ends in 12 poly A residues. Its copy number is estimated at 50–100 per diploid genome. It has a long ORF that codes for 3198 amino acid residues (Schwarz-Sommer and Saedler, 1988).

3.   MECHANISM AND REGULATION OF TRANSPOSITION

Gierl and Saedler (1989) have grouped transposable elements into two classes: those with terminal inverse repeats (TIRs), such as *Ds/Ac* or *Spm/En*, and the others represented by retrotransposons. The latter has been divided into retrovirus-like elements flanked by long terminal repeats (LTRs), as in *Bs*-1 (maize) and nonviral retrotransposons ending in a deoxyadenosine monophosphate residues (dAMP) end on the 3' side as in *Cin* 4. *Bs*-1 and *Cin* 4 have sequences homologous to those of viral reverse transcriptases. The retroposons apparently move by reverse transcription.

An excision-repair model for *Ds* stability in the absence of *Ac* and its high mutability in the presence of *Ac* has been proposed (Mouli and Notani, 1970). It now seems to be established that *Ac* moves by excision and generally leaves a footprint behind. The substrate for the transposase is apparently the TIRs. That it should be coding for its own transposase is apparent from the observation that when *Ac* is moved to tobacco, *Ac* can still transpose (Baker *et al.*, 1987). Sundaresan (1988) has discovered an extrachromosomal *Mu* that is considered an intermediate in transposition. These are most likely covalently closed circular DNA.

4.   ACTIVATION/INACTIVATION OF TRANSPOSABLE ELEMENTS

The *Ac* element at *waxy* locus undergoes cyclic activation and inactivation. This is associated with the modification of its DNA. It has been shown that *Ac* in *wx* m-7 in an active form is hypomethylated and the inactive form is methylated (Chomet *et al.*, 1987). Schwartz and Dennis (1986) also found that the active *Ac* element in *wx*-m9 is hypomethylated and completely methylated in the inactive allele. According to Schwartz (1989), the transposase product of *Ac* causes demethylation of the *Bam* HI site in the promoter region of *Ac*. This remains methylated in *Ds*-9. Chandler and Walbot (1986) also noted that active *Mu* element(s) became inactive and this is correlated with the modification of its DNA. However, both active and inactive elements can be present in a plant and they consider it a reflection of progressive occurrence.

Chen *et al.* (1987) have proposed that *Ac* elements preferentially insert in hypomethylated sites. However, this does not seem to be true with *Ds* elements. Compared to bulk DNA, it is observed that the genes (*A1*, *Adh1*, and *Sh1*) are hypomethylated.

5. CLONING OF GENES BY TRANSPOSON TAGGING AND CLONING OF TRANSPOSABLE ELEMENTS BY GENE TAGGING

Using an *Ac* DNA clone, the *bz1* gene has been isolated from maize (Fedoroff *et al.*, 1984). Similarly, McLaughlin and Walbot (1987) have cloned the *bz2* gene. Essentially this method utilizes mutations at a locus caused by known transposable elements and hybridized with the central part of the transposable element, which is not as abundantly represented as the ends in the genome. Theres *et al.* (1987) have similarly cloned *bz2* locus which had the presence of a *Ds* element. This method is a little more difficult, but the absolute linkage of the transposable element to the gene could eventually yield the clone. Doring (1989) lists no less than 12 alleles of different genes that have been cloned by transposon tagging. Similarly, no less than 13 alleles of different transposable elements have been isolated by the converse method of gene tagging. Most of the genes used were those that affect the endosperm composition or color.

6. DISCUSSION

Transposable elements appear to move either by excision and insertion or by reverse transcription. They can be present as a number of intact copies (e.g., *Ac*) or as derived defective elements (e.g., *Ds*). Their fixation and role in the genome are not clear. That they can rearrange the genome or mutate genes is quite certain, although the extent to which this happens in nature is not known. That Mendel's *wrinkled* allele in pea should contain an *Ac*-like element is a surprise but emphasizes the point that at least part of the natural variability may be due to these elements.

The rules of activation and inactivation of these elements appear to be complex. That this is achieved by methylation is becoming clear. Although genomic stress can clearly activate these elements, it is not known whether regular genes are also induced, as in bacteria in which several damage-inducible gene loci have been found. In terms of long-range adaptation, induction of heritable variation could have value for coping with any new stresses, but for immediate response, something such as enzymatic adaptation may be of greater help. There may also be the question of whether transposable DNA sequences are treated as "foreign" DNA by the plant genome, in which case different rules of methylation may apply to it. At least in some cases (e.g., *Bs*1 and *Cin* 4 in maize), their similarity to retroviral genomes is obvious. Excisable elements, however, are different in that sense, but do find parallels in prokaryotes.

Do these elements have a role in development or are these just a kind of "selfish" DNA, performing no regular function for the cell? There appears at present only circumstantial evidence for a major role for transposable elements in regular plant development. Also, a very small amount of information like coding for a transposase or reverse transcriptase, may circumscribe their role in this regard. On the other hand, as mutators they could play a role in long-range adaptation of a species to changed adverse environment. With new variability, there should be mutants to select that will adjust the organism to new conditions. Under normal circumstances, it would be best for the organism to keep these elements quiescent, otherwise elements like *Mu* might break down the stability of the genome of an organism. Does this mean that such elements are recognized as such by the cell and are inactivated by methylation? There is no evidence on this point. It would be interesting if transposable elements were recognized in this manner.

7. REFERENCES

Allagikar, S. B., Pawar, S. E., Mitra, R. K., and Notani, N. K., 1990, Analysis of a case of somatic instability in maize, in *Proc. D.A.E. Symp. on Advances in Molecular Biology*, Bombay, pp. 487–494.

Baker, B., Coupland, G., Fedoroff, N., Starlinger, P., and Schell, J. 1987, Phenotypic assay for excision of the maize controlling element *Ac* in tobacco, *EMBO J.* **6**:1547–1554.

Bhattacharya, M. K., Smith, A. M., Noel Ellis, T. H., Hedley, C., and Martin, C. 1990, The wrinkled-seed character described by Mendel is caused by a transposon-like insertion in a gene encoding starch-branching enzyme, *Cell* **60**:115–122.

Brink, R. A., and Nilan, R. A., 1952, The relation between light variegated and medium variegated pericarp in maize, *Genetics* **37**:519–544.

Carpenter, R., Hudson, A., Robbins, T., Almeida, J., Martin, C., and Coen, E., 1988, Genetic and molecular analysis of transposable elements in *Antirrhinum majus*, in *Plant Transposable Elements* (O. E. Nelson, ed.), pp. 69–80, Plenum Press, New York.

Chandler, V. L., and Walbot, V., 1986, DNA modification of a maize transposable element correlates with loss of activity, *Proc. Natl. Acad. Sci. USA* **83**:1767–1771.

Chen, C. H., Oishi, K. K., Kloeckener-Gruissen, B., and Freeling, M., 1987, Organ-specific expression of maize *Adh*1 is altered after a *Mu* transposon insertion, *Genetics* **116**:469–477.

Chomet, P. S., Wessler, S., and Dellaporta, S. L., 1987, Inactivation of the maize transposable element *Activator (Ac)* is associated with its DNA modification, *EMBO J.* **6**:295–302.

Dooner, H., English, J., Ralston, E., and Weck, E., 1986, A single genetic unit specifies two transposition functions in the maize element *Activator, Science* **234**:210–211.

Doring, H-P., 1989, Tagging genes with maize elements: An overview, *Maydica* **34**:73–88.

Fedoroff, N., Furtek, D., and Nelson, O. E., 1984, Cloning of the *Bronze* locus in maize by a simple and generalizable procedure using the transposable controlling element *Ac*, *Proc. Natl. Acad. Sci. USA* **81**:3825–3839.

Gierl, A., Schwartz-Sommer, Z., and Saedler, H., 1985, Molecular interactions between the components of the *En*-I transposable elements in *Zea mays*, *EMBO J.* **4**:579–583.

Gierl, A., and Saedler, H., 1989, Maize transposable elements, *Annu. Rev. Genet.* **23**:71–85.

Greenblatt, I. M., 1984, A chromosome replication pattern deduced from pericarp phenotypes resulting from movements of the transposable elements *Modulator* in maize, *Genetics* **108**:471–485.

Kim, H-Y., Schiefelbein, J. W., Raboy, V., Furtek, D. B., and Nelson, O. E., Jr., 1987, RNA splicing permits expression of a maize gene with a defective suppressor-mutator transposable element insertion in an exon, *Proc. Natl. Acad. Sci. USA* **84**:5863–5867.

Kunze, R., Stochaj, U., Laufs, J., and Starlinger, P., 1987, Transcription of transposable element *Activator (Ac)* of *Zea mays* L., *EMBO J.* **6**:1555–1563.

McClintock, B., 1951, Chromosome organization and genic expression, *Cold Spring Harbor Symp.* **16**:13–47.

McLaughlin, M., and Walbot, V., 1987, Cloning of a mutable *bz2* allele of maize by transposon tagging and differential hybridization, *Genetics* **117**:771–776.

Mouli, C., and Notani, N. K., 1970, Absence of a detectable change in *Ds* at the *Al* locus in maize following mutagenic treatments, *Can. J. Genet. Cytol.* **12**:436–442.

Muller-Neumann, M., Yoder, Y. I., and Starlinger, P., 1984, The DNA sequence of the transposable element *Ac* of *Zea mays* L., *Mol. Gen. Genet.* **198**:19–24.

Pereira, A., Cuypers, H., Gierl, A., Schwarz-Sommer, Zs., and Saedler, H., 1986, Molecular analysis of the *En/Spm* transposable element system of *Zea mays*, *EMBO J.* **5**:835–841.

Peterson, P. A., 1961, Mutable *al* of *En* system in maize, *Genetics* **46**:759–771.

Pohlman, R. F., Fedoroff, N. V., and Messing, J., 1984, The nucleotide sequence of the maize controlling element *Activator*, *Cell* **37**:635–643.

Robertson, D. S., 1978, Characterization of a mutator system in maize, *Mutat. Res.* **51**:21–28.

Robertson, D. S., and Stinard, P. S., 1987, Genetic evidence of *Mutator* induced deletions in the short arm of chromosome 9 of maize, *Genetics* **115**:353–361.

Robertson, D. S., Morris, D. W., Stinard, P. S., and Roth, B. A., 1988, Germ line and somatic *Mutator* activity: Are they functionally related? in *Plant Transposable Elements* (O. E. Nelson, ed.), pp. 17–42, Plenum Press, New York.

Schwartz, D., and Dennis, E., 1986, Transposase activity of the *Ac* controlling element in maize is regulated by its degree of methylation, *Mol. Gen. Genet.* **205**:476–482.

Schwartz, D., 1989, Gene-controlled cytosine demethylation on the promoter region of the *Ac* transposable element in maize, *Proc. Natl. Acad. Sci. USA* **86**:2789–2793.

Schwarz-Sommer, Z., and Saedler, H., 1988, Transposition and retrotransposition in plants, in *Plant Transposable Elements* (O. E. Nelson, ed.), pp. 175–187, Plenum Press, New York.

Sommer, H., Carpenter, R., Harrison, B. J., and Saedler, H., 1985, The transposable element *Tam 3* of *Antirrhinum majus* generates a novel type of sequence alterations upon excision, *Mol. Gen. Genet.* **199**:225–231.

Sundaresan, V., 1988, Extrachromosomal *Mu*, in *Plant Transposable Elements* (O. E. Nelson, ed.), pp. 251–259, Plenum Press, New York.

Taylor, L. P., and Walbot, V., 1987, Isolation and characterization of a 1.7 kb transposable element from a mutator line of maize, *Genetics* **117**:297–307.

Theres, N., Scheele, T., and Starlinger, P., 1987, Cloning of the *Bz2* locus of *Zea mays* using the transposable element *Ds* as a gene tag, *Mol. Gen. Genet.* **209**:193–197.

Wessler, S. R., Boran, G., and Varagona, M., 1987, The maize transposable element *Ds* is spliced from RNA, *Science* **237**:916–918.

Chapter 5

Potentials of Woody Plant Transformation

Peter L. Schuerman and Abhaya M. Dandekar

1. INTRODUCTION

Woody plants might well be considered to be one of the more intractable groups of experimental organisms in existence, yet they are some of the most valuable, both aesthetically and commercially. Manipulation of these plants can be difficult for the researcher, breeder, and grower alike, but because of the peculiarities and economic value of the woody plants, success can be particularly rewarding. This chapter deals with the use of genetic engineering to modify woody plants, describing how such modifications can serve the purposes of both researchers and commercial breeders.

2. RATIONALE

Classical genetic analyses of woody plants based on progeny tests are impeded by the long generation time characteristic of many members of this group. Therefore, less information can be obtained from crosses within a given period of

Peter L. Schuerman and Abhaya M. Dandekar   Davis Crown Gall Group, Department of Pomology, University of California, Davis, California 95616-8630.

Subcellular Biochemistry, Volume 17: Plant Genetic Engineering, edited by B. B. Biswas and J. R. Harris. Plenum Press, New York, 1991.

time. Because the study of woody plants by conventional means is restricted by these factors, relatively little is known at the molecular and genetic levels about traits such as alternate-year bearing, overwintering, juvenility, tree architecture, and wood production, traits that can only be studied in woody plants. The ability to genetically transform these plants could circumvent this complication and could possibly improve the understanding of these characters. Transformation could also improve biochemical studies of plants, which are hampered by the existence of the cell wall and the variety of enzymes and phenolic compounds released from the vacuole upon homogenization of plant tissue. The advanced state of cell wall development in some woody plant tissues can make cell lysis particularly difficult. Additionally, woody plants tend to be especially rich in phenolic compounds, which can react with proteins and nucleic acids during extraction (but see Couch and Fritz, 1990).

There is also the economic aspect of research in woody crops. Cereals aside, international trade in fresh, temperate tree fruits rivals that of the major crops of the world (James, 1987). However, conventional genetic manipulation of the crops in this economically important group is not without obstacles. Genetic transformation is an alternative approach that may avoid these difficulties. Transformation procedures allow one to make small, specific changes in the genome of a cell, i.e., the addition of one or a few genes. This is in sharp contrast to conventional breeding, where entire sets of chromosomes are combined. As an illustration of classical breeding, consider backcross breeding, a technique that can be used to produce the equivalent of single-gene transfer. The transfer of a single gene trait, e.g., disease resistance, from a donor species (often a wild, unadapted species) to a well-adapted recipient cultivar, which, though lacking the trait in question, has many desirable characteristics, can be achieved by repeated backcrossing of the F_1 generation to the recipient species, accompanied by progeny selection for the trait of interest. In this manner, progressive elimination of the undesirable donor genes is achieved and the desirable gene is retained. The pace of such an introgression procedure is limited by the generation time of the crop in question. Retention of the recipient's unique genetic constitution is only possible if the plant is true-breeding (completely homozygous). Another point to consider is that selection for the desired allele is equivalent to selection for alleles tightly linked to it, and these alleles may be detrimental to the market value of the plant. This phenomenon is called "linkage drag."

Woody species are a good example of the importance of the limitations of introgression. First, with generation times of years rather than months, an introgressive approach to the transfer of a single trait in woody species becomes temporally demanding. Second, genes can only be introgressed from sexually compatible donors. Third, most woody cultivars are not true-breeding; the levels of outcrossing that occur in such populations ensure a high degree of heterozygosity. For these reasons, production of woody cultivars has resulted mostly from the large-scale asexual propagation of chance seedlings, hybrids, and clonal variants that showed some desirable character. Such processes are somewhat random, in contrast to transformation, with which predictable single-gene modifications are possible. A gene of interest can be inserted, by one of several techniques, into the genome of a single cell. The single cell can then be cultured so that it develops into a plant, all cells of which carry the gene. This strategy is limited not by generation time but by regeneration time, that is, the time it takes to produce a plant from the single cell. Also, the phenomenon of detrimental linkage drag does not occur in transformation. The potential monetary gain from such a modified cultivar to a commercial grower is dependent not only on the nature of the introduced gene (i.e., what problem or problems are solved by the gene), but also on the fact that the unique genetic constitution of the cultivar has not been disturbed, as it would be in an introgression procedure.

3. PRODUCTION OF TRANSGENIC PLANTS

The production of transgenic plants requires (1) recombinant DNA manipulation, which is important in isolating and manipulating the genes to be used, (2) the ability to insert DNA into the genome of a plant cell (transformation), (3) the ability to regenerate a single cell into a whole plant (tissue culture), and (4) the ability to distinguish between transformed and untransformed regenerated plants and tissue.

3.1. DNA Manipulation

A particular sequence of DNA, as it occurs in a cell, is extremely dilute. In order to do work with a desired sequence of DNA, the sequence must be isolated and replicated so that reagent-level quantities can be obtained. This replicative process is specifically what is referred to as gene cloning. The ability to cut and join DNA fragments in a highly precise fashion facilitates this isolation and purification of genes and makes modification of genes possible as well. An example of gene cloning is the introduction of a DNA fragment into a plasmid, which is then introduced into a bacterium so that it can be replicated as the bacterium replicates. The bacteria can be then be lysed and the many copies of the plasmid can be isolated. The replicates of the DNA fragment on each plasmid can be recovered after cleaving each DNA fragment and plasmid apart. Another application of recombinant DNA technology is the production of chimeric genes (Fraley et al., 1983). In short, this is the joining of structural portions of one gene with regulatory portions of another. The regulatory regions of a gene (sequences, such as promoters, enhancers, and polyadenylation signals, which modulate expression) can be joined to the protein-coding (structural) region of another gene. Due to the transcriptional and translational differences between pro-karyotes and eukaryotes, eukaryotic control regions must be joined to the pro-tein-coding region of a prokaryotic gene in order to make expression of the prokaryotic gene product possible in the eukaryote. Also, protein-coding regions may need to be modified in order to govern how the gene is expressed and how' the gene product is targeted and sorted to different cellular compartments.

3.2. Transformation

Several transformation techniques have been described in the literature, some of which are more widely applicable than others. Transformation involves mobilizing a piece of DNA carrying a reporter gene (a gene that gives a scorable and/or selectable phenotype to the cells it enters), and perhaps one or more genes of interest, into a plant cell. This topic will be discussed in greater detail later in this chapter.

3.3. Tissue Culture and *In Vitro* Propagation

The regeneration of a plant from a transformed cell is largely dependent on both the plant and the type of cell transformed. Transformation of germ cells such as spores, pollen, and egg cells might seem to simplify this task, but techniques for transforming such cells have not been developed to a great extent. More often, somatic cells are the target of transformation attempts, so techniques to regenerate whole plants from these somatic cells are required. Tissue culture of somatic cells revolves around the treatment of a piece of excised plant tissue (the explant) with growth regulators, with the intention of stimulating the cells of the explant to grow in a certain fashion. Regeneration is the production of new plants from explant tissue. There are two modes of regeneration, indirect and direct. Indirect regeneration involves an intermediate callus phase (callus being a tumorlike proliferation of cells), while direct regeneration does not. In an indirect system, cells of the callus differentiate to produce adventitious meristems or somatic embryos. In a direct system these are produced directly from explant cells. The resulting organism is referred to as a plantlet. If more than one cell contributes to the production of an adventitious meristem, and there is variation in the genotypes of the cells in the callus, a chimera or mosaic (an organism containing cells with differing genotypes) can result. Somatic embryogenesis differs in that one somatic cell gives rise to the plantlet and does so by undergo-ing a series of changes characteristic of those which a developing embryo under-goes. In this case, single somatic cells behave as if they are zygotes and develop accordingly. Chimerism is unlikely and happens only if genetic changes occur after initiation of the embryogenesis, when the developing embryo is composed of more than one cell (Dandekar *et al.*, 1989). The details of woody plant culture are beyond the scope of this chapter; for more information see Haissig *et al.* (1987), Tulecke (1987), and Wann (1988).

3.4. Screening and Selecting Plantlets

Production of transgenic plants involves the regeneration or propagation of explant tissue or germ cells that have been transformed. Transformation, however, is not completely efficient, and most of the cells in an experiment will escape modification. Plants obtained from subsequent regeneration will not all be transgenic. For this reason, genetic selection and screening are necessary. "Reporter genes" are either used alone as indicators of transformation efficiency or linked to the gene that is being introduced into the plant. Reporter genes can be selectable and/or screenable. Selectable reporter genes confer a selective advantage to the cells in which they are present, while screenable reporter genes do not (ideally) affect cell viability but are easily detected. An example of a selectable marker is the chimeric aminoglycoside phosphotransferase (APT) gene (Herrera-Estrella *et al.*, 1983a) referred to in the literature as APH(3′)II or NPTII, derived from the bacterial Tn5 APT gene (Beck *et al.*, 1983), which confers kanamycin resistance to eukaryotic cells. After transforming with this gene, the explant material is cultured on medium containing kanamycin; thus, only transformed cells are able to grow, divide, and eventually give rise to plants. Other examples of selectable markers include genes for hygromycin resistance and resistance to methotrexate (Fraley *et al.*, 1986). Regulatory elements used to express these coding regions in plants include those obtained from various *Agrobacterium* T-DNA genes (discussed later) that contain eukaryotic promoter and enhancer motifs and those from the 19S and 35S transcripts of the cauliflower mosaic virus (CaMV) (Kuhlemeier *et al.*, 1987).

Examples of screenable markers are the genes for octopine and nopaline syntheses (obtained from *Agrobacterium tumefaciens*), the β-glucuronidase gene (Jefferson *et al.*, 1987), and the chloramphenicol acetyltransferase (*CAT*) gene (Fraley *et al.*, 1986). The former two cause cells carrying them to produce octopine and nopaline, respectively, neither of which is normally found in plants (De Greve *et al.*, 1982; Zambryski *et al.*, 1983). Opines are detectable by a paper-electrophoretic assay carried out on tissue extracts (Otten and Schilperoort, 1978; Petit *et al.*, 1983). β-Glucuronidase is specified by a gene isolated from *Escherichia coli*. This gene has been modified so that it can be expressed in plant cells (Jefferson *et al.*, 1987). The presence of the β-glucuronidase enzyme in tissue can be detected by using colorimetric, fluorometric, and histochemical assay techniques (Jefferson, 1987). *CAT* activity in plant tissue extracts can be assayed by the extract's ability to acetylate radioactively labeled chloramphenicol; the acetylated form is distinguished by a change in electrophoretic mobility (Herrera-Estrella *et al.*, 1983b; De Block *et al.*, 1984).

4. TRANSFORMATION

Several methods of plant transformation have been reported in the literature, including *Agrobacterium*-mediated transformation (AMT) (De Block *et al.*, 1984; Horsch *et al.*, 1985; Ulian *et al.*, 1988; McGranahan *et al.*, 1988; Fillatti *et al.*, 1987; James *et al.*, 1989, 1990a,b), direct DNA uptake into protoplasts (Paszkowski *et al.*, 1984; Hain *et al.*, 1985; Schocher *et al.*, 1986), electroporation (Fromm *et al.*, 1985, 1986), microprojectile transformation (MPT) (Klein *et al.*, 1987; McCabe *et al.*, 1988), microinjection (Crossway *et al.*, 1986), pollentube transformation (Luo and Wu, 1988), and injection of floral tillers (De La Pena *et al.*, 1987). Of these, electroporation, MPT, and AMT will be discussed herein as they are the most generalized and most widely applicable techniques. Of these three, only AMT has been used successfully to transform woody species as of this writing.

4.1. Electroporation

Electroporation can be used only in the context of a highly developed tissue culture regimen wherein whole plants can be regenerated from protoplasts. Protoplasts are cells that have had their cell walls stripped from them by means of hydrolytic enzyme digestion in an osmotically adjusted medium (Revilla *et al.*, 1987). The plant cell wall is a barrier to macromolecules such as DNA and so must be removed. Protoplasts created in this way are first placed in a buffered solution of DNA. Next, electrodes are inserted into the solution and a sharp discharge of electricity is sent through the medium. This causes pores to open on the plasmalemma of the protoplasts, and the DNA in the solution can enter. A fraction of this DNA finds its way to the nucleus and is expressed. Such expression is termed transient and subsides with time. Stable expression occurs when some of the DNA that enters the nucleus integrates into a chromosome or chromosomes of the plant cell. Subsequent treatments of the protoplasts that promote the refabrication of the cell wall and regeneration of whole plants can then yield transgenic plants. A general strategy for the isolation of mesophyll protoplasts from woody plants has been published (Revilla *et al.*, 1987), but the regeneration of such protoplasts is not always possible. Examples of woody plant genera that have been regenerated from protoplasts include *Citrus* (Vardi *et al.*, 1982), *Prunus* (Ochatt *et al.*, 1987; Ochatt and Power, 1988), *Malus* (Patat-Ochatt *et al.*, 1988; Wallin and Johansson, 1989), *Pyrus* (Ochatt and Caso, 1986), *Broussonetia* (Oka and Ohyama, 1985), and *Populus* (Russel and Mc-Cown, 1986, 1988).

4.2. Microprojectile Method

Microprojectile transformation involves the acceleration of small, dense particles (such as gold or tungsten pellets) into plant tissue. The particles are coated with DNA and thus carry DNA into the cells of the tissue, owing to their ability to penetrate the cell wall and plasmalemma. The underlying strategy here is the same as that of electroporation in that both are designed to circumvent the physical barriers presented by cell wall and plasmalemma and deliver a large quantity of DNA in order to overcome the low probability of a single DNA molecule finding its way into the nucleus and integrating into a chromosome. As with electroporation, transient expression is also observed with this technique, and both techniques have been used to reveal the activity of promoters in experiments using chimeric genes (Fromm *et al.*, 1985; Klein *et al.*, 1987).

4.3. *Agrobacterium*-Mediated Method

Because of the success of this particular technique, it will be considered in the greatest detail. *Agrobacterium tumefaciens, A. rhizogenes,* and *A. rubi* are all phytopathogenic bacteria. The mechanism of their pathogenicity is the transfer of a piece of DNA from the bacterium to the plant cell, which changes the metabolism of the target cell, causing, among other things, uncontrolled proliferation of the affected cells. Such an overgrowth is referred to as a hyperplasia or neoplasia. *A. tumefaciens* is responsible for crown gall disease, which produces tumors; *A. rhizogenes* is the agent of hairy root disease, an overgrowth of fine roots; and *A. rubi* causes cane gall on *Rubus* species. These hyperplasias produce opines, novel compounds that can be metabolized by the bacteria. Using recombinant DNA techniques, it has been possible to modify agrobacteria so that alternative genes can be transferred instead of pathogenesis genes. Before proceeding any further, a background on the biology of *Agrobacterium* should be presented.

4.3.1. Biology of *Agrobacterium*

A. tumefaciens is the best understood and most widely studied species of agrobacterium. Accordingly, this discussion will be limited to *A. tumefaciens* with the understanding that other agrobacteria are fundamentally similar. *A. tumefaciens* is a ubiquitous gram-negative, rhizosphere-dwelling plant pathogen. Only a small fraction of agrobacteria isolated from soil are pathogenic (Kerr, 1969), and the lack of pathogenicity is almost always due to the absence of the tumor-inducing plasmid (Ti plasmid; Zaenen *et al.*, 1974; Van Larebeke *et al.*, 1975). Conversely, the Ti plasmid is required for pathogenicity (Watson *et al.*, 1975). Cells of *A. tumefaciens* that harbor a functional Ti plasmid have the ability to cause tumorous growths on a wide variety of dicotyledonous plants and on a few monocotyledonous plants (De Cleene and De Ley, 1976). This transformation event is brought about by the transfer of a segment of the Ti plasmid known as T-DNA into the nuclear genome of the susceptible plant (Chilton *et al.*, 1977, 1982; Bevan and Chilton, 1982). On the Ti plasmid, the T-DNA is flanked by 25 bp direct repeats (Yadav *et al.*, 1982). Encoded within the T-DNA are various genes that are transcribed and translated in plant tissue by virtue of their eukaryotic regulatory elements (promoters, enhancers, polyadenylation signals, etc.) to produce a novel set of proteins. Some of these proteins are enzymes that are involved in the synthesis of auxin, i.e., *IAA* (Schröder *et al.*, 1983; Thom-ashow *et al.*, 1986), cytokinin, i.e., ribosylzeatin (Akiyoshi *et al.*, 1984), and opines (Guyon *et al.*, 1980). The overproduction of auxin and cytokinin leads to cell proliferation; the genes responsible for this can be removed (disarmed Ti plasmid; Zambryski *et al.*, 1983) and replaced with other genes and thus can be transferred to infectible plant tissue. Infection with such "disarmed" strains allows the regeneration of morphologically normal plants that contain the introduced foreign gene (De Block *et al.*, 1984; Horsch *et al.*, 1985; McCormick *et al.*, 1986). The transferred genes are stably incorporated into the plant genome and have been shown to display normal Mendelian inheritance (De Block *et al.*, 1984, Horsch *et al.*, 1984) through sexual reproduction.

The genes responsible for transmission of T-DNA into a host plant cell are located in the virulence (*vir*) region of the Ti plasmid. Based on mutational analyses, the *vir* region can be divided into six complementation groups, designated *vir* A, B, C, D, E, and G. Mutations in *vir* A, B, D, and G arrest virulence, while those in *vir* C and E attenuate virulence. These are all responsible for moving the T-DNA into the plant cell's genome. Because they are the only virulence genes expressed constitutively, *vir* A and *vir* G are thought to regulate the other *vir* genes: B, C, D, and E (Stachel and Zambryski, 1986). Virulence genes are induced by plant phenolic compounds (Bolton *et al.*, 1986) such as acetosyringone and α-hydroxyacetosyringone. Other factors have been reported to influence the induction of virulence genes, such as pH, temperature, sugars (Alt-Mörbe *et al.*, 1988, 1989), opines (Veluthambi *et al.*, 1989), and the osmoprotectant compound glycine betaine (Vernade *et al.*, 1988). The current hypothesis proposes that the *vir* A protein functions in the initial recognition of phenolic inducers (Stachel *et al.*, 1985; Winans *et al.*, 1986). The inducer–*vir* A complex allosterically activates the *vir* G protein by means of a phosphorylation (Jin *et al.*, 1990a,b), which, in turn, activates transcription of the remaining *vir* genes by directly interacting with their promoter sequences (Stachel and Zambryski, 1986). The *vir* D gene codes for an endonuclease (Yanofsky *et al.*, 1986) that recognizes and cleaves the T-DNA border sequences (Wang *et al.*, 1987). The right border sequence has been shown to be essential for T-DNA transfer and functions in a polar fashion (Wang *et al.*, 1984; Shaw *et al.*, 1984). The left border, however, is not necessary for pathogenicity (Shaw *et al.*, 1984).

4.3.2. Virulence Considerations

When deciding which *Agrobacterium* strain to utilize for transformation of a plant species, the virulence of different strains on the plant is a consideration, because a high degree of virulence is likely to improve transformation efficiency. *Agrobacterium* does have host range limitations, and a growing body of evidence indicates wide variations in the rates of infection of single plant species with different *Agrobacterium* strains, such as soybean (Owens and Cress, 1985) and a range of woody plants (Martin, 1987; Dandekar *et al.*, 1988, 1990; Uratsu *et al.*, 1991). The isolation of super virulent strains of *Agrobacterium* that cause numerous large tumors further underscores this fact (Hood *et al.*, 1984; Pythoud *et al.*, 1987). The supervirulent phenotype of *Agrobacterium* strain A281 has been localized to a segment of DNA located in the *vir* region of the Ti plasmid pTiBo542 (Hood *et al.*, 1986; Jin *et al.*, 1987; Pythoud *et al.*, 1987).

4.3.3. Recombinant Vector Plasmids

In order to make use of agrobacteria as vectors for foreign gene transfer, it is necessary to make genetic modifications of the Ti plasmid. A growing body of literature now exists on the construction of different Ti-plasmid vectors. The plasmid vectors used to transform plants using *A. tumefaciens* fall into two broad categories. One class are the "*cis*" or cointegrating vectors where the modified T-DNA and *vir* region are present on the same plasmid. In the second class, known as "binary" vectors, the *vir* and T-DNA segments are contained on two separate, but compatible plasmids within the same bacteria, the *vir* functions being supplied in *trans*. The binary system was developed by Hoekema *et al.* (1983), and today there is a growing number of small, easily manageable binary vectors that have been developed (Bin 19: Bevan, 1984; pEND4K: Klee *et al.*, 1985; pGA471: An *et al.*, 1985). The binary vectors are easier to manipulate as they can replicate in both *E. coli* and *Agrobacterium*, greatly facilitating the transfer of DNA segments between the two microorganisms. The frequency of transformation in a particular plant system is dependent on the type of Ti-plasmid vector chosen.

5.   USEFUL TRAITS FOR WOODY PLANTS

In order to produce superior varieties of a plant species using a transformation approach, the desired trait should be encoded by a few genes, ideally one. Desirable traits for woody plants include pest resistance, disease resistance, drought and cold tolerance, herbicide resistance, improved fruit quality, rootability of mature tissue, reduced juvenility, and alteration of tree form and architecture (e.g., dwarfing or semidwarfing capacity). Very few of these traits have been characterized in sufficient detail to be good candidates for gene transfer technology. Traits such as disease and pest resistance found in plant germplasm can sometimes be shown to be controlled by a single locus, but such genes have not been cloned at present. However, transgenic plants with resistance to disease, herbicides, and insect damage have been produced utilizing genes from nonplant sources.

5.1. Virus Resistance

The development of virus resistance in cultivated crops has been approached through breeding programs that make use of resistance genes in related species. Often these resistance genes are linked with other undesirable genes, as discussed earlier. Because the breeding approach is not always feasible, a variety of management techniques are more generally used to lessen the incidence of disease in the field. These techniques include the production of virus-free plants, the use of chemical pesticides (to kill insects that transmit viruses), and the use of viral cross-protection methods. Viral cross-protection is a phenomenon exhibited by plants which, after infection with a mild strain of a virus, show resistance to more virulent strains. In management schemes, sensitive plants are deliberately infected with a mild, asymptomatic strain of the virus (Fulton, 1986). Advances in the understanding of the structure and function of viral nucleic acid and in the understanding of the host–pathogen relationship have given rise to the development of new strategies for combating viral disease in plants. There are three approaches to developing resistance to a virus: (1) using antisense viral RNA, (2) cross-protection using genes encoding the viral coat protein, and (3) expression of satellite RNA.

5.1.1. Antisense RNA

For reasons not completely understood, expression of a gene can be inhibited by transcription of the corresponding antisense gene within the same cell (Weintraub, 1990). An antisense gene is a gene in which the opposite strand serves as the template during transcription. Chimeric antisense genes derived from viral genes have been used to give plants the ability to inhibit expression of said viral genes during infection, thus inhibiting viral propagation. Transgenic plants have been tested that contain antisense transcripts against cucumber mosaic virus (CMV) (Cuozzo et al., 1988) and potato virus X (PVX) (Hemenway et al., 1988) and are able to protect plants at low inoculum concentrations.

5.1.2. Cross-Protection

The phenomenon of cross-protection spurred the use of chimeric genes specifying viral coat protein as a way of protecting plants from viral disease (Bevan et al., 1985; Powell et al., 1986). This approach has now been successfully used against a number of different viruses, e.g., tobacco mosaic virus (TMV) (Bevan et al., 1985; Powell et al., 1986; Nelson et al., 1988), alfalfa mosaic virus (AlMV) (Tumer *et al.*, 1987; Van Dun *et al.*, 1987), CMV (Cuozzo *et al.*, 1988), and PVX (Hemenway *et al.*, 1988). One hypothesis for the mode of this resistance is that after the infecting virus sheds its coat upon entry into the plant cell, the naked viral genome (RNA) is quickly recoated by the free cytoplasmic coat protein before translation and replication can occur. Translation of disassembled virus *in vitro* can be inhibited by the addition of free coat protein (Tumer *et al.*, 1987), supporting this model. Transgenic plants expressing the coat protein of TMV, AlMV, CMV, or PVX display a delay in appearance of viral symptoms and have a marked reduction in levels of virus after inoculation. Recently, resistance against mixed virus infection (potato viruses X and Y) was engineered in potato (Russet Burbank) through the expression of two coat protein–encoding genes (Lawson *et al.*, 1990). These experiments demonstrate that the coat protein approach can be extended to bestow resistance to mixed viral infections.

5.1.3. Satellite RNA

Satellite RNAs are polynucleotides that are sometimes found associated with RNA viruses. They are propagated and transmitted from one plant to another by the replication and transmission machinery of their companion virus. Satellite RNAs have the ability to modify the virulence phenotype of their companion virus. Satellite RNAs that decrease the virulence of their host virus have been copied into cDNA (using the enzyme reverse transcriptase) and this cDNA has been expressed in transgenic plants that produce satellite RNA (Harrison *et al.*, 1987; Gerlach *et al.*, 1987). Transgenic plants expressing the satellite RNAs of the tobacco ring spot virus (TobRV) and CMV were shown to decrease the severity of the infection caused by the companion virus (Harrison *et al.*, 1987; Gerlach *et al.*, 1987). Amplification of satellite RNA in transgenic plants infected with the companion virus was observed in these studies. The possibility that these amplified sequences could be packaged and released into the environment through viral vectors raises the issue of transgenic release. Further controversy is fueled by the fact that some satellite RNAs can actually increase virulence (Simon *et al.*, 1988) and that variants of unknown pathology might be created through recombination *in planta* with other satellite RNAs (Cascone *et al.*, 1990). More information about the ecology of satellite RNAs must be obtained before the potential risks and benefits can be evaluated.

5.2. Insect Resistance

Resistance to insects has been achieved by employing genes that code for protease inhibitors and for insecticidal crystalline proteins (ICPs). Both work by disturbing the digestion of the feeding insect, though in different ways.

5.2.1. Protease Inhibitors

Protease inhibitors are proteins produced by plants in order to discourage insect feeding. These proteins interfere with protein digestion in the insect gut. Protease inhibitors are produced by a variety of plants and have been found to be toxic to diverse insects (Green and Ryan, 1972; Gatehouse and Boulter, 1983; Ryan *et al.*, 1986). A gene encoding a protease inhibitor, the cowpea trypsin inhibitor (CpTI), has been transferred and expressed in tobacco plants (Hilder *et al.*, 1987). Transgenic plants expressing CpTI were protected when infested with larvae of *Heliothis virescens*.

5.2.2. Insecticidal Crystalline Proteins

ICPs are a class of proteins produced by different strains of the microbe *Bacillus thuringiensis*. Excellent reviews have been published on the biology of *B. thuringiensis* (Aronson *et al.*, 1986; Whiteley and Schnepf, 1986; Höfte and Whiteley, 1989). The ICPs of *B. thuringiensis* are found in the spore (Dulmage, 1981) of this bacterium as a paracrystalline inclusion body. This crystal is composed of proteins previously referred to as the delta endotoxin (Angus, 1954) and now called the insecticidal crystal proteins (ICPs; Höfte and Whiteley, 1989). Different strains of *B. thuringiensis* make ICPs that are toxic to specific groups of insect pests (Goldberg and Margalit, 1977; Yamamoto and McLaughlin, 1981; Krieg *et al.*, 1983; Aronson *et al.*, 1986; Herrnstadt *et al.*, 1986; Fischhoff *et al.*, 1987; Vaeck *et al.*, 1987). Most of the commercially significant insect pests of woody plants belong to the Lepidoptera and may be sensitive to *B. thuringiensis*–encoded ICPs. ICPs are used routinely as a topical insecticide in the form of suspensions of *B. thuringiensis* spores.

When the ICP is ingested by the feeding insect larvae, it is first solubilized in the alkaline midgut of the insect. Here the soluble ICPs are then acted upon by midgut protease(s), which cleave the ICPs and create the highly toxic insecticidal crystal protein fragments (ICPFs) that kill the insect larvae (Lilley *et al.*, 1980). In susceptible insects the ICPF interacts with the epithelial cells that line the midgut, creating pores in the membrane that result in osmotic lysis of these cells (Harvey *et al.*, 1983; Knowles and Ellar, 1987). Recent evidence indicates that the ICPF binds to specific cell receptors on the surface of brush border membranes of midgut epithelium cells of susceptible insects (Hofmann *et al.*, 1988). It is thought that this binding disrupts the cells of the epithelium, killing the insect.

Transgenic tobacco (Vaeck *et al.*, 1987) and tomato (Fischhoff *et al.*, 1987) plants expressing ICPF have been obtained. These transgenic plants expressed sufficient quantities of ICPF to protect the plants from damage caused by the feeding larvae of tobacco and tomato hornworm, both voracious pests. Currently, at least four plant biotechnology companies are field-testing several annual crop species expressing ICPFs. Some important considerations accompany this approach. The ICPF must be made in sufficient quantities to kill the insect pest and must be stable in the plant tissue and not degraded rapidly. The production of the protein must not interfere with the plant's metabolism and must not compromise the quality of the produce in any way.

5.3. Herbicide Resistance

Herbicides, like all toxins, interfere with normal cellular function. Herbicides are distinguished from other toxins in their specific disruption of metabolic pathways unique to plants. This is similar to the classification of antibiotics, which have their greatest disruptive effect on prokaryote-specific metabolism. Toxins interact with cellular structures such as enzymes, cell walls, and ribosomes and inhibit or modify normal cellular processes. Resistance to a toxin can be achieved by three types of mechanisms, as described by Comai and Stalker (1986). These can be referred to as:

1. **Altered target** The normally sensitive enzyme (or other structure) is altered so that, while it retains activity, it does not interact with the toxin (at least not as strongly).
2. **Overexpression** The normally sensitive enzyme may be increased in its cellular concentration, so that even during exposure to the toxin, functional enzymes remain.
3. **Detoxification** The organism may be able to degrade or modify the toxin.

The first strategy would be possible if the gene for such a modified enzyme could be cloned. The second approach would be viable if the gene for the enzyme could be isolated and its protein coding region attached to a more powerful promoter. The third would be possible if genes for detoxification enzymes could be isolated. In each case the resulting gene would be used in a transformation scheme in order to confer resistance to the targeted plant. An example of the altered target mechanism is the creation of glyphosate resistance in plants. Glyphosate is a herbicide, marketed by the Monsanto Company as Roundup, which is toxic because it interferes with the biosythesis of aromatic amino acids. This pathway is common to plants and bacteria, so it is not surprising that glyphosate can also act as an antibiotic. Mutant *Salmonella typhimurium* were obtained that displayed resistance to glyphosate; the gene that had been mutated, *aro*A, encodes 5-enolpyruvyl-shikimate 3-phosphate (EPSP) synthase and was shown to confer glyphosate resistance to *E. coli* (Comai *et al.*, 1983). The gene product of *aro*A confers resistance because glyphosate is unable to bind to and inactivate it, so synthesis of aromatic amino acids is not disturbed. The *aro*A gene was used to make a chimeric gene that could be expressed in plants. The construct was then transformed into tobacco and shown to give resistance to glyphosate (Comai *et al.*, 1985). Resistance to other herbicides has been achieved in plants; a comprehensive review of the subject is that of Padgette *et al.* (1989).

6. EXAMPLES OF WOODY PLANT TRANSFORMATION

Transgenic plants of poplar, walnut, apple, strawberry, plum, and neem have all been successfully produced by means of *Agrobacterium*-mediated transformation. The following is a summary of the strategies used in the transformation of each species.

6.1. Poplar (*Populus alba* × *grandidentata*)

This work was reported by Fillatti *et al.* (1987). Leaf explants of this hybrid poplar were cocultivated with two different strains of *A. tumefaciens*. One of these strains, LBA4404, is a disarmed (nononcogenic) strain (Hoekema *et al.*, 1983), and the other, C58, is an oncogenic, nopaline-producing strain (Hamilton and Fall, 1971). Prior to cocultivation, each strain was transformed with the binary plasmid pPMG85/587, which carries a recombinant T-DNA. The T-DNA on this plasmid carries two copies of the chimeric aminoglycoside phosphotransferase gene and one copy of glyphosate-resistant EPSP synthase. The explants were cultured, and kanamycin-resistant shoots were obtained from adventitious meristems. The particularly interesting feature of the results of this study is that no transformants were obtained from those explants cocultivated with the disarmed strain; only explants that had been cocultivated with C58 plus pPMG85/587 gave rise to transformants. The kanamycin-resistant shoots were separated into two groups, A and B, on the basis of shoot morphology. Type A shoots were visually indistinguishable from normal regenerated shoots, and type B shoots were similar to crown-gall teratoma shoots. The difference between these groups was that the B group had arisen from cells transformed with both the artificial and wild-type T-DNAs, while the A group had arisen from cells that had received only the artificial T-DNA. Western blots revealed that the introduced ESPS synthase gene was expressed in transformed plants of both types. Actual resistance of the plants to glyphosate has yet to be confirmed.

6.2. Walnut (*Juglans regia*)

Walnut is a good example of a crop species in which adventitious bud formation has not been reported to occur from tissue segments, making it impossible to use the leaf disk transformation system described by Horsch *et al.* (1985). Like a number of woody species, walnut undergoes somatic embryogenesis (Tulecke and McGranahan, 1985). Walnut somatic embryos have been successfully used as a target tissue for transformation and regeneration of transgenic plants. Initiated originally from developing zygotic embryos, walnut somatic embryos proliferate in culture, producing numerous secondary embryos from single cells in the epidermal layer (Tulecke and McGranahan, 1985; Polito *et al.*, 1989). The epidermal cells in intact somatic embryos are susceptible to transformation by wild-type and recombinant strains of *A. tumefaciens* and provide a means to regenerate nonchimeric transgenic plants (McGranahan *et al.*, 1988; Dandekar *et al.*, 1989). Although the somatic embryo tissue was highly transformable with *Agrobacterium*, transgenic plants containing the marker gene encoding APT could only be selected at a low frequency (McGranahan *et al.*, 1988). This gene transfer system has now been made more efficient using a combination of a selectable and a screenable marker gene to identify putative transformants (McGranahan *et al.*, 1990). This was accomplished by using binary plasmid vectors containing two marker genes, one encoding β-glucuronidase (GUS), a screenable marker gene, and the other a selectable marker gene encoding APT (Dandekar *et al.*, 1989; McGranahan *et al.*, 1990).

6.3. Apple (*Malus pumila*)

Although the lack of an efficient tissue culture and regeneration system has been one of the major stumbling blocks preventing the development of gene transfer technology in tree crops, this does not hold true for apple, where a wide variety of tissue culture procedures have been successfully carried out. Apple cultivars can be micropropagated efficiently (Jones, 1976; Zimmerman, 1983), and somatic embryogenesis has been reported from leaf explants (James *et al.*, 1984). Several studies have been reported in the literature on regeneration through the formation of adventitous buds and have been reviewed by James (1987). A high-frequency regeneration system has been reported from leaf tissue of different apple cultivars, including McIntosh, Triple Red Delicious (Fasolo *et al.*, 1989), and Greensleeves (James *et al.*, 1988). Transformation of leaf discs of Greensleeves with *Agrobacterium* carrying the binary vector Bin 6 have been reported (James, 1987; James *et al.*, 1989). Transgenic apple plants were subsequently obtained from these leaf discs, and although the frequency of transformation was low, the study clearly indicates that apple can be transformed in this manner. Recent studies indicate that several factors could influence the transformation frequency of apple, including the type of explant tissue used, the strain of *Agrobacterium*, and the design of plasmid vectors (Dandekar *et al.*, 1990). Genetic alterations in an *Agrobacterium* strain by introducing additional copies of the virulence genes *vir* A, B, and G greatly stimulated virulence on apple (Dandekar *et al.*, 1990). Apart from *Agrobacterium*-mediated transformation, it should now be possible to try direct DNA transfer methods because recent reports show that apple plants can be regenerated from protoplasts (Patat-Ochatt *et al.*, 1988; Wallin and Johansson, 1989).

6.4. Strawberry (*Fragaria* × *ananassa*)

James *et al.* (1990a,b) have reported transformation of the strawberry cultivar Rapella using a disarmed binary vector system. *APT* (kanamycin resistance) was used as a selectable marker and nopaline synthase (*NOS*) was used as a screenable marker. The comparatively short generation time of strawberry, compared to other woody plants, was advantageous for the purpose of carrying out inheritance studies. Seeds from the selfed transformants were germinated, and the tissue from these plants was tested for either nopaline content or the ability to produce callus on a medium containing kanamycin. All nopaline-positive plants were also kanamycin resistant, as expected. Nopaline synthase was found to segregate in the F_2 in a 3:1 ratio (3 nopaline-positive to 1 nopaline-negative). Additionally, strawberry plants were transformed with the T-DNA-encoded isopentyltransferase (*ipt*) gene, the enzyme of which catalyzes the production of the cytokinin *trans*-zeatin. Such plants were characterized by a lack of apical dominance, hormone autotrophy, and an inability to form roots.

Another report of strawberry transformation is that of Nehra *et al.* (1990) of the cultivar Redcoat. *Agrobacterium* was used to introduce a recombinant T-DNA carrying three genes: *APT, NOS,* and *GUS*. Assays for *APT* and *GUS* activity and Southern blot analysis all indicated the insertion of the T-DNA and marker genes into the regenerated plants.

Diploid strawberry (*Fragaria vesca*; $2n = 2x = 14$) has been suggested as a model system for woody plant transformation studies (Schuerman, unpublished). *F. vesca* has a generation time of 7 weeks, is highly inbred (isogenic), is easily propagated, and has a small genome, approximately 22 times that of *E. coli* (Ahmadi *et al.*, 1988). It is also regenerable from various tissues (Schuerman and Dandekar, unpublished) and infectable by wild-type strains of *Agrobacterium* (Uratsu *et al.*, 1990).

6.5. Plum (*Prunus domestica*)

An abstract describing the transformation of plum has been presented (Mante *et al.*, 1990). The abstract includes a description of a regeneration/transformation system that uses plum hypocotyl segments. A binary vector was used, the T-DNA of which carried chimeric *APT* and *GUS* genes. Regenerants were selected on the basis of root formation on kanamycin-containing medium and were assayed for *APT* and *GUS* activity. Transformed plants were detected. Southern analyses, indicating insertion of the T-DNA into the genome of the regenerated plants, were also reported.

6.6. Neem (*Azadirachta indica*)

This work was reported by Naina *et al.* (1989). Armed strains of *Agrobacterium* (K12 × 562E and K12 × 167, both octopine strains) were used to mobilize two different recombinant T-DNAs into cells of aseptic *Az. indica* seedlings. The recombinant T-DNAs were carried on binary vectors pCGN562 and pCGN167 (Dandekar *et al.*, 1987). The T-DNA on the former contains an *APT* chimeric construct with octopine synthase control sequences, while the latter contains an *APT* chimeric construct with the 35S CaMV promoter and a transcription termination sequence from the T-DNA of pTiA6. The seedlings were wounded by one of three methods and a suspension of the bacteria was subsequently applied. Infections with strain K12 × 562E gave rise to tumors, which then produced shoots. Strain K12 × 167 produced no tumors, but did give rise to shoots with teratomas at the base. Both types of shoots were rooted and transferred to soil. Arginine was included in the plant culture medium in order to avert possible depletion of this compound in transformed tissues (arginine is a substrate in octopine synthesis). Opine assays revealed the presence of octopine in shoots derived from infections with both strains.

7. ENVIRONMENTAL IMPACT OF TRANSGENIC WOODY PLANTS

For many applications, especially those related to crop improvement, field trials are absolutely necessary, and if this is not an option, transformation becomes an academic exercise only. Perhaps the most frequently considered arguments against the release of genetically modified organisms into the environment are ecological (Gould, 1988; Pimentel *et al.*, 1989). Such arguments stress the importance of thorough consideration of the possible impact of placing a transformed plant in the environment, citing historical ecological disasters that have occurred when newly introduced organisms have infiltrated and dominated existing ecosystems. The release of plants that are resistant to insects, for instance, would place pressure on insect populations to overcome this resistance. Viral populations would respond in a similar fashion when plants with viral resistance are released. The likelihood of resistance development in these populations must be estimated before the risk can be assessed. Extreme reliance on single resistant cultivars could be devastating in the event that the resistance is overcome.

Perhaps the least helpful response to transgenic release comes from those who confuse the issue by portraying man-made taxonomic divisions as sacred barriers or who complain that genetic engineering is simply too dangerous to be done in the first place. Of more assistance are authors who offer guidelines for the evaluation of risk associated with transgenic release (Raffa, 1989). Deployment of new genes in an environment may turn out to be the most critical factor in the success or failure of a genetic engineering program. Gene deployment can be done on a population level, on a temporal level, on an anatomical level, or using a combination of these. Examples would be the incorporation of an insect resistance gene into only a portion of the trees in an orchard (population level), expression of the foreign gene in only some of the tissues of the plant (anatomical level), and only when the tissues are immature (temporal level). This approach would reduce the intensity of the selection pressure on the insect population and would maintain a susceptible population that would not damage the produce of the transgenic trees.

Ecological studies will doubtless be of assistance in risk evaluation. With regard to insect resistance genes, many studies have already been undertaken to estimate the likelihood of susceptibility loss in insect populations (McGaughey, 1985; McGaughey and Johnson, 1987; Stone et al., 1989; Van Rie et al., 1990). Transgenic plants do not need to be created and released in order to unbalance an ecosystem; introductions of natural, foreign species can serve just as well. Transgenic release entails similar risks. By taking a lesson from ecological mistakes, and with a commitment to the type of responsible management that any worthwhile endeavor deserves, the risks of transgenic release can be minimized.

8. REFERENCES

Ahmadi, H., Bringhurst, R. S., Uratsu, S. L., and Dandekar, A. M., 1988, "Alpine" *Fragaria vesca*, a woody plant model for molecular study, *Genome* 30(Suppl. 1): 35.31.22.

Akiyoshi, D. E., Klee, H., Amasino, R. M., Nester, E. W., and Gordon, M. P., 1984, T-DNA of *Agrobacterium tumefaciens* encodes an enzyme of cytokinin biosynthesis, *Proc. Natl. Acad. Sci. USA* 81:5994–5998.

Alt-Mörbe, J., Neddermann, P., von Lintig, J., Weiler, E. W., and Schröder, J., 1988, Temperature-sensitive step in Ti plasmid *vir*-region induction and correlation with cytokinin secretion by *Agrobacteria, Mol. Gen. Genet.* 213:1–8.

Alt-Mörbe, J., Kühlmann, H., and Schröder, J., 1989, Differences in induction of Ti plasmid virulence genes *virG* and *virD*, and continued control of *virD* expression by four external factors, *Mol. Plant-Microbe Inter.* 2:301–308.

An, G., Watson, B., Stachel, S., Gordon, M., and Nester, E. W., 1985, New cloning vehicles for transformation of higher plants, *EMBO J.* 4:277–284.

Angus, T. A., 1954, A bacterial toxin paralyzing silkworm larvae, *Nature* 173:545–546.

Aronson, A. I., Beckman, W., and Dunn, P., 1986, *Bacillus thuringiensis* and related insect pathogens, *Microbiol. Rev.* 50:1–24.

Beck, E., Ludwig, G., Auerswald, E., Reiss, B., and Shaller, H., 1983, Nucleotide sequence and exact localization of the neomycin phosphotransferase gene from transposon Tn5, *Gene* 19:327–336.

Bevan, M., 1984, Binary *Agrobacterium* vectors for plant transformation, *Nucleic Acids Res.* 12:8711–8721.

Bevan, M. W., and Chilton, M-D., 1982, T-DNA of the *Agrobacterium* Ti- and Ri-plasmids, *Annu. Rev. Genet.* 16:357:384.

Bevan, M. W., Mason, S. E., and Goelet, P., 1985, Expression of tobacco mosaic virus coat protein by a cauliflower mosaic virus promoter in plants transformed by *Agrobacterium, EMBO J.* **4:**1921–1926.

Bolton, G. W., Nester, E. W., and Gordon, M. P., 1986, Plant phenolic compounds induce expression of *A. tumefaciens* loci needed for virulence, *Science* **232:**983–985.

Cascone, P. J., Carpenter, C. D., Li, X. H., and Simon, A. E., 1990, Recombination between satellite RNAs of turnip crinkle virus, *EMBO J.* **9:**1709–1715.

Chilton, M-D., Drummond, H. J., Merlo, D. J., Sciaky, D., Montoya, A. L., Gordon, M. P., and Nester, E. W., 1977, Stable incorporation of plasmid DNA into higher plant cells: The molecular basis of crown gall tumorigenesis, *Cell* **11:**263–271.

Chilton, M-D., Tepfer, D. A., Petit, A., David, C., Casse-Delbart, F., and Tempé, J., 1982, *Agrobacterium rhizogenes* inserts T-DNA into the genomes of the host plant root cells, *Nature* **295:**432–434.

Comai, L., and Stalker, D., 1986, Mechanism of action of herbicides and their molecular manipulation, *Oxford Surv. Plant Mol. Cell Biol.* **3:**166–195.

Comai, L., Sen, L., and Stalker, D., 1983, An altered *aro*A gene product confers resistance to the herbicide glyphosate, *Science* **221:**370–371.

Comai, L., Facciotti, D., Hiatt, W. R., Thompson, G., Rose, R. E., and Stalker, D. E., 1985, Expression in plants of a mutant *aro*A gene from *Salmonella typhimurium* confers tolerance to glyphosate, *Nature* **317:**741–744.

Couch, J. A., and Fritz, P. J., 1990, Isolation of DNA from plants high in polyphenolics, *Plant Mol. Biol. Rep.* **8:**8–12.

Crossway, A., Oakes, J. V., Irvine, J. M., Ward, B., Knauf, V. C., and Shewmaker, C. K., 1986, Integration of foreign DNA following microinjection of tobacco mesophyll protoplasts, *Mol. Gen. Genet.* **202:**179–185.

Cuozzo, M., O'Connell, K. M., Kaniewski, W., Fang, R-X., Chua, N-H., and Tumer, N. E., 1988, Viral protection in transgenic plants expressing the cucumber mosaic virus coat protein or its antisense RNA, *Bio/Technology* **6:**549–558.

Dandekar, A. M., Gupta, P. K., Durzan, D. J., and Knauf, V., 1987, Transformation and foreign gene expression in micropropagated Douglas-fir (*Pseudotsuga menziesii*), *Bio/Technology* **5:**587–590.

Dandekar, A. M., Martin, L. A., and McGranahan, G. H., 1988, Genetic transformation and foreign gene expression in walnut tissue, *J. Am. Soc. Hort. Sci.* **113:**945–949.

Dandekar, A. M., McGranahan, G. H., Leslie, C. A., and Uratsu, S. L., 1989, *Agrobacterium*-mediated transformation of somatic embryos as a method for the production of transgenic plants, *J. Tissue Cult. Meth.* **12:**145–150.

Dandekar, A. M., Uratsu, S. L., and Matsuta, N., 1990, *Agrobacterium*-mediated transformation of apple: Factors influencing virulence, *Acta Hort.* **280:**483–494.

De Block, M., Herrera-Estrella, L., Van Montagu, M., Schell, J., and Zambryski, P., 1984, Expression of foreign genes in regenerated plants and their progeny, *EMBO J.* **3:**1681–1689.

De Cleene, M., and De Ley, J., 1976, The host range of crown gall, *Bot. Rev.* **42:**389–466.

De Greve, H., Leemans, J., Hernalsteens, J-P., Thia-toong, L., DeBeuckeleer, M., Willmitzer, L., Otten, L., Van Montagu, M., and Schell, J., 1982, Regeneration of normal and fertile plants that express octopine synthase, from tobacco crown galls after deletion of tumor-controlling functions, *Nature* **300:**752–755.

De La Pena, A., Lorz, H., and Schell, J., 1987, Transgenic rye plants obtained by injecting DNA into young floral tillers, *Nature* **325:**274–276.

Dulmage, H. T., 1981, Insecticidal activity of isolates of *Bacillus thuringiensis* and their potential for pest control, in *Microbial Control of Pests and Plant Diseases, 1970–1980* (H. D. Burges, ed.), pp. 193–222, Academic Press, London.

Fasolo, F., Zimmerman, R. H., and Fordham, I., 1989, Adventitious shoot formation on excised leaves of *in vitro* grown shoots of apple cultivars, *Plant Cell Tissue Organ Cult.* **16:**75–87.

Fillati, J. J., Sellmer, J., McCown, B., Haissig, B., and Comai, L., 1987, *Agrobacterium* mediated transformation and regeneration of *Populus, Mol. Gen. Genet.* **206:**192–199.

Fischhoff, D. A., Bowdish, K. S., Perlak, F. J., Marrone, P. G., McCormick, S. M., Niedermeyer, J. G., Dean, D. A., Kusano-Kretzmer, K., Mayer, E. J., Rochester, D. E., Rogers, S. G., and Fraley, R. T., 1987, Insect tolerant transgenic tomato plants, *Bio/Technology* **5:**807–813.

Fraley, R. T., Rogers, S. G., Horsch, R. B., Sanders, P. R., Flick, J. S., Adams, S. P., Bittner, M. L., Brand, L. A., Fink, C. L., Fry, J. S., Galluppi, G. R., Goldberg, S. B., Hoffmann, N. L., and Woo, S. C., 1983, Expression of bacterial genes in plant cells, *Proc. Natl. Acad. Sci. USA* **80:**4803–4807.

Fraley, R. T., Rogers, S. G., and Horsch, R. B., 1986, Genetic transformation in higher plants, *CRC Rev. Plant Sci.* **4:**1–46.

Fromm, M., Taylor, L. P., and Walbot, V., 1985, Expression of genes transferred into monocot and dicot plant cells by electroporation, *Proc. Natl. Acad. Sci. USA* **82:**5824–5828.

Fromm, M., Taylor, L. P., and Walbot, V., 1986, Stable transformation of maize after gene transfer by electroporation, *Nature* **319:**791–793.

Fulton, R. W., 1986, Practices and precautions in the use of cross protection for plant virus disease control, *Annu. Rev. Phytopathol.* **24:**67–81.

Gatehouse, A. M. R., and Boulter, D., 1983, Assessment of the antimetabolic effects of trypsin inhibitors from cowpea (*Vigna unguiculata*) and other legumes on development of the bruchid beetle *Callosobruchus maculatus, J. Sci. Food Agric.* **34:**345–350.

Gerlach, W. L., Llewellyn, D., and Haseloff, J., 1987, Construction of a plant disease resistance gene from the satellite RNA of tobacco ringspot virus, *Nature* **328:**802–805.

Goldberg, L. J., and Margalit, J., 1977, A bacterial spore demonstrating rapid larvicidal activity against *Anopheles sergentii, Uranotaenia unguiculata, Culex univitattus, Aedes aegypti,* and *Culex pipiens, Mosquito News* **37:**355–358.

Gould, F., 1988, Evolutionary biology and genetically engineered crops, *BioScience* **38:**26–33.

Green, T. R., and Ryan, C. A., 1972, Wound induced proteinase inhibitor in plant leaves: A possible defense mechanism against insects, *Science* **175:**776–777.

Guyon, P., Chilton, M-D., Petit, A., and Tempé, J., 1980, Agropine in "null type" crown gall tumors: Evidence for the generality of the opine concept, *Proc. Natl. Acad. Sci. USA* **77:**2693–2697.

Hain, R., Stable, P., Czernilofsky, A. P., Steinbiß, H. H., Herrera-Estrella, L., and Schell, J., 1985, Uptake, integration, expression and genetic transmission of a selectable chimeric gene by plant protoplasts, *Mol. Gen. Genet.* **199:**161–168.

Haissig, B. E., Nelson, N. D., and Kidd, G. H., 1987, Trends in the use of tissue culture in forest improvement, *Bio/Technology* **5:**52–59.

Hamilton, R. H., and Fall, M. Z., 1971, The loss of tumor-initiating ability in *Agrobacterium tumefaciens* by incubation at high temperature, *Experientia* **27:**229–230.

Harrison, B. D., Mayo, M. A., and Baulcombe, D. C., 1987, Virus resistance in transgenic plants that express cucumber mosaic virus satellite RNA, *Nature* **328:**799–802.

Harvey, W. R., Cioffi, M., Dow, J. A. T., and Wolfersberger, M. G., 1983, Potassium ion transport ATPase in insect epithelium, *J. Exp. Biol.* **106:**91–117.

Hemenway, C., Fang, R-X., Kaniewski, W. K., Chua, N-H., and Tumer, N. E., 1988, Analysis of the mechanism of protection in transgenic plants expressing the potato virus X coat protein or its antisense RNA, *EMBO J.* **7:**1273–1280.

Herrera-Estrella, L., De Block, M., Messens, E., Hernalsteens, J-P., Van Montagu, M., and Schell, J., 1983a, Chimeric genes as dominant selectable markers in plant cells, *EMBO J.* **2:**987–995.

Herrera-Estrella, L., Depicker, A., Van Montagu, M., and Schell, J., 1983b, Expression of chimeric genes transferred into plant cells using a Ti-plasmid–derived vector, *Nature* **303:**209–213.

Herrnstadt, C., Soares, G. G., Wilcox, E. R., and Edwards, D. L., 1986, A new strain of *Bacillus thuringiensis* with activity against coleopteran insects, *Bio/Technology* **4:**305–308.

Hilder, V. A., Gatehouse, A. M. R., Sheerman, S. E., Barker, R. F., and Boulter, D., 1987, A novel mechanism of insect resistance engineered into tobacco, *Nature* **330:**160–163.

Hoekema, A., Hirsh, P. R., Hooykaas, P. J. J., and Schilperoort, R. A., 1983, A binary plant vector strategy based on separation of *vir* and T-region of the *Agrobacterium tumefaciens* Ti-plasmid, *Nature* **303:**179–181.

Hofmann, C., Vanderbruggen, H., Höfte, H., Van Rie, J., Jansens, S., and Van Mellaert, H., 1988, Specificity of *Bacillus thuringiensis* delta endotoxins is correlated with the presence of high affinity binding sites in brush border membrane of target insect midguts, *Proc. Natl. Acad. Sci. USA* **85:**7844–7848.

Höfte, H., and Whiteley, H. R., 1989, Insecticidal crystal proteins of *Bacillus thuringiensis*, *Microbiol. Rev.* **53:**242–255.

Hood, E. A., Jen, G., Kayes, L., Kramer, J., Fraley, R. T., and Childton, M-D., 1984, Restriction endonuclease map of pTiBo542, a potential Ti plasmid vector for genetic engineering of plants, *Bio/Technology* **2:**702–709.

Hood, E. A., Helmer, G. L., Fraley, R. T., and Chilton, M-D., 1986, The hypervirulence of *Agrobacterium tumefaciens* A281 is encoded in a region of pTiBo542 outside T-DNA, *J. Bacteriol.* **168:**1291–1301.

Horsch, R. B., Fraley, R. T., Rogers, S. G., Sanders, P. R., Lloyd, A., and Hoffmann, N. L., 1984, Inheritance of functional foreign genes, *Science* **223:**496–498.

Horsch, R. B., Fry, J. E., Hoffmann, N. L., Eichholtz, D., Rogers, S. G., and Fraley, R. T., 1985, A simple and general method for transferring genes into plants, *Science* **227:**1229–1231.

James, D. J., 1987, Cell and tissue culture technology for the genetic manipulation of temperate fruit trees, in *Biotechnology and Genetic Engineering Reviews*, Vol. 5 (G. E. Russell, ed.), pp. 33–79, Intercept Ltd., Newcastle-upon-Tyne.

James, D. J., Passey, A. J., and Deeming, D. C., 1984, Adventitious embryogenesis and the *in vitro* culture of apple seed parts, *J. Plant Physiol.* **115:**217–229.

James, D. J., Passey, A. J., and Rugini, E., 1988, Factors affecting high frequency plant regeneration from apple leaf tissues cultured *in vitro*, *J. Plant Physiol.* **132:**148–154.

James, D. J., Passey, A. J., Barbara, D. J., and Bevan, M. W., 1989, Genetic transformation of apple (*Malus pumila* Mill.) using a disarmed Ti-binary vector, *Plant Cell Rep.* **7:**658–661.

James, D. J., Passey, A. J., and Barbara, D. J., 1990a, *Agrobacterium*-mediated transformation of the cultivated strawberry (*Fragaria x anannassa* Duch.) using disarmed binary vectors, *Plant Sci.* **69:**79–94.

James, D. J., Passey, A. J., and Barbara, D. J., 1990b, Regeneration and transformation of apple and strawberry using disarmed Ti-binary vectors, in *Genetic Engineering of Crop Plants. 49th Nottingham Easter School* (G. Lycett and D. Grierson, eds.), pp. 239–248, Sutton Bonington, University of Nottingham, Butterworths, UK.

Jefferson, R. A., 1987, Assaying chimeric genes in plants: The *GUS* gene fusion system, *Plant Mol. Biol. Rep.* **5:**387–405.

Jefferson, R. A., Kavanagh, T. A., and Bevan, M. W., 1987, *GUS* fusion: β-Glucuronidase as a sensitive and versatile gene marker in higher plants, *EMBO J.* **6:**3901–3907.

Jin, S., Komari, T., Gordon, M. P., and Nester, E. W., 1987, Genes responsible for the super-virulence phenotype of *Agrobacterium* tumefaciens A281, *J. Bacteriol.* **169:**4417–4425.

Jin, S., Roitsch, T., Ankenbauer, R. G., Gordon, M. P., and Nèster, E. W., 1990a, The VirA protein of *Agrobacterium tumefaciens* is autophosphorylated and is essential for *vir* gene regulation, *J. Bacteriol.* **172:**525–530.

Jin, S., Roitsch, T., Christie, P. J., and Nester, E. W., 1990b, The regulatory *VirG* protein specifically binds to a *cis*-acting regulatory sequence involved in transcriptional activation of *Agrobacterium tumefaciens* virulence genes, *J. Bacteriol.* **172:**531–537.

Jones, O. P., 1976, Effect of phloridzin and phloroglucinol on apple shoots, *Nature* **262:**392–393.

Kerr, A., 1969, Crown gall of stone fruit. I. Isolation of *Agrobacterium* isolates, *Aust. J. Biol. Sci.* **22**:111–116.

Klee, J. H., Yanofsky, M. F., and Nester, E. W., 1985, Vectors for transformation of higher plants, *Bio/Technology* **3**:637–642.

Klein, T. M., Wolf, E. D., Wu, R., and Sanford, J. C., 1987, High-velocity microprojectiles for delivering nucleic acids into living cells, *Nature* **327**:70–73.

Knowles, B. H., and Ellar, D. J., 1987, Colloid-osmotic lysis is a general feature of the mechanism of action of *Bacillus thruingiensis* δ-endotoxins with different insect specificities, *Biochim. Biophys. Acta* **924**:509–518.

Krieg, V. A., Huger, A. M., Langenbruch, G. A., and Schnetter, W. Z., 1983, *Bacillus thuringiensis* var. *tenebrionis:* A new pathotype effective against larvae of Coleoptera, *Z. Angew. Entomol.* **96**:500–508.

Kuhlemeier, C., Green, P. J., and Chua, N-H., 1987, Regulation of gene expression in higher plants, *Annu. Rev. Plant Physiol.* **38**:221–257.

Lawson, C., Kaniewski, W., Haley, L., Rozman, R., Newell, C., Sanders, P., and Tumer, N., 1990, Engineering resistance to mixed virus infection in a commercial potato cultivar: Resistance to potato virus X and potato virus Y in transgenic Russet Burbank, *Bio/Technology* **8**:127–134.

Lilley, M., Ruffell, R. N., and Sommerville, H., 1980, Purification of the insecticidal toxin in crystals of *Bacillus thuringiensis, J. Gen. Microbiol.* **118**:1–11.

Luo, Z-X., and Wu, R., 1988, A simple method for the transformation of rice via the pollen-tube pathway, *Plant Mol. Biol. Rep.* **6**:165–174.

Mante, S., Cornell, U., Morgens, P., Scorza, R., Cordts, J., and Callahan, A., 1990, *Agrobacterium*-mediated transformation of plum (*Prunus domestica*) hypocotyl segments and regeneration of transgenic plants, *In Vitro* **26**:44A.

Martin, L., 1987, Genetic transformation and foreign gene expression in tissue of different woody species, M.S. thesis. University of California, Davis.

McCabe, D. E., Swain, W. F., Martinell, B. J., and Christou, P., 1988, Stable transformation of soybean by particle acceleration, *Bio/Technology* **6**:923–926.

McCormick, S., Niedermeyer, J., Fry, J., Barnason, A., Horsch, R., and Fraley, R., 1986, Leaf disc transformation of cultivated tomato (*L. esculentum*) using *Agrobacterium tumefaciens, Plant Cell Rep.* **5**:81–84.

McGaughey, W. H., 1985, Insect resistance to the biological insecticide *Bacillus thuringiensis, Science* **229**:193–195.

McGaughey, W. H., and Johnson, D. E., 1987, Toxicity of different serotypes and toxins of *Bacillus thuringiensis* to resistant and susceptible Indianmeal moths (Lepidoptera: Pyralidae), *J. Econ. Entomol.* **80**:1122–1126.

McGranahan, G. H., Leslie, C. A., Uratsu, S. L., Martin, L. A., and Dandekar, A. M., 1988, *Agrobacterium*-mediated transformation of walnut somatic embryos and regeneration of transgenic plants, *Bio/Technology* **6**:800–804.

McGranahan, G. H., Leslie, C. A., Uratsu, S. L., and Dandekar, A. M., 1990, Improved efficiency of the walnut somatic embryo gene transfer system, *Plant Cell Rep.* **8**:512–516.

Naina, N. S., Gupta, P. K., and Mascarenhas, J. P., 1989, Genetic transformation and regeneration of transgenic neem (*Azadirachta indica*) plants using *Agrobacterium tumefaciens, Curr. Sci.* **58**:184–187.

Nehra, N. S., Chibbar, R. N., Kartha, K. K., Datla, R. S. S., Crosby, W. L., and Stushnoff, C., 1990, *Agrobacterium*-mediated transformation of strawberry calli and recovery of transgenic plants, *Plant Cell Rep.* **9**:10–13.

Nelson, R. S., McCormic, S. M., Delanney, X., Dube, P., Layton, J., Anderson, E. J., Kaniewska, M., Proksch, R. K., Horsch, R. B., Rogers, S. G., Fraley, R. T., and Beachy, R. N., 1988, Virus tolerance, plant growth, and field performance of transgenic tomato plants expressing coat protein from tobacco mosaic virus, *Bio/Technology* **6**:403–409.

Ochatt, S. J., and Caso, O. H., 1986, Shoot regeneration from leaf mesophyll protoplasts of wild pear (*Pyrus communis* var, *pyraster* L.), *J. Plant Physiol.* **122**:243–249.

Ochatt, S. J., and Power, J. B., 1988, An alternative approach to plant regeneration from protoplasts of sour cherry (*Prunus cerasus* L.), *Plant Sci.* **56**:75–79.

Ochatt, S. J., Cocking, E. C., and Power, J. B., 1987, Isolation, culture and plant regeneration of Colt cherry (*Prunus avium* × *pseudocerasus*) protoplasts, *Plant Sci.* **50**:139–143.

Oka, S., and Ohyama, K., 1985, Plant regeneration from leaf mesophyll protoplasts of *Broussonetia kazinoki* Sieb. (paper mulberry), *J. Plant Physiol.* **119**:455–460.

Otten, L. A. M. B., and Schilperoort, R. A., 1978, A rapid microscale method for the detection of lysopine and nopaline dehydrogenase activities, *Biochem. Biophys. Acta* **527**:497–500.

Owens, L. D., and Cress, D. E., 1985, Genotypic variability of soybean response to *Agrobacterium* strains harboring the Ti or Ri plasmids, *Plant Physiol.* **77**:87–94.

Padgette, S. R., Della-Cioppa, G., Shah, D. M., Fraley, R. T., and Kishore, G. M., 1989, Selective herbicide tolerance through protein engineering, in *Cell Culture and Somatic Cell Genetics of Plants*, Vol. 6 (I. K. Vasil, ed.), pp. 441–476, Academic Press, Inc., San Diego.

Paszkowski, J., Shillito, R. D., Saul, M., Mandak, V., Hohn, T., Hohn, B., and Potrykus, I., 1984, Direct gene transfer to plants, *EMBO J.* **3**:2717–2722.

Patat-Ochatt, E. M., Ochatt, S. J., and Power, J. B., 1988, Plant regeneration from protoplasts of apple rootstocks and scion varieties (*Malus* × *domestica* Borkh.), *J. Plant Physiol.* **133**:460–465.

Petit, A., David, C., Dahl, G. A., Ellis, J. G., Guyon, P., Casse-Delbart, F., and Tempé, J., 1983, Further extension of the opine concept: plasmids in *Agrobacterium rhizogenes* cooperate for opine degradation, *Mol. Gen. Genet.* **190**:204–214.

Pimentel, D., Hunter, M. S., LaGro, J. A., Efroymson, R. A., Landers, J. C., Mervis, F. T., McCarthy, C. A., and Boyd, A. E., 1989, Benefits and risks of genetic engineering in agriculture, *BioScience* **39**:606–614.

Polito, V. S., McGranahan, G. H., Pinney, K., and Leslie, C., 1989, Origin of somatic embryos from repetitively embryogenic cultures of walnut (*Juglans regia* L.): Implications for *Agrobacterium*-mediated transformation, *Plant Cell Rep.* **8**:219–221.

Powell, A. P., Nelson, R. S., De, B., Hoffmann, N., Rogers, S. G., Fraley, R. T., and Beachy, R. N., 1986, Delay of disease development in transgenic plants that express the tobacco mosaic virus coat protein gene, *Science* **232**:738–743.

Pythoud, F., Sinkar, V. P., Nester, E. W., and Gordon, M. P., 1987, Increased virulence of *Agrobacterium rhizogenes* conferred by the *vir* region of pTiBO542: Application to genetic engineering of poplar, *Bio/Technology* **5**:1323–1327.

Raffa, K. F., 1989, Genetic engineering of trees to enhance resistance to insects—Evaluating the risks of biotype evolution and secondary pest outbreak, *BioScience* **39**:524–534.

Revilla, M. A., Ochatt, S. J., Doughty, S., and Power, J. B., 1987, A general strategy for the isolation of mesophyll protoplasts from deciduous fruit and nut tree species, *Plant Sci.* **50**:133–137.

Russel, J. A., and McCown, B. H., 1986, Culture and regeneration of *Populus* leaf protoplasts isolated from non-seedling tissue, *Plant Sci.* **46**:133–142.

Russel, J. A., and McCown, B. H., 1988, Recovery of plants from leaf protoplasts of hybrid-poplar and aspen clones, *Plant Cell Rep.* **7**:59–62.

Ryan, C. A., Bishop, P. D., Graham, J. S., Broadway, R. M., and Duffey, S. S., 1986, Plant and fungal cell wall fragments activate expression of proteinase inhibitor genes for plant defense, *J. Chem. Ecol.* **12**:1025–1035.

Schocher, R. J., Shillito, R. D., Saul, M. W., Paszkowski, J., and Potrykus, I., 1986, Co-transformation of unlinked foreign genes into plants by direct gene transfer, *Bio/Technology* **4**:1093–1096.

Schröder, G., Waffenschmidt, S., Weiler, E. W., and Schröder, J., 1983, The T-region of Ti-plasmids codes for an enzyme synthesizing indole-3-acetic acid, *EMBO J.* **2**:403–409.

Shaw, C. H., Watson, M. D., Carter, G. H., and Shaw, C. H., 1984, The right hand copy of the nopaline Ti-plasmid 25 bp repeat is required for tumour formation, *Nucleic Acids. Res.* **12**:6031–6041.

Simon, A. E., H. Engel, H., Johnson, R. P., and Howell, S. H., 1988, Identification of regions affecting virulence, RNA processing and infectivity in the virulent satellite of turnip crinkle virus, *EMBO J.* **7**:2645–2651.

Stachel, S., and Zambryski, P., 1986, *vir*A and *vir*G control the plant-induced activation of the T-DNA transfer process of *A. tumefaciens, Cell* **46**:325–333.

Stachel, S. E., Messens, E., Van Montagu, M., and Zambryski, P., 1985, Identification of the signal molecules produced by wounded plant cells that activate T-DNA transfer in *Agrobacterium tumefaciens, Nature* **318**:624–629.

Stone, T. B., Sims, S. R., and Marrone, P. G., 1989, Selection of tobacco budworm for resistance to a genetically engineered *Pseudomonas fluorescens* containing the delta-endotoxin of *Bacillus thuringiensis* subsp. kurstaki, *J. Invertebr. Pathol.* **53**:228–234.

Thomashow, M. F., Hugly, S., Buchholz, W. G., and Thomashow, L. S., 1986, Molecular basis for the auxin independent phenotype of crown gall tumor tissue, *Science* **231**:616–618.

Tulecke, W., 1987, Somatic embryogenesis in woody perennials, in *Cell and Tissue culture in Forestry*, Vol. 2, (J. M. Bonga and D. J. Durzan, eds.), pp. 61–91. Martinus Nijhoff, Boston.

Tulecke, W., and McGranahan, G. H., 1985, Somatic embryogenesis and plant regeneration from cotyledon tissue of walnut, *Juglans regia* L., *Plant Sci.* **40**:53–67.

Tumer, N. E., O'Connell, K. M., Nelson, R. S., Sanders, P. R., Beachy, R. N., Fraley, R. T., and Shah, D. M., 1987, Expression of alfalfa mosaic virus coat protein gene confers cross-protection in transgenic tobacco and tomato plants, *EMBO J.* **6**:1181–1189.

Ulian, E. C., Smith, R. H., Gould, J. H., and McKnight, T. D., 1988, Transformation of plants via the shoot apex, *In Vitro Cellul. Dev. Biol.* **24**:951–954.

Uratsu, S. L., Ahmadi, H., Bringhurst, R. S., and Dandekar, A. M., 1991, Relative virulence of *Agrobacterium* strains on strawberry (*Fragaria vesca*), *Hort. Sci.* **26**:196–199.

Vaeck, M., Reynaerts, A., Hofte, H., Jansens, S., DeBeuckeleer, M., Dean, C., Zabeau, M., Van Montagu, M., and Leemans, J., 1987, Transgenic plants protected from insect attack, *Nature* **328**:33–37.

Van Dun, C. M. P., Bol, J. F., and van Vloten-Doting, L., 1987, Expression of alfalfa mosaic virus and tobacco rattle virus coat protein genes in transgenic tobacco plants, *Virology* **159**:299–305.

Van Larebeke, N., Genetello, C., Schell, J., Schilperoort, R. A., Hermans, A. K., Hernalsteens, J. P., and Van Montagu, M., 1975, Acquisition of tumour-inducing ability by non-oncogenic agrobacteria as a result of plasmid transfer, *Nature* **255**:742–743.

Van Rie, J., McGaughey, W. H., Johnson, D. E., Barnett, B. D., and Van Mellaert, H., 1990, Mechanism of insect resistance to the microbial insecticide *Bacillus thuringiensis, Science* **247**:72–74.

Vardi, A., Spiegel-Roy, P., and Galun, E., 1982, Plant regeneration from *Citrus* protoplasts, variability in methodological requirements among cultivars and species, *Theor. Appl. Genet.* **62**:171–176.

Veluthambi, K., Krishnan, M., Gould, J. H., Smith, R. H., and Gelvin, S. B., 1989, Opines stimulate induction of the *vir* genes of the *Agrobacterium tumefaciens* Ti plasmid, *J. Bacteriol.* **171**:3969–3703.

Vernade, D., Herrera-Estrella, A., Wang, K., and Van Montagu, M., 1988, Glycine betaine allows enhanced induction of the *Agrobacterium tumefaciens vir* genes by acetosyringone at low pH, *J. Bacteriol.* **170**:5822–5829.

Wallin, A., and Johansson, L., 1989, Plant regeneration from leaf mesophyll protoplasts of *in vitro* cultured shoots of a columnar apple, *J. Plant Physiol.* **135**:565–570.

Wang, K., Herrera-Estrella, L., Van Montagu, M., and Zambryski, P., 1984, Right 25 bp terminus sequence of the nopaline T-DNA is essential for and determines the direction of DNA transfer from *Agrobacterium* to the plant genome, *Cell* **38:**455–462.

Wang, K., Stachel, S. E., Timmerman, B., Van Montagu, M., and Zambryski, P. C., 1987, Site-specific nick in the T-DNA border sequence as a result of *Agrobacterium vir* gene expression, *Science* **235:**587–591.

Wann, S. R., 1988, Somatic embryogenesis in woody species, in *Horticultural Reviews*, Vol. 10 (J. Janick, ed.), pp. 153–181, Timber Press, Portland, OR.

Watson, B., Currier, T. C., Gordon, M. P., Chilton, M-D., and Nester, E. W., 1975, Plasmid required for virulence of *Agrobacterium tumefaciens*, *J. Bacteriol.* **123:**255–264.

Weintraub, H. M., 1990, Antisense RNA and DNA, *Sci. Am.* **262:**40–46.

Whiteley, H. R., and Schnepf, H. E., 1986, The molecular biology of parasporal crystal body formation in *Bacillus thuringiensis, Annu. Rev. Microbiol.* **40:**549–576.

Winans, S., Ebert, P., Stachel, S., Gordon, M., and Nester, E. W., 1986, A gene essential for *Agrobacterium* virulence is homologous to a family of positive regulatory loci, *Proc. Natl. Acad. Sci. USA* **83:**8278–8282.

Yadav, N. S., Vanderleyden, J., Bennett, D. R., Barnes, W. M., and Chilton, M. D., 1982, Short direct repeats flank the T-DNA on a nopaline Ti-plasmid, *Proc. Natl. Acad. Sci. USA* **79:**6322–6326.

Yamamoto, T., and McLaughlin, R. E., 1981, Isolation of a protein from the parasporal crystal of *Bacillus thuringiensis* var *kurstaki* toxic to mosquito larvae, *Aedes taeniorhynchus, Biochem. Biophys. Res. Commun.* **103:**414–421.

Yanofsky, M. F., Porter, S. G., Young, C., Albright, L. M., Gordon, M. P., and Nester, E. W., 1986, The *vir*D operon of *Agrobacterium tumefaciens* encodes a site-specific nuclease, *Cell* **47:**471–477.

Zaenen, I., Van Larebeke, N., Teuchy, H., Van Montagu, M., and Schell, J., 1974, Supercoiled circular DNA in crown gall inducing *Agrobacterium* strains, *J. Mol. Biol.* **86:**109–127.

Zambryski, P., Joos, H., Genetello, C., Leemans, J., Van Montagu, M., and Schell, J., 1983, Ti plasmid vector for the introduction of DNA into plant cells without alteration of their normal regeneration capacity, *EMBO J.* **2:**2143–2150.

Zimmerman, R. H., 1983, Tissue Culture, in *Methods in Fruit Breeding* (J. N. Moore and J. Janick, eds.), pp. 124–135, Purdue University Press, West Lafayette, IN.

Chapter 6

Genetic Manipulation of Male Gametophytic Generation in Higher Plants

Ercole Ottaviano†, M. Enrico Pè, and Giorgio Binelli

1. POLLEN BIOTECHNOLOGY AND PLANT BREEDING

The success of a plant breeding program largely depends on the genetic composition of the base population used to select improved varieties and on the use of efficient selection procedures. The conventional approach offers a large spectrum of methods allowing increase of genetic variability and efficiency of selection. Practical results are well documented by the remarkable genetic gains obtained for many crops (Russell, 1974; Duvick, 1981; Borlaug, 1983). However, recent advances in biotechnology offer the possibility of increasing the efficiency of the conventional selection procedures and of enriching the genetic pool of a crop species by adding genes from alien species or obtained from *in vitro* manipulation. In this context, pollen technology can play an important role.

A promising method for obtaining high efficiency in selection, and one which can easily be integrated in the conventional methods, is based on pollen (male

†Deceased

Ercole Ottaviano, M. Enrico Pè, and Giorgio Binelli   Department of Genetics and Microbiology, University of Milan, 20133 Milan, Italy.

Subcellular Biochemistry, Volume 17: Plant Genetic Engineering, edited by B. B. Biswas and J. R. Harris. Plenum Press, New York, 1991.

gametophyte) selection (MGS). It consists in the application of selection pressures during the male gametophytic phase, so as to increase the frequency of fertilization by pollen grains carrying genes that confer better adaptability. Recent experimental results show that this selection operates in nature and can play an important role in the evolution of wild (Mulcahy, 1979; Ottaviano and Sari Gorla, 1979) and cultivated (Ottaviano and Mulcahy, 1986) plants and in the control of the genetic structure of plant populations (Ottaviano *et al.*, 1988b; Ottaviano, 1988).

Hybrid genotypes obtained by crossing inbred lines having high combining ability are the best system for maximizing heterosis and for combining positive traits of two different genotypes. Hybrid seed production can be obtained by means of the genetic control of pollen development or pollen function (cytoplasmic or chromosome male sterility, self-incompatibility, gametophytic factors) or by means of chemical treatments (see Duvick, 1965; Edwardson, 1970; Frankel and Galun, 1977; Bianchi and Lorenzoni, 1975; McRae, 1985; for review). All the genetic mechanisms described are found in nature and can also be achieved by standard genetic manipulation. However, the isolation of genes controlling pollen development or the interaction between pollen and style and of genes specifically expressed in the male gametophytic generation offers the possibility of transferring efficient genetic mechanisms between alien species and of developing new strategies for the production of hybrid varieties.

Because of its simple structure and the possibility of *in vitro* germination and *in vitro* culture in the late stages of development, the pollen grain is an ideal vector for genetic transformation: if foreign DNA is integrated in the sperm cells, it would be transmitted to the progeny by means of normal sexual reproduction, without the difficulties encountered in other transformation systems, which call for plant regeneration from cell or tissue cultures.

The discussion of these topics in the following sections of this article calls for specific reference to developmental stages of the male gametophytic generation. For this reason the main features of the process are briefly described. Detailed information can be found in specialized reviews (Heslop-Harrison, 1971a, 1987; Shivanna and Johri, 1985).

Sexual reproduction in angiosperm plants gives rise to two alternate generations: (1) the gametophyte, which starts from haploid cells (spores) produced by the meiotic division and leads to the production of gamete cells, and (2) the sporophyte, which develops from the zygote, differentiates female (pistil) and male (anther) flower structures, and produces, respectively, female and male meiocytes. The female gametophyte (the embryo sac) is a microscopic structure enclosed in a specialized flower structure, the pistil (Figure 1). The embryo-sac mother cell (meiocyte) undergoes female meiosis, producing four megaspores, one of which undergoes three mitoses, resulting in eight haploid cells. These develop into the egg cell, two synergids, and three antipodals, and the remaining two (polar *nuclei*) fuse to form a diploid central cell. The male gametophyte (pollen) develops inside the anther (Figure 2), is shed at maturation, and, after

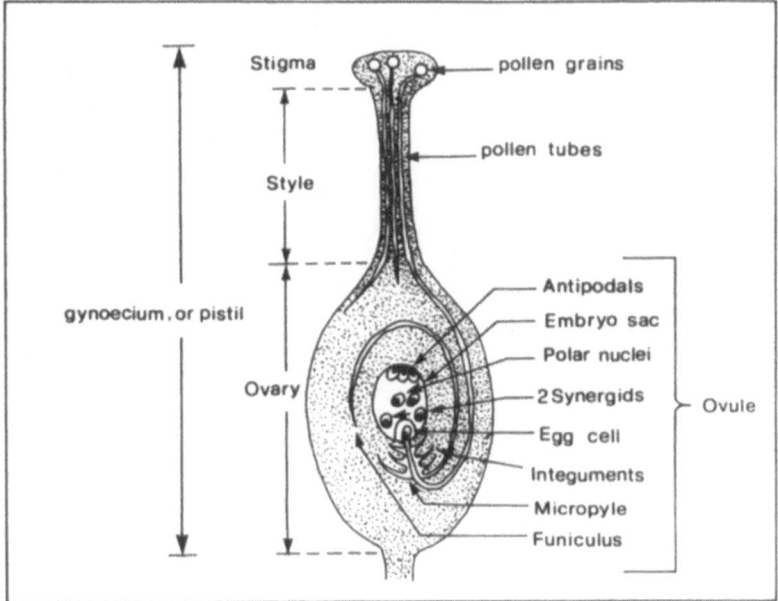

FIGURE 1. Longitudinal cross-section diagram of pistil showing also pollen tube growth.

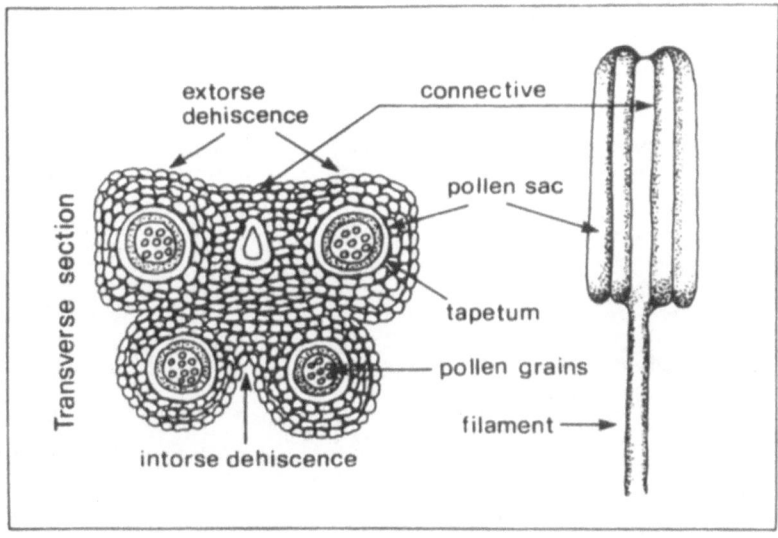

FIGURE 2. Anther and its transverse cross-section diagram.

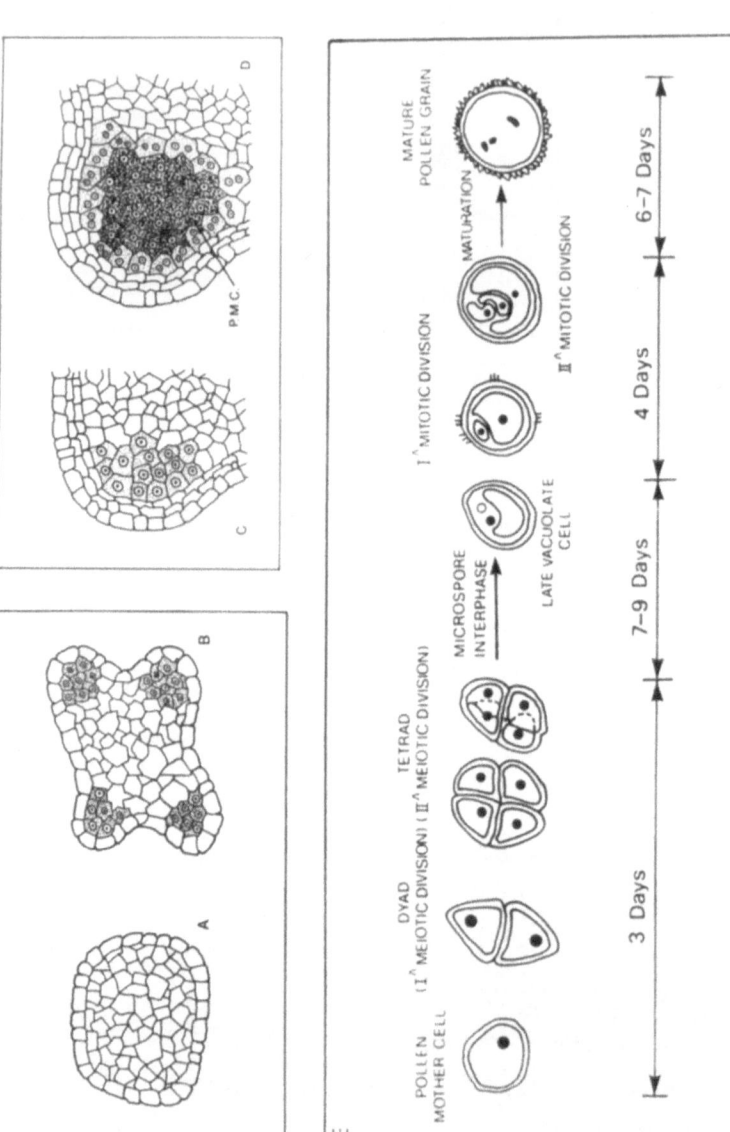

FIGURE 3. (A, B) Cross-section of a developing anther: (A) undifferentiated meristematic tissue; (B) with differentiated archesporial cells; (C, D) with wall layer and melocytes. (E) Microsporogenesis and pollen formation (time expressed in days is referred to maize male gametophyte).

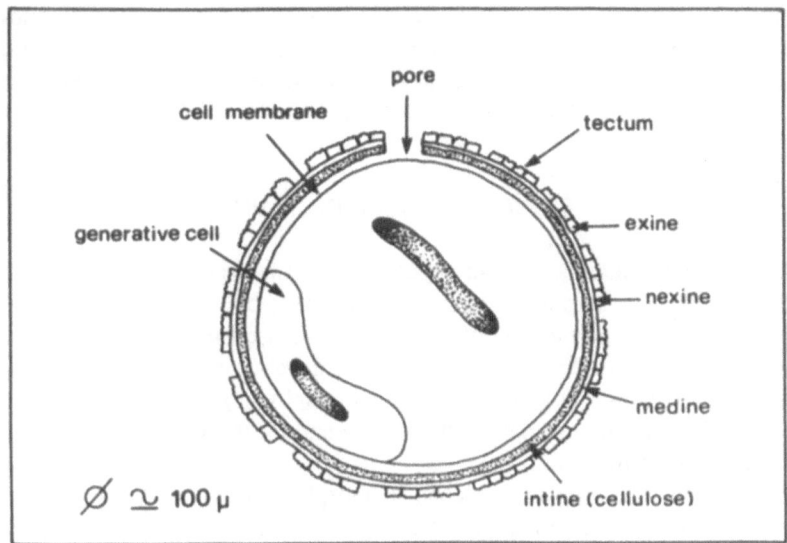

FIGURE 4. Cross-section diagram of a pollen grain.

germination on the stigma, develops a pollen tube which, passing through the transmitting tissue of the style, reaches the ovule and discharges the two male gametophytic cells. One of these fuses with the egg cell to produce the zygote and the other with the diploid central cell to produce the triploid endosperm.

The origin and development of male gametophytic generation are illustrated in Figure 3. Four groups of archesporial cells differentiate from the meristematic tissue contained in the young anthers. Archesporial cells divide to form the wall layers and the microspore mother cells. These undergo meiosis, producing a tetrad of microspores. Prior to meiosis, sporogeneous cells are interconnected by normal plasmodesmata, each approximately 350 Å in diameter. At the first stage of meiosis (leptotene), other and larger (2.5 nm) channels are formed, so that the meiocytes function as a single syncytium. Deposition of callose (B-1.3-glucane) in the wall of individual meiocytes begins during zygotene, leading to complete isolation of the meiocytes. During meiosis, the formation of the callosic wall is accomplished after each division, as in monocots, or postponed until the end of meiosis, as in dicots. After tetrad release, each microspore develops and undergoes the first haploid mitotic division, forming the vegetative and the generative cells. During development in the anther (trinucleate pollen species) or during pollen tube growth in the style (binucleate pollen species), the sperm cell undergoes the second haploid mitotic division, giving rise to the two spermatic cells. In the short time between shedding and germination, the mature pollen grain (Figure 4) lives as an independent organism. Germination consists of extrusion of the pollen tube through a germ pore. Vegetative nucleus and sperm cells move from the pollen grain into the pollen tube. During growth in the style-transmitting tissue, the tip of the tube, which encloses the sperm cells and the vegetative nucleus, is separated from the distal part of the tube by callose plugs.

2. THE EXTENT OF GENE EXPRESSION IN POLLEN

Genetic manipulations of plant populations by means of selection pressures applied to the male gametophytic generation are based on a number of biological assumptions: (1) a large number of genes are expressed in male gametophytic generation; (2) genes expressed in gametophytic generation are also expressed in the sporophytic tissues, (3) genes showing gametophytic gene expression have a significant effect on male gametophytic fitness and on the variability of characters of agronomical interest.

Observations showing the absence of morphological variability of pollen grains produced by a single highly heterozygous plant and theoretical considerations based on the positive aspects of diploidy support the classical view that gene expression in the gametophytic phase constitutes an independent domain, specific for the control of pollen functions (see Heslop-Harrison, 1971b, for a review of the topic). On the·other hand, the literature reports a large amount of data concerning inheritance of chromosome deficiencies, distorted segregations, genetic analysis of single pollen grains based on staining techniques, or protein electrophoresis, which support the idea of a large genetic domain common to sporophytic and gametophytic generation (see Ottaviano and Mulcahy, 1989, for a review of the topic).

However, recently a number of studies have produced quantitative estimates of the amount of gene expression in gametophytic generation and of the extent of gametophytic–sporophytic gene expression. In these studies three basic experimental approaches have been used: analysis of the isoenzymatic profiles by electrophoresis, heterologous hybridization of pollen cDNA to polyA $^+$ RNA from sporophytic tissues, and study of the gametophytic expression of genes affecting endosperm development.

Isozyme analysis to study gene expression in pollen was first proposed by Brewbaker (1971) and by Mascarenhas (1975). The test for gametophytic gene expression is simply based on the study of the electrophoretic pattern of dimeric or multimeric enzymes. The monomers forming these enzymes are codified by different alleles, which are classified "fast" or "slow" according to the electrophoretic mobility of their products: plants heterozygous for fast and slow alleles produce hybrid bands in the sporophytic tissues, while the zymogram of the pollen sample will show only the two parental bands if the protein is codified by the product of postmeiotic gene expression: single pollen grains contain only one of the two alleles and consequently the hybrid protein cannot be formed. The validity of this approach has been confirmed in maize by the use of partially diploid pollen, which allows heterozygous genotypes to be obtained also in this tissue (Frova et al., 1983). In tomato, Tanksley et al. (1981) assayed nine enzyme systems resulting from the expression of 28 structural genes and estimated that 62% of these were expressed in pollen, 58% in both pollen and sporophytic tissues (genetic overlap), and 3% in pollen only. In maize, where 15 enzyme systems for 34 structural genes have been analyzed (Sari-Gorla et al., 1986), the estimates were 86%, 72%, and 6%, respectively. In the *Populus* species, where 15 systems for 45–51 genes were analyzed, the results revealed similar values: genetic overlap was from 74 to 80%, and for genes expressed only in pollen, from 11 to 17% (Rajora and Zsuffa, 1986). Comparable results, although based on the study of a smaller number of enzyme systems, have been found in barley by Pedersen et al. (1987), who reckoned the amount of genetic overlap at 60%.

Quantitative estimates of gene expression in pollen of the genetic overlap based on a large number of genes have been obtained by means of mRNA analysis. In *Tradescantia paludosa,* Willing and Mascarenhas (1984) found that mature pollen grains and sporophytic tissues contain 20,000 and 30,000 different mRNAs, respectively. Moreover, mRNAs found in pollen have been shown to code for proteins similar to those synthesized during pollen germination and tube growth (Mascarenhas, 1984). Heterologous hybridization between cDNA from sporophytic tissues and $poly(A)^+RNA$ from pollen and the reciprocal hybridization have led to the estimate that 60% of the sequences analyzed are found in pollen, 54% are expressed in both gametophytic and sporophytic tissues, and 15% are specific to pollen. In corn, the same analysis shows that 24,000 different sequences are found in pollen and 31,000 in shoots, while hybridization between cDNA from pollen and $polyA^+RNA$ from shoots revealed 65% of genetic overlap (Willing et al., 1988).

In maize, we have studied 32 *de* (defective endosperm) mutants representing 32 different genes controlling endosperm development. Analysis of the F_2 segregation revealed that in 22 cases the gene affecting endosperm development also affects microspore development or pollen tube growth (Ottaviano et al., 1988a).

The findings relating to the estimates of gametophytic gene expression and gametophytic–sporophytic genetic overlap are summarized in Table I. The similarity of the results obtained by these studies carried out on different plant species and different classes of genes strongly supports the idea that the large extent of gametophytic–sporophytic gene expression is a general phenomenon in higher plants. However, most of the information relates to mature or germinating pollen. In the male gametophytic phase, mature pollen is a relatively quiescent stage in which metabolic activities are reduced to a very low level. Therefore, the

Table I
Estimates of the Percentage of Genes Expressed in Pollen and in the Sporophyte

Expressed in pollen only	Expressed in both pollen and sporophyte[a] (%)		Expressed in sporophyte only (%)	Assay	Reference
3%	58	(59)	39	Isozyme (tomato)	Tanksley et al., 1981
6%	73	(78)	21	Isozyme (corn)	Sari-Goria et al., 1986
11–17%	74–78		8–10	Isozyme (*Populus*)	Rajora and Zsuffa, 1986
10%		(60)	31	Isozyme (barley)	Pedersen et al., 1987
15%	54	(60)	31	mRNA (*Tradescantia*)	Willing and Mascarenhas, 1984
	65			mRNA (corn)	Willing et al., 1988
		(65)	35	*de* mutants (corn)	Ottaviano et al., 1988a

[a]Figures in parentheses indicate the percentage of sporophytically expressed genes expressed in pollen. (From Ottaviano and Mulcahy, 1989, updated.)

extent of gene expression evaluated at this stage is probably an underestimate. In fact, pollen has a quite complex developmental pattern from meiosis, through the phases of microspore mitosis, to dehiscence: the most important developmental processes of the microspore include wall morphogenesis by means of intine and exine formation, haploid cell division(s), and synthesis of storage substances. All these stages are characterized by intense metabolic activity, in terms of both mRNA and protein synthesis (Mascarenhas *et al.*, 1984, Stinson *et al.*, 1987).

At the molecular level, information on stage-specific gene expression in male gametophytic generation has been obtained in maize for genes codifying isozymes and for genes whose expression is induced by temperature stresses. With regard to microsporogenesis, the isozyme approach has revealed: (1) genes expressed just after tetrad release and all through microspore development (*Adh-1, Got-2,* β-*glu-1*), (2) genes expressed only in the later stages (*Cat-1*), and (3) genes showing sporophytic control in early microspore stages and gametophytic control in later stages (*Got-1*). Heat-shock proteins, although not found in mature pollen, are synthesized during microspore development, two polypeptides (84, 72 kDa) being detected all through microsporogenesis, while a group (94, 74, 58, 46 kDa) is synthesized during the uninucleate stage (before the first haploid mitotic division) and one band appears only after this stage in the binucleated and trinucleated microspores (Frova *et al.*, 1987, 1989). Postpollination gene expression studies have been based on comparative analysis of non-pollinated and pollinated silks, using as female and male sources genotypes differing in electrophoretic mobility, so that the source of the enzyme activity can be recognized in pollinated silks. Of the six isozymes analyzed (*Adh-1,* β-*glu-2,*

Cat-1,-3,-4, Got-1), male gametophytic activity during pollen tube growth was detected only for *Got-1* (Frova *et al.*, 1989).

3. GAMETOPHYTIC SELECTION AND PLANT BREEDING

High efficiency of selection in the male gametophytic generation is related to the haploid state and the large population size. In fact, because of the haploid state: (1) the evolution rate in the gametophytic phase is much higher than in diploids (Ottaviano and Sari Gorla, 1979; Pfahler, 1983), and (2) the probability of selecting complex allele combinations is greatly increased (number of genotypic combinations for *n* genes is much lower in haploids than in diploids, i.e., 2^n versus 4^n). On the other hand, a large population size allows application of a great intensity of selection.

In the possibility of operating on very large populations, which can be handled in a highly controlled environment, and the relatively simple structure of the grains, MGS reflects the basic characteristic of *in vitro* cell and tissue selection. However, MGS offers some specific advantages: (1) it can operate on a large spectrum of genetic variability produced by recombination and on the haploid genome; the only *in vitro* cultures offering the same advantages are the double haploid lines from anther cultures (Faraughi-Wehr *et al.*, 1986); (2) it is not dependent on plant regeneration; (3) because it is applied independently of the genotypes, it can easily be incorporated in a plant breeding program; and (4) it does not require sophisticated technologies.

Quantitative estimates of sporophytic–gametophytic gene expression indicate that MGS can operate on a very large number of genes, which are also involved in the control of sporophytic traits. However, the potential of MGS can be established only on the basis of knowledge concerning traits of agronomical interest that can be efficiently assayed and selected in the male gametophyte. In this regard, the possibility of pollen assays being useful in plant breeding has been proved by a number of studies. The most representative associations found between pollen and sporophytic characters are shown in Table II. Correlations between pollen competitive ability due to tube growth rate and sporophytic growth and vigor have been found in maize (Mulcahy, 1971, 1974), in which it has also been proved that the gametophytic character can be used to predict combining ability (Ottaviano *et al.*, 1980). Associations between pollen and sporophyte have been reported for tolerance to herbicide ethofumesate in *Beta vulgaris* (Smith, 1986), for resistance to the pathotoxin produced by *Helmintosporium maydis* (Laughnan and Gabay, 1973), and for tolerance to heavy metals in *Silene dioica* and *S. alba* (Searcy and Mulcahy, 1985a,b). Bino *et al.* (1988) have demonstrated that phytotoxic compounds produced by *Alternaria alternata* f.sp. lycopersici, eliciting disease symptoms in detached leaf bioassay,

Table II
Sporophytic-Gametophytic Associations

Traits		Species	Study
Gametophyte	Sporophyte		
Tube growth rate	Kernel weight	Maize	Mulcahy, 1971
	Seedling weight	Maize	Mulcahy, 1974
	Combining ability	Maize	Ottaviano *et al.*, 1980
Competitive ability	dry weight	Maize	Yamada 1983
Helminthosporium maydis resistance	Helminthosporium maydis resistance	Maize	Laughnan & Gabay, 1973
Alternaria brassicicola resistance	Alternaria brassicicola resistance	Brassica	Hodgkin & McDonald, 1986
Alternaria alternata resistance	Alternaria alternata resistance	Tomato	Bino *et al.*, 1988
Kanamycin resistance	Kanamycin resistance	Tomato	Bino *et al.*, 1987
Ethofumesate resistance	Ethofumesate resistance	Sugarbeet	Smith, 1986
Heavy metal tolerance	Heavy metal tolerance	Silene dioica	Searcy & Mulcahy, 1985b
High temperature tolerance	High temperature tolerance	Maize	Binelli *et al.*, 1985
Ozone tolerance	Ozone tolerance	Nicotiana Petunia	Feder, 1986
Glucosinate content	Glucosinate content	Brassica	Dungey *et al.*, 1988
Fatty acid quality	Fatty acid quality	Brassica	Evans *et al.*, 1987, 1988

inhibited both germination and tube growth *in vitro;* pollen from susceptible genotypes was more sensitive than pollen from resistant plants. More recently Bino *et al.* (1987) have demonstrated that tomato pollen from plants transgenic for a chimeric gene conferring resistance to kanamycin is resistant to the antibiotic.

The correlations between sporophytic and gametophytic tissues indicate that the traits to be studied can be evaluated at pollen level and that efficient pollen assays can be developed for use in plant breeding. Moreover, a high selection efficiency is obtained when the trait is controlled by the gametophyte and selection can consequently be applied within populations produced by a single heterozygous plant; i.e., the unit on which selection is applied is the pollen grain and not the single plant.

The efficiency of MGS has been tested in a number of cases (Table III); two different selective criteria have been adopted: the first consists in selecting for pollen competitive ability, the second in selecting for tolerance to environmental stresses.

Two methods have been devised to select for pollen competitive ability: the first is based on the variation of pollination intensity (number of pollen grains per

Table III
Sporophytic Effects of Gametophytic Selection Growth and Fertility

Characters	Species	Reference
Kernel and seedling weight root growth, dry matters product	Maize	Ottaviano et al., 1983, 1988a Landi et al., 1989
Seedling weight and vigor	Dianthus	McKenna and Mulcahy, 1983
Kernel weight, plant height	Cotton, Vigna, wheat	Ter-Avanesian, 1978
Fertility	Vigna	Ter-Avanesian, 1978
	Petunia	Mulcahy et al., 1975
Seedling weight and height	Turnera ulmifolia	McKenna, 1986
Plant vigor	Cucurbita pepo	Stephenson et al., 1986 Lee and Hartgerink, 1986
Seedling weight and germinability	Dianthus chinensis Anchusa officinalis	McKenna and Mulcahy, 1983 McKenna, 1986
Stern length, leaf number	Lotus corniculatus	Schlichting et al., 1987

flower) and the second on the variability of the distance that competing pollen has to cover in the style. The first method is based on the assumption that the intensity of competition between pollen tubes growing in the same style is related to the number of pollen grains on that style. With the increase of selection intensity, one expects a reduction of variances and an increase (or reduction) of mean values in the sporophytic progeny for those characters affected by genes showing overlap with pollen competitive ability. To avoid confusion between gametophytic selection and sporophytic selection, pollen from a single plant or from clones of single plants must be used.

The first account of pollination density affecting both mean and variance was provided by Ter-Avanesian (1949, 1978) in Vigna, cotton, and wheat. The characters affected were those expressing plant vigor and fertility. Similar results have been reported for a number of different species: Turnera ulmifolia (McKenna, 1986), Cucurbita pepo (Stephenson et al., 1986), Lotus corniculatus (Schlichting et al., 1987), and Cassia fasciculata (Lee and Hartgerink, 1986).

The second method, first suggested by Correns (1928), is based on the assumption that in competitive conditions the probability of fertilization by the faster-growing pollen tubes increases with the length of the style. Maize is very suitable for application of this method: silk length varies according to the position of the flower on the ear, increasing from the top to the base. This structure allows selection of gametophytes according to silk length: basal kernels are produced under high intensity of selection and apical kernels under low intensity. The most comprehensive results of this approach have been obtained in a study of a maize population (Ottaviano et al., 1982, 1988b): a selection program was carried out for high- and low-pollen competitive ability according to a recurrent selection scheme. The results obtained showed that: (1) for the gametophytic traits a very clear response to selection was obtained for pollen tube growth rate *in vivo:* mean values of the population obtained at high intensity of selection were higher than those of the population obtained at low intensity; (2) correlated responses were obtained for sporophytic traits, i.e., kernel weight, root tip growth *in vitro,* and seedling weight. Quantitative analysis allowed the partitioning of the variability of both characters. For pollen tube growth rate, 88% of the variability between families was due to genetic effects. The portion released by MGS accounted for 19% of total variability. The same partitioning for the sporophytic traits showed 60% and 16%, respectively. Considering that the procedure adopted was strictly based on within-plant selection, the response observed demonstrates that the pollen tube growth rate is affected by genes expressed in the postmeiotic phase and that these genes are also involved in the control of kernel and seedling weight and root tip growth. Data confirming these conclusions have been produced by Landi *et al.* (1989). The same method adapted to different flower structures has revealed the effect of MGS on sporophytic generation in *Dianthus chinensis* (Mulcahy and Mulcahy, 1975) and in *Anchusa officinalis* (McKenna, 1986). All these studies based on selection for pollen competitive ability produced correlated responses in sporophytic generation for traits that are the expression of plant growth and vigor. It may seem difficult to understand why genes controlling pollen tube growth rate should also affect these traits. However, taking into account the results of isozyme analysis showing genetic overlap for factors involved in the control of basic metabolic processes, such as starch metabolism, energy production, and wall component synthesis, it can be hypothesized that the correlated response is produced by selection acting on the variability produced by these genes.

A number of experimental results show that MGS can be effective for improving tolerance to environmental stresses (Table IV). The first report of this effect was provided by Zamir *et al.* (1982) and Zamir and Vallejos (1983) in tomato for response to low temperature stresses. The material used was the F_1 from an interspecific cross between *Lycopersicon hirsutum,* a species adapted to low temperatures, and a sensitive species, *L. esculentum.* The response to selection applied to developing microspores and especially to pollen tubes growing in the style was detected by isozyme analysis. The progeny from stressed gametophytes received a higher proportion of the genome from *L. hirsutum,* as shown by the differential transmission of isoenzyme alleles; the genes controlling the character are localized on two different chromosomes and expressed in the postmeiotic phase. Moreover, root elongation of seedlings from low-temperature backcrosses ($F_1 \times L.$ *esculentum*) was less inhibited by cold than that of the other progeny (Zamir and Gadish, 1987). The same approach was used to test the efficiency of MGS for improving tolerance to salinity: the stress applied to

Table IV
Sporophytic Effects of Gametophytic Selection Adaptability
to Environmental Stresses

Selected character for tolerance to	Selection stage[a]	Species	Reference
Temperature (low)	MS, PF	Tomato	Zamir et al., 1981, 1982, 1983, 1987
Salinity	MS, PF	S. pennellii × L. esculentum	Sacher et al., 1983 Sari-Goria et al., 1987
Heavy metals	MS, PF	Silene dioica	Searcy and Mulcahy, 1985a,b
Storage effects	MP	Potato	Pallais et al., 1986
Fusarium	(PF)	Tomato	Rabinowitch et al., 1978
Herbicide (Chlorsulfuron)	MS, PF	Maize	Sari et al., 1989

[a]MS, microsporogenesis; PF, pollen function; MP, mature pollen.

developing microspores of F_1 plants produced by crossing *Solanum pennellii* (salt tolerant) and *S. esculentum* (sensitive) gave rise to a preferential male gametophytic transmission of the genome from the tolerant parent (Sari-Gorla *et la.*, 1988a) and a majority of tolerant progeny (Sacher *et al.*, 1983). Heavy metal tolerance has been studied in *Silene dioica* and *Mimulus guttatus* (Searcy and Mulcahy, 1985a,b). It was demonstrated that the expression of metal tolerance in the two species is significantly influenced by the pollen genome, and that selection during pollen development is effective in increasing tolerance in the sporophyte. In maize, Sari-Gorla *et al.* (1988b) demonstrated a significant response to pollen selection for herbicide (Chlorsulfuron) tolerance. Selection was applied during microspore maturation (tassels were cut 2 weeks before anthesis and grown in artificial medium) or during pollen tube growth (pollinated silks were sprayed with the herbicide solution). Treated pollen produced from F_1 genotypes obtained by crossing a sensitive with a tolerant line was backcrossed to the sensitive parent. Response to selection was evaluated by growing the segregating seedling progeny on a substrate containing the herbicide. Plant and root elongation of seedlings from the backcross produced by treated pollen were less inhibited than in the case of the backcross obtained from untreated pollen.

The results of the studies reported in this section indicate that a large number of characters generally expressed at the cellular level can be efficiently assayed and selected in gametophytic generation. The spectrum would be even larger if selective pressure could be applied to genes whose expression could be experimentally induced during microsporogenesis or tube growth. It has been demonstrated in developing corn pollen that genes for the synthesis of heat-shock proteins (hsp) are induced by a temperature of 38°C from the first stages after meiosis until a few days before anthesis (Frova *et al.*, 1987). Although the function(s) of hsp's is far from clear, some data suggest that they may be involved in the mechanisms of protection against temperature stresses; therefore hsp's potentially represent a suitable marker for the detection and selection of heat-resistant genotypes.

The main problems to be solved for an efficient use of MGS in plant breeding relate to the development of suitable selection procedures. These are quite easy to apply in the case of abiotic stresses, which can be reproduced in growth chambers. Resistance to phytotoxins, however, cannot be easily assayed *in vivo,* because the anthers or the pistils would be damaged and the pollen killed by the treatment.

Selection of developing microspores can be obtained by means of *in vitro* culture techniques. As mentioned above, *in vitro* cultures of detached inflorescences have provided an efficient system of selection for herbicide tolerance in corn. A very promising technique could be based on *in vitro* cultures of isolated microspores: in tobacco. Heberle-Bors *et al.* (1988) obtained maturation *in vitro* of pollen grains that can be used for pollination to produce a seed set.

Selection of a pollen population prior to pollination can also be obtained by incubating the pollen grains in a medium suitable for germination but containing selective chemicals. In *Brassica napus,* Hodgkin (1988) developed this technique to apply selective pressure for resistance to phytotoxic compounds of the fungal pathogen *Alternaria brassicicola.* The results obtained showed that the treated pollen can be used successfully in pollination and that resistance to pathotoxins is increased in the gametophyte of the progeny.

4. MALE GAMETOPHYTIC MUTANTS

Manipulation of the reproduction system by means of pollen technologies requires identification and isolation of genes controlling pollen function. The more direct approach to this problem is the production and molecular characterization of male gametophytic mutants.

Microsporogenesis, germination, and pollen tube growth inside the stylar tissues are under the control of a very complex genetic system. The well-documented existence of postmeiotic gene expression and the fact that, especially during its early development, the male gametophyte is connected with the sporophytic tapetal cells within the anthers can account for the coordinated genetic control over pollen development exhibited by either the sporophyte or the gametophyte. The genotypes of pollen and of the plant, along with environmental effects, also play a fundamental role during pollen tube extrusion and growth in the pistil, when complex interactions between growing pollen grains and stylar tissues take place.

A lot of information is now available on the ultrastructural and physiological characteristics of pollen development and functions (see Mascarenhas, 1975;

Sunderland and Huang, 1987), whereas many aspects of genetic control of the male gametophyte still remain to be elucidated.

In our opinion, the lack of available mutants affecting microsporogenesis and pollen function and the poor biochemical and molecular characterization of the known mutants are a major problem still to be solved in order to obtain an understanding of the genetics of plant reproductive systems and which can be solved only partially by the application of recombinant DNA technology. Four classes of gametophytic mutants are necessary in order to dissect the genetic regulation of the male gametophytic phase: (1) sporophytic mutants affecting microsporogenesis, (2) gametophytic mutants affecting microsporogenesis, (3) gametophytic mutants influencing germination and pollen tube growth, (4) mutants controlling the pollen–style interaction. A comprehensive review of the genetics of angiosperm pollen has recently been published (Ottaviano and Mulcahy, 1989). We shall focus on analysis of the types of mutants mentioned above. Meiotic mutants will not be discussed in this chapter; for extensive reviews see Peloquin (1986) and Golubovskaya (1989).

4.1. Sporophytic Mutants for Pollen Development

A class of cytoplasmic mutants that cause pollen abortion and therefore male sterility exists in over than 140 plant species (Kaul, 1988). Cytoplasmic male sterility (CMS) is widely used in many crop species for the production of hybrid seed because male fertility can be restored by the action of one or a few nuclear restorer genes. CMS has been extensively studied in maize, where four major cytoplasms are known, the normal cytoplasm (N) and cytoplasm (types C, T, and S) that causes male sterility (Buchert, 1961; Laughnan and Gabay-Laughnan, 1983; Hanson and Conde, 1985). It has been demonstrated that rearrangements in the mitochondrial DNA are responsible for CMS (Pring et al., 1977; Pring and Levings, 1978; Abbott and Fauron, 1986; Bailey-Serres et al., 1986; for a review on plant mitochondria genome and its implication in CMS, see Levings and Brown, 1989). Dewey et al. (1986) found two open reading frames, called *ORF25* and *T-urf 13*, in the mitochondrial genome of *cms-T*. *ORF25* is transcribed in all four of the major maize cytoplasms, while *T-urf13* is cotrascribed with *ORF25* only in *cms-T*, giving rise to up to six different transcripts (Rocheford and Pring, 1990). A 13-kDa polypeptide was associated with *T-urf13* and is present in all *cms-T* plant tissues (Dewey et al., 1987; Wise et al., 1987). There is strong evidence that *T-urf13* is responsible for *cms-T* and for the severe susceptibility to *Bipolaris majdis* that *cms-T* exhibits. The joint action of the two nuclear genes *Rf1* and *Rf2* can cure the male sterility of *cms-T*. Both genes are sporophytically expressed: a plant with *cms-T* cytoplasm carrying *Rf1* and *Rf2* in the heterozygous state is fully fertile, producing 100% of viable pollen grains. *Rf1* interferes with *T-urf13*, reducing the amount of the 13-kDa polypeptide and altering the *T-urf13* (Dewey *et al.*, 1987; Kennell *et al.*, 1987). The observation that inbred lines show differences in the *T-urf13/ORF25* transcripts suggests a more complex interaction between the mitochondrial and nuclear genome in addition to the interaction with the restorer genes (Rocheford and Pring, 1990).

In *cms-S* cytoplasm the sterility has been associated with the presence of the two episomes *S1* and *S2*, molecules of self-replicating, double-stranded DNA (Levings *et al.*, 1980). Studying naturally occurring restoration events, Laughnan and colleagues have demonstrated that the genetic background influences reversion and that not every mutant loses the two episomes (Escote *et al.*, 1985; Zabala *et al.*, 1989). Mitochondria genomes in *N* and *S* cytoplasms contain sequences homologous to the *S* episomes (McNay *et al.*, 1983), suggesting that male sterility in *cms-S* cytoplasm is probably due to rearrangements in the mitochondrial DNA. *Cms-S* cytoplasm is restored by a single dominant restorer allele, Rf_3, the only gametophytic restorer gene so far known, which will be considered in the next section.

Cms-C is also restored to fertility by a dominant sporophytic gene, Rf_4. Recently a new maize cytoplasm causing CMS, designated *cms-Y*, has been discovered in China; to regain fertility it requires the presence of three nuclear restorer genes, Rf_4, Rf_5, and Rf_7, to regain fertility (Quin and Dun, 1990).

The other category of sporophytic mutants affecting microsporogenesis is represented by the genomic male sterile mutants (*ms*); *ms* mutants have been reported in several plant species (Gotteschalk and Kaul, 1974; Albertsen and Phillips, 1981; Kaul and Murthy, 1985). Biochemical or physiological alterations of the tapetal cells can often alter their trophic functions in favor of developing pollen grains and can bring about microspore degenerations. Although biochemical and molecular basis of this type of male sterility is not known, genetic data—degeneration of the microspores not revealing gametophytic control—indicate that the character has sporophytic basis and is very likely due to alterations of the trophic functions of the tapetal cells.

4.2. Gametophytic Mutants for Pollen Development and Function

Not many gametophytic mutants are known to influence pollen development. Mangelsdorf (1932) described a gametophytic mutant, *sp small pollen,* that reduces the size of pollen grain in maize. The finding that wild-type pollen from a heterozygous plant that segregates +/sp is bigger than wild-type pollen from a +/+ homozygous plant is a clear demonstration of intergametophytic competition for nutrients between the pollen grains developing in the same anther (Ottaviano and Mulcahy, 1989).

Genetic analyses showing distorted Mendelian segregations due to genes affecting pollen development, pollen germination, or tube growth have been reported in several plant species and recently reviewed (Ottaviano and Mulcahy, 1989). Particularly interesting is Rf_3, a nuclear gene that restores fertility of *cms-S* cytoplasm and represents the only known case of a restorer-of-fertility gene under gametophytic control. *Cms-S* Rf_3/rf_3 plants are fertile but segregate normal and sterile pollen grains in a 1:1 ratio (Buchert, 1961; Laughnan and Gabay, 1973). Studying naturally occurring restorers of *cms-S* cytoplasm, Laughnan *et al.* (1989a,b) have found that the restorer genes in different inbred lines map in different chromosomal positions. The results obtained from elegant genetic experiments seem to suggest that Rf_3 is a transposable element. This could represent a striking example of controlling activity exhibited by a transposon-like element on a trait so important for plant fitness.

A difference of fertilization ability between Rf_3 and rf_3 alleles is suggested by some data obtained in our laboratory (unpublished data): we crossed *cms-S* sterile plants derived from the inbred line Ny 821 with Rf_3/rf_3 plants of the same line; the expected 1:1 ratio of fertile versus male sterile plants in the progeny was not observed. Instead we detected a dramatic deviation in favor of fertile plants. Similar results were obtained using WF9 inbred lines. This suggests that the Rf_3 allele has a greater competitive ability than rf_3.

4.3. Sporophytic–Gametophytic Mutants Affecting Pollen Development and/or Function

This class of mutants is particularly useful for the dissection of the complex genetic program that regulates the male gametophyte. Furthermore, their isolation and characterization could be useful in an attempt to genetically manipulate important sporophytic traits by intervention in the gametophytic phase. The extensive gametophytic–sporophytic gene expression (see preceding section) clearly accounts for the existence of this type of mutant.

Meinke and colleagues (1982, 1986) have demonstrated the gametophytic expression of 10 embryo-lethal mutants in *Arabidopsis thaliana*. Upon selfing plants heterozygous for embryo-lethal mutations, they observed a nonrandom distribution of aborted seeds along the siliques. It is interesting to note that temperature affected the distribution of aborted seeds in several of the mutants analyzed.

The *Arabidopsis* embryo-lethal mutants—the observation that components of pollen development and function show positive correlation with endosperm development (Mulcahy, 1971; Ottaviano *et al.*, 1980) and that some alleles causing defective endosperm mutants in maize are expressed in the male gametophyte—suggested an indirect approach for the selection of viable mutant alleles with sporophytic–gametophytic effect.

A set of 34 defective endosperm (*de*) mutants of maize was tested for gametophytic effect. Three classes of defective endosperm/gametophytic effect (*de-ga*) were revealed, based on deviation from the expected Mendelian segregation: (1) *de-ga* mutants that affected microsporogenesis; (2) mutants showing effects on both pollen development and function; (3) mutants where only the pollen tube growth rate was affected. Direct observation of pollen grains showed that some *de-ga* mutants have an influence on pollen size and percentage of sterility in the anthers of plants heterozygous for the *de-ga* mutation (Ottaviano *et al.*, 1988a).

As for the sporophytic effect of such mutants, they influence endosperm development, reducing the dry matter accumulation rate and in some instances also plant development. The kinetics of dry matter accumulation appears to be mutant-specific (Manzocchi *et al.*, 1980a,b). A startling reduction of indolacetic acid synthesis has been observed in two mutants: the cells were smaller, but their number was not reduced.

4.4. Genes Controlling Pollen–Style Interaction

Physiological and genetic analyses have demonstrated that pollen germination and pollen tube growth are characterized by a complex network of interactions between pollen grains and pistil tissues and deeply influenced by the environment (Heslop-Harrison, 1987; Ottaviano and Mulcahy, 1989).

Pollen tube growth is independent of stylar nutrients only during the first phase of growth; signals of unknown molecular and chemical composition are probably sent from the pollen tube toward the ovary and from the ovary throughout the style (Bergamini-Mulcahy and Mulcahy, 1986). Analysis of the alcohol dehydrogenase, catalase, and glutamic-oxalacetic transaminase enzyme systems in pollinated and nonpollinated maize styles has shown that even if no pollen–style hybrid enzymatic molecules were produced and no additional gene expression was detected in response to pollination, mature silks were still quite active. The production of additional *Adh* and *Cat* enzymatic activity was observed, suggesting a functional support given to the growing pollen tubes of the styles (Frova, 1990).

The genetic control of pollen–pistil interactions is fundamental in the shaping of the reproductive system of the plants: it can determine the extreme situation of exclusive cross-fertilization as well as the opposite condition of reproductive isolation. The first type of action is typical of the self-incompatibility systems; the second is that exhibited by the gametophytic factors (*Ga*) found in maize, rice, and other species.

Self-incompatibility systems have been developed by several plant species and their specific reaction is controlled by simple multiallelic genes, the *S* genes (de Nettancourt, 1977). Pollen carrying the same *S* allele as the pistil cannot perform fertilization. Self-incompatibility can be sporophytic or gametophytic and has been thoroughly studied at the physiological and molecular level: these studies have led to the isolation and cloning of *S* genes in *Brassica oleracea* (Nasrallah *et al.*, 1985, 1987) and in *Nicotiana alata* (Anderson *et al.*, 1986). A comprehensive review of the genetic and molecular aspects of self-incompatibility has recently been published (Ebert *et al.*, 1989). On the basis of nucleotide sequence analysis of three different *S* alleles of *N. alata* and biochemical data, it has been postulated that the *S* gene codifies for a ribonuclease (McClure *et al.*, 1989, 1990).

As for the gametophytic factors, in maize nine different genes have been identified that confer high competitive ability on *Ga* pollen growing on *Ga/Ga* or *Ga/ga* styles (Nelson, 1952; Schwartz, 1960; Bianchi and Lorenzoni, 1975). Nelson has demonstrated that *ga* pollen germinates on *Ga/Ga* silks as much as *Ga* pollen, but the pollen tube growth rate rapidly diminishes and ceases after about 8 hr. The result is an evident poor fertilization ability of *ga* pollen that is outgrown by *Ga* pollen in competition on *Ga/Ga* silks.

A particularly strong effect is that of the allele *Ga^s*, which totally inhibits the growth of *ga* pollen on *Ga/Ga* styles. These observations can only be explained by specific interactions between the *Ga*-controlled functions in the pollen and in the style.

The existence of a genetic component of the pollen–pistil interaction was also revealed in maize by experiments based on mixed pollination procedure. Pollen–style interaction was evaluated analyzing differences in the proportion of uncolored kernels on ears obtained by mixed pollination with pollen from different lines (uncolored kernel) mixed in equal amounts with pollen from a standard line carrying alleles for colored aleurone.

It has been demonstrated that the pollen tube growth rate of five inbred lines depends on the genotype of the stylar tissue (Ottaviano *et al.*, 1980). An interesting observation was the positive correlation between specific combining ability (SCA) evaluated on the basis of pollen tube growth rate (an estimate of pollen–style interaction) and SCA evaluated on the basis of seedling dry weight.

Pollen–style interaction was studied by means of mixed pollination techniques, as previously described also in seven *de-ga* mutants (Ottaviano *et al.*, 1988a). Pollen collected from *de-ga/de-ga* plants was mixed with pollen from an unrelated standard line and used to pollinate *de-ga/de-ga* and *de-ga/+* plants of the same *de-ga* mutant. A positive interaction was observed for three mutants when *de-ga* pollen was growing on styles carrying the normal allele, whereas three other mutants showed the opposite situation, with a significantly higher percentage of uncolored kernels on *de-ga/de-ga* styles than on *de-ga/+* styles.

4.5. Manipulation of Pollen Development

In the past decade molecular biology has provided powerful tools for identification of cell- and/or tissue-specific genes and for their characterization. Of the possible applications to genetic manipulation of pollen development and, more in general, of plant reproduction that molecular biology could permit, particularly important are the isolation of genes expressed during microsporogenesis and floral development and the identification of regulatory DNA sequences that account for their spatial and temporal specificity.

Paramount in this respect are the studies carried out in the R. Goldberg and J. Mascarenhas laboratories leading to the first isolation of anther- and pollen-specific genes. Anther-specific genes were isolated by screening a cDNA library made up of mRNAs isolated from tobacco anthers (Goldberg, 1988) and used to study their regulation during the development of the anther (Koltunow *et al.*, 1990). The analysis of recombinant cDNA libraries obtained from mature pollen of *Tradescantia paludosa* and maize allowed the isolation of several cDNA clones espressed either in pollen or in pollen and in sporophytic tissues (Stinson *et al.*, 1987). Recently anther- and pollen-specific clones have been isolated in tomato (Gasser *et al.*, 1988), *Oenothera organensis* (Brown and Crouch, 1990), and *Brassica napus* (Albani *et al.*, 1990).

Some corresponding genomic clones have been isolated and the first DNA sequencing data are now available (Hanson *et al.*, 1989; Hamilton *et al.*, 1989; Twell *et al.*, 1989a; Brown and Crouch, 1990).

It is beyond the purpose of this review to go into the details of the molecular characteristics ascertained by these works; for these, reference should be made to the extensive reviews recently published on the topic (Mascarenhas, 1989, 1990). The possible applications of molecular biology to the genetic manipulation of pollen development could be exemplified by the strategy for the controlled induction of male sterile-line plants by means of self-killing genes. This approach implies the stable transformation of plants with a gene, such as the gene coding for a ribonuclease, fused to the promoter region of an anther on a pollen-specific gene. By means of a correct temporal and spatial control, the ribonuclease activity triggered by the specific promoter should prevent anther development and/or pollen grain formation, giving rise to sterile plants. The feasibility of this approach has been recently demonstrated by an elegant work in *Brassica napus* by Mariani *et al.* (1990). Although problems such as the restoration of fertility in the progeny of transformed plants are still to be solved, this or similar strategies could be of great interest in hybrid production for the plant breeding of the near future.

5.   GENE TRANSFER BY MEANS OF POLLEN TRANSFORMATION

Efficient methods for delivering exogenous DNA into plant cells are currently available and are powerful tools in plant molecular biology. Transient or stable transformation with molecularly engineered DNA sequences allows detection and fine analysis of regulatory regions responsible for temporal and/or spatial gene expression and can provide a better knowledge of the structure and regulation of the plant genome. Furthermore, taking advantage of the totipotency of plant cells to differentiate and give rise to whole fertile plants, plant gene transformation offers the possibility of introduction of specific genes conferring agriculturally interesting traits. *Agrobacterium tumefaciens*–mediated gene transfer has become the method of choice of people working with plants, mainly of the dicotyledonous family, suitable for infection (Horsch *et al.*, 1985). Although there is evidence of *A. tumefaciens* infecting some monocotyledonous species (Hooykaas-Van Slogteren *et al.*, 1984; Grimsley *et al.*, 1987), most of the agronomically important species, such as cereals, are not susceptible to infection from *A. tumefaciens*. Methods of direct DNA transfer to protoplasts, such as chemically mediated DNA uptake (Paszkowski *et al.*, 1984), liposome-mediated gene transfer (Deshayes *et al.*, 1985), electroporation (Fromm *et al.*, 1985), and microinjection (Crossway *et al.*, 1986), are being used with various degrees of efficiency for genetic transformation of all species, without any host range limitation. For a review of cereal transformation, see Potrykus (1990).

Unfortunately, these methods have in common a serious drawback: the possibility of the stable transfer of an introduced gene to the following generations depends on the regeneration via tissue-culture manipulations of transgenic fertile plants. First, not every plant species is easy to regenerate, the cereals again being the most difficult to handle. Furthermore, *in vitro* culture induces an unpredictable and often deleterious somaclonal variation.

Transformation of the male gametophyte or its use as carrier of exogenous DNA into the embryo sac can circumvent all these limitations by exploiting the natural fertilization ability of pollen. Far-sighted in this respect were the observations that led Hess *et al.* (1974) to suggest the use of pollen as a "supervector" to introduce foreign genetic material into plants. This was in 1974, when plant transformation attempts were just beginning. Since then serious efforts have been made in this direction by several researchers, but although pollen transformation has been successful, it soon became clear that it was not so easy as had been thought and hoped, and in fact, a simple, general, efficient way to obtain transgenic plants by means of pollen manipulation has not yet been found. Nevertheless the observations mentioned still justify the attempt to improve the pollen system for producing transgenic plants.

5.1. Developmental Stages as Targets for Gametophytic Transformation

One aspect of pollen transformation systems that has to be carefully considered relates to the stage of the male gametophyte cycle that is most suitable for genetic manipulation. In order to obtain stable transformation, the exogenous DNA has to overcome all the external barriers, get into the cells, and be integrated into the generative nucleus. Mature pollen grains have thick walls of several layers of chemically different materials; at anthesis they are dehydrated and metabolically inactive. In both binucleated and trinucleated species no DNA synthesis takes place, whether in mature pollen or in growing pollen tubes (Mascarenhas, 1975). Nevertheless, DNA polymerase activity has been detected in both vegetative and generative nuclei (Wever and Takats, 1971; Takats and Wever, 1971), and DNA repair activity has been demonstrated to occur in *Petunia* mature pollen after rehydration in the generative nucleus (Jackson, 1987).

A favorable stage for DNA uptake could be during rehydration of the pollen grains. Heslop-Harrison has extensively studied the process, showing that water movement in and out of the pollen grain is controlled by the pollen walls. The regulatory action is associated with changes of the pollen surface, in general an increase of the area of exposed intine, the inner layer of pollen wall (Heslop-Harrison, 1979, 1987). Since intine is more permeable to water, this exposed area could serve for exogenous DNA uptake. The growing pollen tube seems to be a less protected structure than mature pollen grain. Pollen tubes have thinner cell walls than most plant cells, and this characteristic has been exploited in microinjection experiments, where the rigid cell wall has been the limiting factor impeding the extension of use of injection transfers successful in animal cell systems to plant cell systems. More accessible appear to be the pollen tube tips, which lack a true cell wall. Macromolecule uptake has been demonstrated in rehydrated and germinating pollen grains, but whether it occurs through pollen pores or through pollen tube tips is not quite clear (Hess *et al.*, 1974; de Wet *et al.*, 1986).

Mature pollen grains and developing pollen tubes share the great advantage of being easily accessible for genetic manipulations, but in general, either their metabolic activity is absent or very slight or is mainly dedicated to synthesis of the pollen tube wall. The DNA replication machinery, which could facilitate DNA exchange and interaction between DNA molecules, is repressed, thus limiting the possibility of stable integration of foreign DNA in the genome.

Quite different is the situation in the floral tissues engaged in producing the gametes. Cells that undergo meiosis are characterized by intense metabolic activities, DNA replication, and there is extensive cytoplasmic communication between the sporogenous cells. A network of cytoplasmic bridges, the plasmodesmata, connects all the microsporocytes within an anther locule to the tapetal cells. During prophase I of meiosis the plasmodesmata disappear, but large cytoplasmic connections between the microgametocytes remain throughout meiosis, giving rise to a syncytial structure. These will disappear at the tetrad stage (Heslop-Harrison, 1966; Mascarenhas, 1975).

Studying the development of male germline in rye, Puertas *et al.* (1984) demonstrated that before the first meiotic metaphase, archesporial cells were permeable to caffeine and colchicine, as shown by the appearance of meiotic abnormalities. This observation suggested a possible successful approach for obtaining transgenic plants (de la Peña *et al.*, 1987).

Microsporogenesis, as we have seen, is a very complex process and sperm cell formation involves two DNA replications during which exogenous DNA integration should be favored. Pollen mother cells and microspores cannot be directly manipulated, and the only way to transform them is by injection into sporophytic tissues connected to the microsporocytes or microspores. Experiments involving these techniques will be discussed in the next section.

5.2. Transformation Strategies

Several pollen systems for plant gene transformation have been proposed, most of them being modifications of transformation techniques employed with success in animal cell systems or in other plant cell systems, such as protoplast transformation. DNA of various origins, from plasmid and phage DNA to total eukaryotic DNA, has been used as donor and several marker genes have been used to monitor successful transformation. As in every transformation attempt, pollen-mediated gene transfer has to face the connected problems of selection of the transformant and confirmation of the transformation.

As stated by Hess (1987), proof of gene transfer has to be given at the phenotypical level, at the genetic level, through the transmission of the transferred character to the following progenies; at the biochemical level, through the detection of specific enzymatic activity; and eventually at the molecular level, showing the insertion of foreign DNA into the genome of the recipient organism. The different strategies applied so far in a number of experiments have provided various degrees of proof, most of which have been extensively reviewed (Hess, 1987, 1988; Heberle-Bors *et al.*, 1990).

5.2.1. Pollen Treated with X- and γ-Rays

The observation that pollen grains treated with high doses of ionizing radiation can perform their function of releasing the sperm cells into the embryo sac and that the eggs after pseudofertilization by irradiated pollen admit a second normal fertilization led Pandey (1975, 1980) to suggest the introduction of genes into plants by means of what he described as "egg transformation." Highly fragmented pollen DNA could be directly incorporated in the egg nucleus, which might undergo doubling and give rise to transformed parthenogenic plants or could be integrated along with the DNA brought into the embryo sac by nontreated pollen. Successful results have been claimed in the *Nicotiana* genus although molecular proof of exogenous DNA integration was never given (Pandey, 1983; Jinks *et al.*, 1981; Caligari *et al.*, 1981).

Attempts to induce egg transformation in maize have not obtained favorable results (Sanford *et al.*, 1984; Sari Gorla *et al.*, 1987). The method has been more or less abandoned.

5.2.2. Direct DNA Transfer

This method has been the most frequently used and most successful system for transforming mature and germinating pollen. Bacteriophages carrying *Escherichia coli* β-galactosidase or galactose-1-phosphate uridyl transferase genes have been incubated with *Petunia hybrida* pollen in the appropriate germinating media. There is clear evidence of the stable integrations of these two genes; the β-galactosidase gene has also been demonstrated to be sexually transmitted over several offspring generations (Hess and Dressler, 1984; Hess *et al.*, 1985; Hess, 1987; Hess, 1988). Proof at the molecular level of male transformation by this technique has been obtained (deWet *et al.*, 1988). Plasmids containing the *APH-2* gene coding for a protein that confers resistance to kanamycin, along with the appropriate sequences for correct expression of the gene in the plant cells, were incubated with germinating pollen of maize. Recombinants were selected by screening for kanamycin resistance and the transformation efficiency was estimated to be about 2%.

A surprisingly high transformation efficiency using the direct DNA feeding method has been obtained in maize (Ohta, 1986). Plants of an inbred line, characterized by having several marker genes in the appropriate homozygous state, were used as recipient in a transformation experiment where pollen of the recipient line was mixed with total DNA of a donor line. Phenotypic analysis of the progeny showed transformation efficiency of 9.29%. Gene expression of exogenous DNA was also detected in the following generation, although the frequency of transformed plants was lower, indicating that if transformation was obtained, it was often unstable. No molecular proof of stable gene transfer has so far been given.

Other attempts to obtain transformed plants by the direct DNA method have failed (Sanford *et al.*, 1985; Negrutiu *et al.*, 1986). In his review Hess (1987) stressed that a serious problem was represented by the presence of quite intense nuclease activity in the incubation media, detected when pollen was extracted from the anthers, and not ruled out when other pollen preparations were used. This could be particularly true when donor DNA is supplied as linear or circular plasmid DNA. Total genomic DNA, where enough carrier DNA is present, thus providing a buffered situation, and protected DNA, such as phage DNA, are probably freer from this problem.

Treatment of pollen with exogenous DNA has unexpected mutagenic properties (Hess *et al.*, 1976; Ohta, 1986) and this observation could account for the high frequency of transformation in Ohta's experiments (Hess, 1987).

5.2.3.   Coculture Pollen–*A. tumefaciens*

This method has been proposed in order to exploit the high transformation efficiency of *A. tumefaciens* in pollen transformation experiments in the effort to combine the properties and the advantages of the two systems (Hess, 1986; Sanford and Skubik, 1986). Hess (1987) demonstrated that *Petunia* pollen cocultured with an *A. tumefaciens* strain carrying kanamycin resistance was able to grow better in culture medium with kanamycin than pollen cocultured in culture medium alone or with an *A. tumefaciens* strain sensitive to kanamycin. Furthermore, he showed that activation of the *vir* region on the Ti plasmid, needed to trigger DNA transfer into plant cells, was promoted by pollen extracts but not by stigmatic preparations or pollen and stigma exudates (Hess, 1988).

In vitro culture analysis of the offspring obtained by pollination of *Petunia* flowers with a mixture of *Petunia* pollen and *A. tumefaciens* has demonstrated that undifferentiated growth was favored and that some of the obtained calli could be maintained in hormone-free medium. This strongly suggests gene transfer from the Ti plasmid to pollen (Hess, 1987). In their experiments Sanford and Skubik (1986) observed that *A. tumefaciens* colonized the surface of pollen tubes of tobacco but not of maize or lily. But, despite the high rate of abnormal tobacco seedlings obtained after coculture with *Agrobacterium* and/or naked Ti-plasmid DNA, they detected neither nopaline synthase activity nor T-DNA in the analyzed plants that appeared transformed. The authors suggested that temporal activation of oncogenic gene expression was a possible cause of the abnormal progeny found after co-culture and that, if transformation had occurred, T-DNA could have been selectively lost during plant development.

5.2.4.   Macro- and Microinjection

These methods are based on the physical introduction of foreign DNA solution into plant cells by pressure or by ionophoresis with the aid of normal syringes (macroinjection) or micromanipulators and micropipettes (microinjection). The advantage lies in the possibility of treating plant organs or plant structures whose transformation could be favored by the plant metabolism or plant developmental stage, where a more direct approach would be difficult, if not impossible.

Macroinjection of plasmid DNA carrying the *APH II* gene conferring kanamycin resistance into rye floral tillers led to successful production of transgenic rye plants (de la Peña *et al.*, 1987). DNA was injected using a tuberculin syringe above each tiller node about 14 days before meiosis. Putative transgenic plants were selected by germinating seeds of the following generation on medium with kanamycin and growing seedlings, also in the presence of kanamycin, for 16 days. Seedlings from untreated plants could be easily detected because they were totally white after 10 days of culture in the selective medium. Out of 3023 seeds obtained by cross-pollination of 97 macroinjected plants, two seeds gave rise to seedlings that were able to grow in the selective medium, revealed *APH II* enzymatic activity, and, when analyzed by Southern blot hybridization, revealed the incorporation of plasmid DNA in their genome. The authors were unable to indicate whether the transformation had occurred in the male, in the female, or in both germlines, but, on the basis of the membrane characteristics and the timing of the treatment, were inclined to think it was the male cells that were transformed. Despite the low efficiency, these results aroused a lot of interest in this transformation system, above all because of its simplicity, but to our knowledge, it remains the only example reported in the literature.

A similar approach was followed in maize using the *Adh-1* gene as reporter gene (Bennetzen *et al.*, 1988). Spikelets of immature tassels from *Adh-1* null pollen plants, approximately at pachitene stage, were treated by injecting an aqueous solution containing the *Adh-1* genomic sequence. Pollen shed by the anthers of treated spikelets was subject to colorimetric assay to detect *Adh-1* activity. In some experiments up to 0.1% of pollen grains appeared to be putative transformants, but these grains were not recovered and used for fertilization. Thus the authors do not present any genetic evidence of transformation and cannot rule out the possibility of contamination. The difficulty of interpretation and recovery of transformed pollen grains was one of the reasons why the authors abandoned these experiments.

Microinjection has been unsuccessfully used in attempts to transform the generative nuclei of growing pollen tubes. Furthermore, of the several plant species tested, Hepher *et al.* (1985) found that only in a few was the generative nucleus large enough to be a suitable target.

5.2.5. More Radical Approaches

The presence of the cell wall has always been the major obstacle in every attempt to introduce exogenous DNA into plant cells. To complete our overview of the several different methods used in the attempt to overcome this problem, we have to mention two proposed methods based on crude force. A few years ago Sanford (1983) presented the results of his treatment of mature and germinating pollen of *Vinca minor* with a microlaser. His idea was to perforate the pollen walls with the aid of a laser beam, producing holes that could become entrance gates for exogenous DNA uptake. By adjusting the intensity of the laser blasts and the pollen-germinating medium, and by increasing the external osmotic pressure, he was able to prevent lethal bursting of the treated pollen. Pollen showed a good capacity to recover, sealing the lesions a few seconds after the treatment and thus limiting the cytoplasmic extrusion. However, the observed unidirectional outward flow of cytoplasm led Sanford to conclude that DNA uptake was not likely to occur in these experimental conditions.

This approach was recently resumed. Microspores obtained from immature anthers of *Brassica napus* were placed in hypotonic buffer containing plasmid pBR322 DNA and perforated by a laser pulse. Uptake of the plasmid DNA through the holes, which remained open for about 5 sec, was visualized by ultraviolet illumination. No data were reported concerning either the stability of DNA uptake or the viability and pollination capacity of the treated microspores (Weber *et al.*, 1988).

Plant cell and plant tissue transformation by high-velocity microprojectiles coated with DNA was first proposed a few years ago (Klein *et al.*, 1987). This technique allows the direct delivery of DNA into intact plant cells or tissues. Since then it has been quite extensively used on different plant tissues and plant species, either in stable plant transformation experiments, in attempts to introduce foreign DNA into cell organelles, or for the purpose of detection of transient gene expression of specific genes in promoter studies (see Klein *et al.*, 1990, for a review).

Microprojectile bombardment of hydrated tobacco pollen, tobacco anthers, and intact flowers was adopted as the DNA delivery system in studies of pollen promoter (Twell *et al.*, 1989b). The ability of 600 bp of 5′ flanking DNA of a tomato pollen–specific gene to promote gene expression in pollen was tested by comparison with the cauliflower mosaic virus (CAMV) 35S promoter. The β-glucunomidase (*GUS*) gene was used as reporter gene, and it was either placed under the control of the pollen promoter or fused to the CAMV 35S promoter. Transient *GUS* activity was detected for both chimeric genes in tobacco hydrated pollen grains, but the *GUS* activity driven by the pollen promoter was approximately 1000 times greater than that driven by the CAMV 35S promoter. On the other hand, the pollen promoter was not able to drive *GUS* activity in transformed tobacco leaves.

These results have shown that microprojectile bombardment can be a useful approach for the study of transient gene expression in pollen. It is interesting to note that the treated tobacco pollen did not lose its germination capacity and this is of great importance in the context of an attempt to transform stably pollen grains by the microprojectile technique.

6. REFERENCES

Abbott, A. G., and Fauron, C. M. R. 1986, Structural alteration in a transcribed region of T type cytoplasmic male sterile maize mitochondrial genome, *Curr. Genet.* **10**:777–783.

Albani, D., Robert, L. S., Donaldson, P. A., Altosaar, I., Arnison, P. G., and Fabijanski, S. F., 1990, Characterization of a pollen specific gene family from *Brassica napus* which is activated during early microspore development, *Plant Mol. Biol.* **15**:605–622.

Albertsen, M. C., and Phillips, R. L., 1981, Developmental cytology of 13 male sterile loci in maize, *Can. J. Genet. Cytol.* **23:**195–208.

Anderson, M. A., Cornish, E.C., Mau, S. L., Williams, E. G., Hoggart, R. D., Atkinson, A., Bonig, I., Grego, B., Simpson, R., Roche, P. J., Haley, J. D., Niall, H. D., Tregear, G. W., Coughlan, J. P., Crowford, R. J., and Clarke, A. E., 1986, Cloning of cDNA for a stylar glycoprotein associated with expression of self-incompatibility in *Nicotiana alata, Nature* **321:**38–44.

Bailey-Serres, J., Dixon, L. K., Liddell, A. D., and Leaver, C. J., 1986, Nuclear-mitochondrial interactions in cytoplasmic male sterile sorghum, *Theor. Appl. Genet.* **73:**252–260.

Benito Moreno, R. H., Macke, F., Hauser, M. T., Alwen, A., Heberle-Bors, E., 1988, Sporophytes and male gametophytes from *in vitro* cultured, immature tobacco pollen, in *Sexual Reproduction in Higher Plants* (M. Cresti, P. Gon, and E. Pacini, eds.) pp. 137–142, Springer-Verlag, Berlin.

Bennetzen, J. L., Lin, C., McCormick, S., and Staskawicz, B. J., 1988, Transformation of *Adh* null pollen to Adh+ by macroinjection, *Maize Genet. Coop. Newslett.* **62:**113–114.

Bergamini-Mulcahy, G., and Mulcahy, D. L., 1986, Pollen-pistil interaction, in *Biotechnology and Ecology of Pollen* (D. L. Mulcahy, G. Bergamini-Mulcahy, and E. Ottaviano, eds.), Springer-Verlag, New York.

Bianchi, A., and Lorenzoni, C., 1975, Gametophytic factors in *Zea mays,* in *Gamete Competition in Plants and Animals* (D.L. Mulcahy, ed.), pp. 257–264, Elsevier, Amsterdam.

Binelli, G., Vieira de Manincor, E., and Ottaviano, E., 1985, Temperature effects on pollen germination and pollen tube growth in maize, *Genet. Agri.* **39:**269–281.

Bino, R. J., Hille, J., and Franken, J., 1987, Kanamycin resistance during *in vitro* development of pollen from transgenic tomato plants, *Plant Cell Rep.* **6:**333–336.

Bino, R. J., Franken, J., Witsenboer, H. M. A., Hille, J., and Dons, J. J. M., 1988, Effects of *Alternaria alternata* f.sp. *Lycopersici* toxins on pollen, *Theor. Appl. Genet.* **76:**204–208.

Borlaug, N. E., 1983, Contributions of conventional plant breeding to food production, *Science* **219:**689–693.

Brewbaker, J. L., 1971, Pollen enzymes and isoenzymes, in *Pollen: Development and Physiology* (J. Heslop-Harrison, ed.), pp. 156–170, Butterworth, London.

Brown, S. M., and Crouch, M. L., 1990, Characterization of a gene family abundantly expressed in *Oenothera organensis* pollen that shows sequence similarity to polygalacturonase, *Plant Cell* **2:**263–274.

Buchert, J. G., 1961, The stage of genome-plasmon interaction in the restoration of fertility to cytoplasmically pollen-sterile maize, *Proc. Natl. Acad. Sci. USA* **47:**1426–1440.

Caligari, P. S. D., Ingram, N. R., and Jinks, J. L., 1981, Gene transfer in *Nicotiana rustica* by means of irradiated pollen. I. Unselected progeny, *Heredity* **47:**12–26.

Correns, C., 1928, Bestimmung, Vererbung und Verteilung des Geschlechtes bei den hoheren Pflanzen, *Hanbuch Vererbungswissensch.* **2:**1–138.

Crossway, A., Oakes, J., Iavine, J., Ward, B., Knauf, V., and Shewmaker, C., 1986, Integration of foreign DNA following microinjection of tobacco mesophyll protoplasts, *Mol. Gen. Genet.* **202:**179–185.

de la Pena, A., Lorz, H., and Schell, J., 1987, Transgenic rye plants obtained by injecting DNA into young floral tillers, *Nature* **325:**274–276.

de Nettancourt, D., 1977, Incompatibility in Angiosperms, in *Monographs on Theoretical and Applied Genetics N°3* (R. Frankel, G. A. E. Gall, and H. F. Linskens, eds.), Springer-Verlag, Berlin.

de Wet, J. M. J., Berthand, J., Cubero, J. I., and Hepburn, A., 1988, Genetic transformation of cereals, in *Crop Plant Biotechnology in Tropical Crop Improvement,* Proc. Int. Biotechnol. Workshop, pp. 27–32, Hyderabad.

de Wet, J. M. M., de Wet, A. E., Brink, D. E., Hepburn, A. G., and Woods, J. A., 1986,

Gametophyte transformation in maize, in *Biotechnology and Ecology of Pollen* (D. L. Mulcahy and E. Ottaviano, eds.), pp. 59–64, Springer-Verlag, New York.

Deshayes, A., Herrera-Estrella, L., and Caboshe, M., 1985, Liposome-mediated transformation of tobacco mesophyll protoplasts by Escherichia coli plasmid, *EMBO Jour.* **4**:2731–2737.

Dewey, R. E., Levings III, C. S., and Timothy, D. H., 1986, Novel recombinations in the maize mitochondrial genome produce a unique transcriptional unit in the Texas male sterile cytoplasm, *Cell* **44**:439–449.

Dewey, R. E., Timothy, D. H., and Levings III, C. S., 1987, A mitochondrial protein associated with cytoplasmic male sterility in the T cytoplasm of maize, *Proc. Natl. Acad. Sci. USA* **84**:5374–5378.

Dungey, S. G., Sang, J. P., Rothnie, N. E., Palmer, M. V., Burke, D. G., Knox, R. B., Williams, E. G., Hilliard, E. P., and Salisbury, P. A., 1988, Glucosinolates in the pollen of rapeseed and Indian mustard, *Phytochemistry* **27**:815–817.

Duvick, D. N., 1965, Cytoplasmic pollen sterility in corn, *Adv. Genet.* **13**:1–56.

Duvick, D. N., 1981, Progress in conventional plant breeding, in *Gene Manipulation in Plant Improvement* (J. P. Gustavson, ed.), pp. 17–31, Plenum Publ., New York.

Ebert, P. R., Anderson, M. A., Bernatzky, R., Altschuler, M., and Clarke, A. E., 1989, Genetic polymorphism of self-incompatibility in flowering plants, *Cell* **56**:255–262.

Edwardson, J. R., 1970, Cytoplasmic male sterility, *The Botanical Review*, **36**:341–420.

Escote, L. J., Gabay-Laughnan, S., and Laughnan, J. R., 1985, Cytoplasmic reversion to fertility in cms-S maize need not involve loss of linear mitochondrial plasmids, *Plasmid* **14**:264–267.

Evans, D. E., Rothnie, N. E., Palmer, M. V., Burke, D. G., Sang, J. P., Knox, R. B., Williams, E. G., Hilliard, E. P., and Salisbury, P. A., 1987, Comparative analysis of fatty acids in pollen and seed of rapeseed, *Phytochemistry* **26**:1895–1898.

Evans, D. E., Rothnie, N. E., Sang, J. P., Palmer, M. V., Mulcahy, D. L., Singh, M. B., and Knox, R. B., 1988, Correlations between gametophytic (pollen) and sporophytic (seed) generations for polyunsaturated fatty acids in oilseed rape *Brassica napus* L., *Theor. Appl. Genet.* **76**:411–419.

Faraughi-Wehr, B., Fiendt, W., Scuchmann, R., Kohler, F., and Wenzel, G., 1986, *In vitro* selection for resistance, in *Somaclonal Variation and Plant Improvement* (J. Semal, ed.), pp. 35–44, Nijhoff M./Junk, W., The Hague, The Netherlands.

Feder, W. A., 1986, Predicting species response to ozone using a pollen screen, in *Biotechnology and Ecology of Pollen* (D. L. Mulcahy, G. Bergamini-Mulcahy, and E. Ottaviano, eds.), pp. 89–94, Springer-Verlag, New York.

Frankel, R., and Galun, E., 1977, *Pollination Mechanisms, Reproduction and Plant Breeding*, 281 pp. Springer-Verlag, New York.

Fromm, M., Taylor, L. P., and Walbot, V., 1986, Stable transformation of maize after gene transfer by electroporation, *Nature* **319**:791–793.

Frova, C., 1990, Analysis of gene expression in microspores, pollen and silks of *Zea mays* L. *Sex. Plant Reprod.* **3**:200–206.

Frova, C., Sari Gorla, M., Ottaviano, E., and Pella, C., 1983, Haplo-diploid gene expression in maize and its detection, *Biochem. Genet.* **21**:923–931.

Frova, C., Binelli, G., and Ottaviano, E., 1987, Isozyme and *hsp* gene expression during male gametophyte development in maize, in *Isozymes, Genetics; Development and Evolution* (M. C. Rattazzi, J. G., Scandalios, and G. S. Whitt, eds.), pp. 97–120, Alan R. Liss, New York.

Frova, C., Taramino, G., and Binelli, G., 1989, Heat-shock proteins during pollen development in maize, *Dev. Genet.* **10**:324–332.

Gasser, C. S., Smith, A. G., Budelier, K. A., Hinchee, M. A., McCormick, S., *et al.*, 1988, Isolation of differentially expressed genes from tomato flowers, in *Temporal and Spacial Regulation of Plant Genes* (D. P. S. Verma and B. Goldberg, eds.), pp. 83–96, Springer-Verlag, New York.

Goldberg, R. B., 1988, Plants: Novel developmental processes, *Science* **240:**1460–1466.

Golubovskavya, I. N., 1989, Meiotic mutants in maize: *mei* genes and conception of genetic control of meiosis, *Adv. Genet.* **26:**149–192.

Gotteschalk, W., and Kaul, M. L. H., 1974, The genetic control of microsporogenesis in higher plants, *Nucleus* **17:**133–166.

Grimsley, N., Hohn, T., Davies, T. W., and Hohn, B., 1987, *Agrobacterium*-mediated delivery of infectious maize streak virus into maize plants, *Nature* **325:**177–179.

Hamilton, D. A., Bashe, D. M., Stinson, J. R., and Mascarenhas, J. P., 1989, Characterization of a pollen-specific genomic clone from maize, *Sex. Plant Reprod.* **2:**208–212.

Hanson, D. D., Hamilton, D. A., Travis, J. L., Bashe, D. M., and Mascarenhas, J. P., 1989, Characterization of a pollen-specific cDNA clone from *Zea mays* and its expression, *Plant Cell* **1:**173–179.

Hanson, M. R., and Conde, M. F., 1985, Functioning and variation of cytoplasmic genomes: Lessons from cytoplasmic-nuclear interactions affecting male sterility in plants, *Int. Rev. Cytol.* **94:**213–267.

Heberle-Bors, E., Benito Moreno, R. M., Alwen, A., Stoger, E., and Vicente, O., 1990, Transformation of pollen, in *Current Plant Science and Biotechnology in Agriculture,* Progress in Plant Cellular and Molecular Biology, Proc. VII Int. Congress in Plant Tissue and Plant Cell Culture (H. J. J. Nijkamp, L. H. W. Van Der Plas, and J. Van Aartrijk, eds.), pp. 244–251, Kluwer Academic Publisher, Dordrecht.

Hepher, A., Sherman, A., Gates, P., and Boulter, D., 1985, Microinjection of gene vectors and ovaries as a potential means of transforming whole plants, in *Experimental Manipulation of Ovule Tissues* (G. P. Chapman, S. H. Mantell, and R. W. Daniels, eds.), pp. 52–63, Longman, New York.

Heslop-Harrison, J., 1966, Cytoplasmic connexions between angiosperm meiocytes, *Ann. Bot.* **30:**221–229.

Heslop-Harrison, J., 1971a, *Pollen: Development and Physiology,* pp. 338, Butterworth, London.

Heslop-Harrison, J., 1971b, The pollen wall: structure and development, in *Pollen: Development and Physiology* (J. Heslop-Harrison, ed.), pp. 75–98, Butterworth, London.

Heslop-Harrison, J., 1979, The forgotten generation: Some thoughts on the genetics and physiology of angiosperm gametophytes, in *The Plant Genome: 4TH Innes Symposium* (D. R. Davies and D. A. Hopwood, eds.), pp. 1–14, John Innes Institute, Norwich.

Heslop-Harrison, J., 1987, Pollen germination and pollen tube growth, *Int. Rev. Cytol.* **107:**1–70.

Heslop-Harrison, J., Heslop-Harrison, Y., Knox, R. B., and Howlett, B., 1973, Pollen wall proteins: Gametophytic and sporophytic fraction in pollen wall of Malvaceae, *Ann. Bot.* **37:**403–412.

Hess, D., 1986, The pollen system of gene transfer, in *Genetic Manipulation in Plant Breeding* (W. Horn, C. J. Jensen, W. Odenbach, and O. Scheider, eds.), pp. 803–811, W. de Gruyter, Berlin.

Hess, D., 1987, Pollen-based techniques in genetic manipulation, *Int. Rev. Cytol.* **107:**367–395.

Hess, D., 1988, Direct and indirect gene transfer using pollen as carriers of exogenous DNA, in *Crop Plant Biotechnology in Tropical Crop Improvement,* pp. 19–26, Proc. Int. Biotechnol. Workshop, Hyderabad, India, ICRISAT.

Hess, D., and Dressler, K., 1984, Bacterial transferase activity expressed, in *Investigations on the Tumor Induction in Nicotiana glauca by pollen transfer of DNA isolated from* Nicotiana langsdorfii (D. Hess, G. Schneider, H. Lorz, and G. Blaich, eds.), *Z. Pflanzenphysiol.* **77:**247–254.

Hess, D., Gresshoff, P. M., Fielitz, U., and Gleiss, D., 1974, Uptake of protein and bacteriophage into swelling and germinating pollen of *Petunia hybrida, Z. Pflanzenphysiol.* **74:**371–376.

Hess, D., Dressler, K., and Konle, S., 1985, Gene transfer in higher plants using pollen as vectors: Bacterial transferase activity expressed in Petunia progenies, in *Experimental Manipulation of Ovule Tissue* (G. P. Chapman, S. H. Mantell, and R. W. Daniels, eds.), pp. 224–239, Longman, New York.

Hodgkin, T., 1988, *In vitro* pollen selection in *Brassica napus* L., in *Sexual Reproduction in Higher Plants* (M. Cresti, P. Gori, and E. Pacini, eds.), pp. 57–62, Springer-Verlag, Berlin.

Hodgkin, T., MacDonald, M. V., 1986, The effect of phytotoxin from *Alternaria brassicicola* on *Brassica* pollen, *New Phytol.* **104:**631–636.

Hooykaas-Van Slogteren, G., Hooykaas, P. J., and Schilperoort, R. A., 1984, Expression of Ti plasmid genes in monocotyledonous plants infected with *Agrobacterium tumefaciens*, *Nature* **311:**763–764.

Horsch, R. B., Fry, J. E., Hoffman, N. L., Eichholtz, D., Rogers, S. G., and Fraley, R. T., 1985, A simple and general method for transferring genes into plants, *Science* **227:**1229–1231.

Jackson, J. F., 1988, DNA repair in *Petunia* hybrida pollen, in *Sexual Reproduction in Higher Plants* (M. Cresti, P. Gori, and E. Pacini, eds.), pp. 81–86, Springer-Verlag, Berlin, Heidelberg.

Jinks, J. L., Caligari, P. S. D., and Ingram, N. R., 1981, Gene transfer in *Nicotiana rustica* using irradiated pollen, *Nature* **291:**586–588.

Kaul, M. L. H., 1988, Male sterility in higher plants, 1005 pp., Springer Verlag, Berlin.

Kaul, M. L. H., and Murthy, T. G. K., 1985, Mutant genes affecting higher plant meiosis, *Theor. Appl. Genet.* **70:**449–466.

Kennell, J. C., Wise, R. P., and Pring, D. R., 1987, Influence of nuclear background on transcription of a maize mitochondrial region associated with Texas male sterile cytoplasm, *Mol. Gen. Genet.* **210:**399–406.

Klein, T. M., Wolf, E. D., Wu, R., and Sanford, J. C., 1987, High-velocity microprojectiles for delivering nucleic acids into living cells, *Nature* **327:**70–73.

Klein, T. M., Goff, S. A., Roth, B. A., and Fromm, M. E., 1990, Applications of the particle gun in plant biology, in *Current Plant Science and Biotechnology in Agriculture*, Progress in Plant Cellular and Molecular Biology, Proc. VII Int. Congress in Plant Tissue and Plant Cell Culture, (H. J. J. Nijkamp, L. H. W. Van Der Plas, and J. Van Aartijk, eds.), pp. 56–66, Kluwer, Dordrecht.

Koltunow, A. M., Truettner, J., Cox, K. H., Wallroth, M., and Goldberg, R. B., 1990, Different temporal and spatial gene expression patterns occur during anther development, *The Plant Cell* **2:**1201–1224.

Landi, P., Frascaroli, E., Tuberosa, R., and Conti, S., 1989, Comparison between responses to gametophytic and sporophytic recurrent selection in maize (*Zea mays* L.), *Theor. Appl. Genet.* **77:**761–767.

Laughnan, J. R., and Gabay, S. J., 1973, Reaction of germinating maize pollen to *Helminthosporium maydis* pathotoxins, *Crop Sci.* **43:**681–684.

Laughnan, J. R., and Gabay-Laughnan, S., 1983, Cytoplasmic male sterility in maize, *Annu. Rev. Genet.* **17:**27–48.

Laughnan, J. R., Gabay-Laughnan, S., and Day, J. M., 1989a, Evidence for transposition of the naturally occurring *cms-S* restorer in inbred line CE1, *Maize Genet. Coop. Newslett.* **63:**121.

Laughnan, J. R., Gabay-Laughnan, S., and Day, J. M., 1989b, Naturally occurring restorers of *cms-S* are located at various chromosomal sites in different inbred lines and appear to be transposable, *Maize Genet. Coop. Newslett.* **63:**120–121.

Lee, T. D., and Hartgerink, A. P., 1986, Pollination intensity, fruit maturation pattern, and offspring quality in *Cassia fasciculata* (Leguminosae), in *Biotechnology and Ecology of Pollen* (D. L. Mulcahy, G. Bergamini Mulcahy, and E. Ottaviano, eds.), pp. 417–422, Springer-Verlag, New York.

Levings, C. S., III, and Brown, G., 1989, Molecular biology of plant mitochondria, *Cell* **56:**171–179.

Levings, C. S., III, Kim, B. D., Pring, D. R., Conde, M. F., Mans, R. J., Laughnan, J. R., and Gabay-Laughnan, S. J., 1980, Cytoplasmic reversion of *cms-S* in maize: Association with a traspositional event, *Science* **209:**1021–1023.

Mangelsdorf, P. C., 1932, Mechanical separation of gametes in maize, *J. Hered.* **23:**288–295.

Manzocchi, L. A., Daminati, M., Gentinetta, E., and Salamini, F., 1980a, Viable defective endo-sperm mutants in maize. Kernel weight, protein fraction and zein subunits in mature endosperm, *Maydica* **25**:105–116.

Manzocchi, L. A., Daminati, M. G., and Gentinetta, E., 1980b, Viable defective endosperm mutants in maize. II. Kernel weight, nitrogen and zein accumulation during endosperm develop-ment, *Maydica* **25**:199–210.

Mariani, C., De Beuckeleer, M., Treuttner, J., Leemans, J., and Goldberg, R. B., 1990, Induction of male sterility in plants by a chimaeric ribonuclease gene, *Nature* **347**:737–741.

Mascarenhas, J. P., 1975, The biochemistry of Angiosperm pollen development, *Bot. Rev.* **41**:259–314.

Mascarenhas, J. P., 1984, Molecular mechanisms of heat stress tolerance, in *Applications of Genetic Engineering to Crop Improvement* (G. B. Collins, and J. G. Petolino, eds.), pp. 391–425, M. Nijhoff/Dr W. Junk, Dordrecht.

Mascarenhas, J. P., 1989, The male gametophyte of flowering plants, *Plant Cell* **1**:657–664.

Mascarenhas, J. P., 1990, Gene activity during pollen development, *Annu. Rev. Plant Physiol. Plant Mol. Biol.* **41**:317–338.

Mascarenhas, N. T., Bashe, D., Eisenberg, A., Willing, R. P., Xiao, C. M., and Mascarenhas, J. P., 1984, Messenger RNAs in corn pollen and protein synthesis during germination and pollen tube growth, *Theor. Appl. Genet.* **68**:323–326.

McClure, B. A., Haring, V., Ebert, P. R., Anderson, M. A., Simpson, R. J., Sakiyama, F., and Clarke, A. E., 1989, Style self-incompatibility gene products of *Nicotiana alata* are ribonucleases, *Nature* **342**:955–957.

McClure, B. A., Gray, J. E., Anderson, M. A., and Clarke, A. E., 1990, Self-incompatibility in *Nicotiana alata* involves degradation of pollen rRNA, *Nature* **347**:757–760.

McKenna, M. A., 1986, Heterostyly and microgametophytic selection: The effect of pollen competi-tion on sporophytic vigor in two distylous species, in *Biotechnology and Ecology of Pollen* (D. L. Mulcahy, G. Bergamini Mulcahy, and E. Ottaviano, eds.), pp. 443–448, Springer-Verlag, New York.

McKenna, M., and Mulcahy, D. L., 1983, Ecological aspects of gametophytic competition in *Dianthus chinensis*, in *Pollen: Biology and Implications in Plant Breeding* (D. L. Mulcahy and E. Ottaviano, eds.), pp. 419–424, Elsevier Biomedical, New York.

McNay, J. W., Pring, D. R., and Lonsdale, D. H., 1983, Polymorphism of mitochondrial DNA "S" regions among normal cytoplasm of maize, *Plant Mol. Biol.* **12**:177–189.

McRae, D. H., 1985, Advances in chemical hybridization, *Plant Breeding Rev.* **3**:169–191.

Meinke, D. W., 1982, Embryo-lethal mutants of *Arabidopsis thaliana*: Evidence for gametophytic expression of the mutant genes, *Theor. Appl. Genet.* **63**:381–386.

Meinke, D. W., and Baus, A. D., 1986, Gametophytic gene expression in embryo-lethal mutants of *Arabidopsis thaliana*, in *Biotechnology and Ecology of Pollen* (D. L. Mulcahy, G. Bergamini Mulcahy, and E. Ottaviano, eds.), pp. 15–20, Springer-Verlag, New York.

Mulcahy, D. L., 1971, A correlation between gametophytic and sporophytic characteristics in *Zea mays* L., *Science* **171**:1155–1156.

Mulcahy, D. L., 1974, Correlation between speed of pollen tube growth and seedling weight in *Zea mays* L., *Nature* **249**:491–492.

Mulcahy, D. L., 1979, The rise of the angiosperms: A genecological factor, *Science* **206**:20–23.

Mulcahy, D. L., and Mulcahy, G. B., 1975, The influence of gametophytic competition on spo-rophytic quality in *Dianthus chinensis*, *Theor. App. Genet.* **46**:277–280.

Mulcahy, D. L., Mulcahy, G. B., and Ottaviano, E., 1975, Sporophytic expression of gametophytic competition in *Petunia hybrida*, in *Gamete Competition in Plants and Animals* (D. L. Mulcahy, ed.), pp. 227–232, North Holland Publ. Co., Amsterdam.

Nasrallah, J. B., Kao, T. H., Goldberg, M. L., and Nasrallah, M. E., 1985, A cDNA clone encoding an S locus-specific glycoprotein from *Brassica oleracea*, *Nature* **318**:263–267.

Nasrallah, J. B., Kao, T. H., Chen, C. H., Goldberg, M. L., and Nasrallah, M. E., 1987, Amino-acid sequence of glycoproteins encoded by three alleles of the S locus of *Brassica oleracea*, *Nature* **326**:617–619.

Negrutiu, K., Heberle-Bors, E., and Potrykus, I., 1986, Attempts to transform for kamamycin-resistance in mature pollen of tobacco, in *Biotechnology and Ecology of Pollen* (D. L. Mulcahy, G. Bergamini-Mulcahy, and E. Ottaviano, eds.), pp. 65–70, Springer-Verlag, New York.

Nelson, O. E., 1952, Non reciprocal cross sterility in maize, *Genetics* **37**:101–124.

Ohta, Y., 1986, High efficiency genetic transformation of maize by a mixture of pollen and exogenous DNA, *Proc. Natl. Acad. Sci. USA* **83**:715–719.

Ottaviano, E., 1990, Selection pressure on pollen and its relevance to plant breeding (Sinha, S. K., Sane, P. V., Bhargava, S. C., and Agrawal, P. K., eds.) pp. 1315–1321, Society for Plant Physiology and Biochemistry, New Delhi.

Ottaviano, E., and Mulcahy, D. L., 1986, Gametophytic selection as a factor of crop plant evaluation, in *The Origin and Domestication of Cultivated Plants* (C. Barigozzi, ed.), pp. 101–120, Elsevier, Amsterdam.

Ottaviano, E., and Mulcahy, D. L., 1989, Genetics of Angiosperm Pollen, *Advances in Genetics* **26**:1–64.

Ottaviano, E., and Sari Gorla, M., 1979, Genetic variability of male gametophyte in maize. Pollen genotype and pollen–style interaction, in *Israeli–Italian Joint Meeting on Genetics and Breeding of Crop Plants*, pp. 89–106, Monogr. Genet. Agraria IV, Rome.

Ottaviano, E., Sari Gorla, M., and Mulcahy, D. L., 1980, Pollen tube growth rate in *Zea mays*: Implications for genetic improvement of crops, *Science* **210**:437–438.

Ottaviano, E., Sari-Gorla, M., and Pè, E., 1982, Male gametophytic selection in maize, *Theor. Appl. Genet.* **63**:249–254.

Ottaviano, E., Sari Gorla, M., and Arenari, I., 1983, Male gametophytic competitive ability in maize. Selection and implications with regard to the breeding system, in *Pollen: Biology and Implications for Plant Breeding* (D. L. Mulcahy, and E. Ottaviano, eds.), pp. 367–373, Elsevier Biomedical, New York.

Ottaviano, E., Petroni, D., and Pè, E., 1988a, Gametophytic expression of genes controlling endosperm development in maize, *Theor. Appl. Genet.* **75**:252–258.

Ottaviano, E., Sari Gorla, M., and Villa, M., 1988b, Pollen competitive ability in maize: Within population variability and response to selection, *Theor. Appl. Genet.* **76**:601–608.

Pallais, N., Malagamba, P., Fong, N., Garcia, R., and Scmiediche, P., 1986, Pollen selection through storage: a tool for improving true potato seed quality? in *Biotechnology and Ecology of Pollen* (D. L. Mulcahy, G. Bergamini-Mulcahy, and E. Ottaviano, eds.), pp. 153–158. Springer-Verlag, New York.

Pandey, K. K., 1975, Sexual transfer of specific genes without gametic fusion, *Nature* **256**:310–313.

Pandey, K. K., 1980, Further evidence for egg transformation in *Nicotiana, Heredity* **45**:15–29.

Pandey, K. K., 1983, Evidence for gene transfer by the use of sublethally irradiated pollen in *Zea mays* and theory of occurrence by chromosome repair through somatic recombination and gene conversion, *Mol. Gen. Genet.* **191**:358–365.

Paszkowski, J., Shillito, R. D., Saul, M., Mandak, V., Hohn, T., Hohn, B., and Potrykus, I., 1984, Direct gene transfer to plants, *EMBO J.* **3**:2717–2722.

Pedersen, S., Simonsen, V., Loeschke, V., 1987, Overlap of gametophytic and sporophytic gene expression in barley, *Theor. Appl. Genet.* **75**:200–206.

Peloquin, S. J., 1986, Genetic engineering with meiotic mutants, in *Biotechnology and Ecology of Pollen* (D. L. Mulcahy, and E. Ottaviano, eds.), pp. 361–367, Elsevier Biomedical, New York.

Pfahler, P. L., 1983, Comparative effectiveness of pollen genotype selection in higher plants, in *Biotechnology and Ecology of Pollen* (D. L. Mulcahy, and E. Ottaviano, eds.), pp. 361–367, Elsevier Biomedical, New York.

Potrykus, I., 1990, Gene transfer to cereal: an assessment, *Bio/Technology* **6**:531–542.

Pring, D. R., and Levings, C. S., III, 1978, Heterogeneity of maize cytoplasmic genomes among male-sterile cytoplasms, *Genetics* **89**:121–136.

Pring, D. R., Levings, C. S., III, Hu, W. W., and Timothy, D. H., 1977, Unique DNA associated with mitochondria in "S" type cytoplasm of male sterile maize, *Proc. Natl. Acad. Sci. USA* **74**:2904–2908.

Puertas, M. J., de le Pena, A., Estades, B., and Merino, F., 1984, Early sensitivity to colchicine in developing anthers of rye, *Chromosoma* **89**:121–126.

Quin, T., and Dun, D., 1990, Character and inheritance of a new Y-type cytoplasmic male-sterile line, *Maize Genet. Coop. Newslett.* **64**:61–62.

Rabinowitch, H. D., Reting, N., and Kedar, N., 1978, The mechanism of preferential fertilization in tomatoes carrying the I-allele for *Fusarium* resistance, *Euphytica* **27**:219–224.

Rajora, O. P., and Zsuffa, L., 1986, Sporophytic and gametophytic gene expression in *Populus deltoides* marsh., *P. nigra* L., and *P. maximowiczii* henry, *Can. J. Genet. Cytol.* **28**:476–482.

Rocheford, T. R., and Pring, D. R., 1990, Nuclear-mitochondrial interactions affecting transcription of mitochondrial open reading frames, *Maize Genet. Coop. Newslett.* **64**:61–62.

Russell, W. A., 1974, Comparative performance of maize hybrids representing different eras of maize breeding, in *Proc. 29th Annual Corn and Sorghum Conference* (D. Walkinson, ed.), pp. 81–101, Am. Seed Trade Assoc., Washington, DC.

Sacher, R., Mulcahy, D. L., and Staples, R., 1983, Developmental selection for salt tolerance during self pollination of Licopersicon x Solanum F_1 for salt tolerance of F_2, in *Pollen: Biology and Implications for Plant Breeding* (D. L. Mulcahy, and E. Ottaviano, eds.), pp. 329–334, Elsevier Biomedical, New York.

Sanford, J. C., 1983, Pollen studies using a laser microbeam, in *Pollen: Biotechnology and Implications for Plant Breeding* (D.L. Mulcahy, and E. Ottaviano, eds.), pp. 107–116, Elsevier Biomedical, New York.

Sanford, J. C., and Skubik, K. A., 1986, Attempted pollen-mediated transformation using Ti plasmids, in *Biotechnology and Ecology of Pollen* (D. L. Mulcahy, G. Bergamini-Mulcahy, and E. Ottaviano, eds.), pp. 71–76, Springer-Verlag, New York.

Sanford, J. C., Chyi, Y. S., and Reish, B. I., 1984, Attempted "egg transformation" in *Zea mays* L. using irradiated pollen, *Theor. Appl. Genet.* **68**:269–275.

Sanford, J. C., Skubik, K. A., and Reisch, B. I., 1985, Attempted pollen-mediated plant transformation employing genomic donor DNA, *Theor. Appl. Genet.* **69**:571–575.

Sari-Gorla, M., Frova, C., Binelli, G., and Ottaviano, E., 1986, The extent of gametophytic–sporophytic gene expression in maize, *Theor. Appl. Genet.* **72**:42–47.

Sari-Gorla, M., Villa, M., and Ottaviano, E., 1987, Pollen irradiation and gene transfer in maize, *Maydica* **32**:239–248.

Sari-Gorla, M., Mulcahy, D. L., Gianfranceschi, L., and Ottaviano, E., 1988a, Gametophytic selection for salt tolerance, *Genet. Agri.* **42**:92–93.

Sari-Gorla, M., Ottaviano, E., Frascaroli, E., and Landi, P., 1989, Herbicide-tolerant corn by pollen selection, *Sex Plant Reprod.* **2**:65–69.

Schlichting, C. D., Stephenson, A. G., Davis, L. E., and Winsor, J. A., 1987, Pollen competition and offspring variance, *Evol. Trends Plants* **1**:35–39.

Schwartz, D., 1960, The analysis of a case of cross-sterility in maize, *Proc. Natl. Acad. Sci. USA* **36**:719–724.

Searcy, K. B., and Mulcahy, D. L., 1985a, Pollen selection and the gametophytic expression of metal tolerance in *Silene dioica* (Caryophyllaceae) and *Mimulus guttatus* (Scrophulariaceae), *Am. J. Bot.* **72**:1700–1706.

Searcy, K. B., and Mulcahy, D. L., 1985b, The parallel expression of metal tolerance in pollen and sporophytes of *Silene dioica* (L.) Clairv., *S. alba* (Mill.) Krause and *Mimulus guttatus* DC, *Theor. Appl. Genet.* **69**:597–602.

Shivanna, K. R., and Johri, B. M., 1985, *The Angiosperm Pollen. Structure and Function,* Wiley Eastern Limited, New Delhi.

Smith, G. A., 1986, Sporophytic screening and gametophytic verification of phytotoxin tolerance in sugarbeet (*Beta vulgaris* L.), in *Biotechnology and Ecology of Pollen* (D. L. Mulcahy, G. Bergamini Mulcahy, and E. Ottaviano, eds.), pp. 83–88, Springer-Verlag, New York.

Stephenson, A. C., Winsor, J. A., and Davis, L. E., 1986, Effects of pollen load size on fruit maturation and sporophyte quality in zucchini, in *Biotechnology and Ecology of Pollen* (D.L. Mulcahy, G. Bergamini Mulcahy, and E. Ottaviano, eds.), pp. 429–434, Springer-Verlag, New York.

Stinson, J. R., Eisenberg, A. J., Willing, R. P., Pè, M. E., Hanson, D. D., and Mascarenhas, J. P., 1987, Gene expressed in the male gametophyte of flowering plants and their isolation, *Plant Physiol.* **83**:442–447.

Sunderland, N., and Huang, B., 1987, Ultrastructural aspects of pollen dimorphism, *Int. Rev. Cytol.* **107**:175–219.

Takats, S. T., and Wever, G. H., 1971, DNA polymerase and DNA nuclease activities in S-competent and S-incompetent nuclei from *Tradescantia* pollen grains, *Exp. Cell Res.* **69**:25–28.

Tanksley, S. D., Zamir, D., and Rick, C. M., 1981, Evidence for extensive overlap of sporophytic and gametophytic gene expression in *Lycopersicon esculentum, Science* **213**:453–455.

Ter-Avanesian, D. V., 1949, The role of the number of pollen grains per flower in plant breeding, *Bull. Appl. Bot. Plant Breed., Russian,* **28**:19–33.

Ter-Avanesian, D. V., 1978, The effect of varying the number of pollen grains used in fertilization, *Theor. Appl. Genet.* **52**:77–79.

Torti, G., Manzocchi, L., and Salamini, F., 1986, Free and bound indole-acetic acid is low in the endosperm of the maize mutant defective endosperm-B18, *Theor. Appl. Genet.* **72**:602–605.

Twell, D., Wing, R. A., Yamaguchi, J., and McCormick, S., 1989a, Isolation and expression of an anther-specific gene from tomato, *Mol. Gen. Genet.* **217**:240–245.

Twell, D., Klein, T. M., Fromm, M. E., and McCormick, S., 1989b, Transient expression of chimeric genes delivered into pollen by microprojectile bombardment, *Plant Physiol.* **91**:1270–1274.

Weber, G., Monajembashi, S., Grenlich, K. O., and Wolfrum, J., 1988, Genetic manipulation of plant cells and organelles with a laser microbeam, *Plant Cell Tissue Organ Culture* **12**:219–222.

Wever, G. H., and Takats, S. T., 1971, Isolation and separation of S-competent and S-incompetent nuclei from *Tradescantia* pollen, *Exp. Cell Res.* **69**:29–32.

Willing, R. P., and Mascarenhas, J. P., 1984, Analysis of complexity and diversity of mRNAs from pollen shoots of *Tradescantia, Plant Physiol.* **75**:865–868.

Willing, R. P., Bashe, D., and Mascarenhas, J. P., 1988, An analysis of the quantity and diversity of messenger RNAs from pollen and shoots of *Zea mays, Theor. Appl. Genet.* **75**:751–753.

Wise, R. P., Fliss, A. E., Pring, D. R., and Gegenbach, B. G., 1987, *Urf13-T* of T cytoplasm maize mitrochondria encodes a 13Kd polypeptide, *Plant Mol. Biol.* **9**:121–126.

Yamada, M., 1983, Superiority of pollen from F_1 plants of maize in selective fertilization, *JARQ* **17**:166–172.

Zabala, G., Gabay-Laughnan, S., and Laughnan, J. R., 1989, Nuclear control over molecular characteristics of cms-S male-fertile cytoplasmic revertants, *Maize Genet. Coop. Newslett.* **63**:118–119.

Zamir, D., and Gadish, I., 1987, Pollen selection for low temperature adaptation in tomato, *Theor, Appl. Genet.* **74**:545–548.

Zamir, D., and Vallejos, E. C., 1983, Temperature effects on haploid selection of tomato microspores and pollen grains, in *Pollen: Biology and Implication for Plant Breeding* (D. L. Mulcahy and E. Ottaviano, eds.), pp. 335–342, Elsevier Scientific, New York.

Zamir, D., Tanksley, S. D., and Jones, R. A., 1981, Low temperature effect on selective fertilization
    by pollen mixtures of wild and cultivated tomato species, *Theor. Appl. Genet.* **59:**235–238.
Zamir, D., Tanksley, S. D., and Jones, A. J., 1982, Haploid selection for low temperature tolerance
    of tomato pollen, *Genetics* **101:**129–137.

Chapter 7

Transient Gene Expression of Chimeric Genes in Cells and Tissues of Crops

Hans-Henning Steinbiss and Andrew Davidson

1. INTRODUCTION

The recent progress in understanding plant biology at the molecular level has been due, in part, to the development of efficient gene transfer systems for many plant species. Without doubt, the most efficient way to transfer genes to crops is with *Agrobacterium tumefaciens*. Fraley *et al.* (1984) found that nopaline synthase gene expression could be detected as early as 12 hr after cocultivation of *Agrobacterium* and *Petunia* protoplasts. Whether the state of the introduced DNA is extrachromosomal or integrated was not pertinent, but clearly, rapid analysis of gene expression is possible soon after gene transfer (Janssen and Gardner, 1989; Vancanneyt *et al.*, 1990).

Techniques for transferring DNA into cells often introduce much more DNA into the cell nucleus than the amount that becomes stably incorporated into the host chromosomes. This extrachromosomal DNA is lost over a period of 1–2 weeks as a result of dilution by cell division and susceptibility to intracellular degradation. However, during its transient existence in the cell, the extrachromosomal DNA is known to be transcriptionally active. Measurement of this

Hans-Henning Steinbiss and Andrew Davidson   Max-Planck Institut für Züchtungsforschung, Abt. Genetische Grundlagen der Pflanzenzüchtung, 5000 Köln-30, Germany.

Subcellular Biochemistry, Volume 17: Plant Genetic Engineering, edited by B. B. Biswas and J. R. Harris. Plenum Press, New York, 1991.

transcriptional activity is the basis of the extremely useful transient gene expression assay.

Transient gene expression assays are of special importance when studying plant systems that are not currently amenable to routine transformation procedures using either *Agrobacterium*-mediated or direct gene transfer methods. Such plant systems include many important crop species, in particular the cereals, which do not belong in the host range of *Agrobacterium* (De Cleene, 1985). In addition, transient gene expression is of great value for the rapid evaluation of gene expression and gene transfer methods before lengthy stable transformation investigations are carried out.

The aim of this review is to demonstrate the utility of transient gene expression studies for advancing the understanding of plant molecular biology in general and for improving gene transfer systems so that in the near future it may be possible to stably transform the most economically important crop species.

2. METHODOLOGY

A general consideration of any gene transfer technique is the potential of the recipient cell to express the introduced gene. For transient assays the recipient cell should be capable of the metabolic and physiological functions of experimental interest and, optimally, should respond to external stimuli as cells *in planta*. This is often not the case for protoplast systems, which are the easiest to use for gene transfer techniques. Thus, increasingly, research is concentrated on gene transfer to plant tissues, organs, and organelles, which may more accurately reflect the functions of the plant cell. Additionally, the use of these systems may circumvent tissue culture steps, as regeneration of many plant species from tissue culture has proven difficult.

It should be emphasized that expression levels in all systems can be misleading because of weekly variations in the condition of plant tissues or suspension cultures. Therefore, any comparisons should be carried out with care using the same preparations and at different time intervals.

2.1. Designing Plant Expression Vectors

Vectors for use in transient gene expression studies are comprised of three basic, interchangeable components within a replicating selectable bacterial plasmid: a promoter, a reporter gene, and a 3' noncoding region. The promoter is usually a strong promoter of plant or viral origin, e.g., the cauliflower mosaic virus (CaMV) 35S RNA promoter, the promoter of ribulose bisphosphate 1', 5-carboxylase (RUBISCO), or the promoter of the alcohol dehydrogenase (*Adh1*) gene of maize, but may be interchanged for promoter analysis studies with the promoter of choice.

The reporter gene may be either a selectable marker, such as the bacterial antibiotic resistance genes neomycin phosphotransferase II (*NPTII*), chloramphenicolacetyl transferase (*CAT*), and hygromycin phosphotransferase (*HYG*), or a gene whose protein product is readily detectable, e.g., *Escherichia coli* β-glucuronidase (*GUS*) or luciferase (*LUX*). The 3' noncoding region contains the poly A$^+$ adenylation signal and perhaps a transcriptional terminator.

Examples of these types of convenient vectors have been created by Pietrzak *et al.* (1986), Rogers *et al.* (1987), and Töpfer *et al.* (1987, 1988a,b).

Additionally, the inclusion of other elements, such as the enhancer sequence formed by duplication of the CaMV35S promoter (Kay *et al.*, 1987; Hobbs *et al.*, 1990) or the intron of the maize *Adh1* gene, which may increase the transcriptional activity of a promoter 10- to 100-fold (Callis *et al.*, 1987a), may be desirable. More detailed information concerning the components of vectors for transient gene expression studies are given in a special section below.

In principle, any combination of a promoter–reporter gene–3' noncoding sequence may work for a given plant system. However, in practice, a specific combination of the three basic components often gives a higher expression in a given system. It cannot be excluded that this may be due to the handling of the recipient cells or tissues, and it is strongly recommended that all constructions be compared with the same cell system and under the same laboratory conditions (e.g., Pröls *et al.*, 1988, 1989; Töpfer *et al.*, 1988b).

2.2. Plant Protoplasts—An Ideal Single-Cell System

One major distinction between gene transfer techniques is whether they can be applied to cells with cell walls (e.g., suspension cultured cells, embryos, meristems) or require protoplasts as the recipients. Protoplasts are the simplest plant cells for gene transfer because removing the cell wall eliminates the major barrier to DNA transfer. This advantage is offset by the finding that protoplasts often do not retain the patterns of gene expression present in the cells/tissue from which they were derived.

Several very efficient DNA transfer techniques for obtaining stably transformed plants cells from protoplasts have been described in the past (e.g., Krens *et al.*, 1982; Deshayes *et al.*, 1985; Hain *et al.*, 1985; Potrykus *et al.*, 1985; Shillito *et al.*, 1985; Negrutiu *et al.*, 1987; Krüger-Lebus and Potrykus, 1987). These methods also lead to transient gene expression and thus may be used for transient gene expression studies independently from stable transformation.

Although, most of transient gene expression studies have been performed with tobacco, protoplasts from other important crops, such as sugarbeet (Lindsey and Jones, 1987), potato (Jones *et al.*, 1989), maize (Fromm *et al.*, 1985), as well as trees (Gupta *et al.*, 1988; Seguin and Lalonde, 1988; Bekkaoui *et al.*, 1988, 1990; Wilson *et al.*, 1989; Tautorus *et al.*, 1989; Kobayashi and Uchimiya, 1989), have also been used.

To date, the majority of transient expression studies have been performed using single-cell systems. However, very recently studies using particle bombardment for gene transfer have shown that these effects can also be obtained in tissues and organs.

2.3. Tissues as Tools for Transient Gene Expression Studies

Recent advances in gene transfer techniques have focused on DNA delivery to intact cells and tissues. Transfer to intact cells bypasses the problems associated with protoplasts, namely, that regeneration into whole plants has proven difficult in many cases, especially for cereals, the most economically important crops (for reviews, see Lörz et al., 1988; Vasil, 1988; Steinbiss and Davidson, 1989). However, efficient plant regeneration from protoplasts of rice (Abdullah et al., 1986; Yamada et al., 1986; Terada et al., 1987; Kyozuka et al., 1987, 1988; Yang et al., 1989), maize (Cai et al., 1987; Rhodes et al., 1988a), and very recently even wheat (Vasil et al., 1990), *Sorghum* (Wei and Xu, 1990), and barley (Lazzeri and Lörz, 1990) has been achieved. Successful transformation of rice (Shimamoto et al., 1988; Toriyama et al., 1988; Zhang and Wu, 1988; Zhang et al., 1988) and maize (Rhodes et al., 1988b) has been reported. Aside from these promising results using tissue culture methods, several alternative approaches for the transformation of crops have been suggested and are in the early stages of development.

"High-velocity microprojectile bombardment" (biolistics) represents the major advance for gene transfer into plants. Simple in concept, the technique involves accelerating micron-sized metal particles to velocities sufficient to penetrate intact cells or tissues. DNA either bound to the microprojectiles or present in a drop of fluid surrounding the particles is thereby carried into the cell (Sanford et al., 1987; Klein et al., 1987; Sanford, 1988). At least some of the DNA will arrive in the nucleus and other organelles, where it is biologically active. It is not known whether the DNA is deposited directly in the nucleus or transported there from impacts in other parts of the cell. However, the size of the particles, as well as their speed, makes it possible that the target cell will be damaged and lose its regeneration capacity.

The utility of microprojectiles for gene transfer into plant cells was first demonstrated with transient expression assays in onion epidermal cells (Klein et al., 1987), which are very large in comparison to meristem cells and easy to handle as they are derived from a single layer of tissue. The relevance to other crop plants has been demonstrated by transient assays in maize (Klein et al., 1988a,b; Oard et al., 1990; Cao et al., 1990), rice (Wang et al., 1988; Oard et al., 1990; Cao et al., 1990), barley (Mendel et al., 1989; Kartha et al., 1989; Lee et al., 1989), soybean (McCabe et al., 1988; Christou et al., 1988) and last but not least, wheat (Lee et al., 1989; Oard et al., 1990). Furthermore, use of the glucuronidase marker gene has allowed visualization of the number and distribution of bombarded cells that transiently express the introduced DNA. As demonstrated by the authors cited above, gene transfer by particle bombardment is probably applicable to most plant cells and tissues and, in combination with suitable regeneration frequencies of the recipient cells, will have a promising future.

Embryonic cultures are now available for most of the cereals (Lörz et al., 1988; Vasil, 1988). Such cultures contain a large percentage of cells that are capable of regeneration into fertile plants. However, only a small percentage of the recipient embryogenic cell cluster will become chimeric after gene transfer as transformation efficiencies are usually low. A single transgenic cell in a large cell cluster or in a tissue is generally difficult to recover by metabolic selection, as the inhibited or dying cells often will inhibit its growth. In such a case the ratio of transgenic to nontransgenic cells is important in detecting and recovering stable transformants. Transient expression studies could be helpful in the improvement of this ratio, assuming there is a direct correlation between the number of cells expressing a foreign gene transiently and their ability to give rise to transgenic cells.

For many years it has been known that tissue culture stress induces somaclonal variation (for reviews, see Lee and Phillips, 1988; Scowcroft and Larkin, 1988; Evans, 1989). Thus, more recently, many scientists have started to transform cells of certain intact tissues during the normal life-cycle of the plant, avoiding any tissue culture stress and the problem of regeneration to whole plants. Some potentially useful targets are: pollen (Twell et al., 1989), meristems (Oard et al., 1990) and embryos (Klein et al., 1988a,b; McCabe et al., 1988; Kartha et al., 1989; Cao et al., 1990). In all the cited cases, transient gene expression studies have been used to demonstrate successful gene delivery, a prerequisite for obtaining transgenic cells.

2.4. Plant Cell Organelles as Targets

An additional exciting prospect for the use of microprojectile bombardment is the direct transfer of genes to plant chloroplasts and mitochondria. Stable transformation of chloroplasts has been reported, as monitored by the correction of deletions in the chloroplast genome (Daniell and McFadden, 1987; Boynton et al., 1988; Blowers et al., 1989). Similarly, a deletion in the mitochondrial genome of yeast was corrected by particle bombardment using microprojectiles carrying a gene spanning the deleted area (Johnston et al., 1988; Fox et al., 1988). Although gene transfer to higher plant organelles resulting in transient expression has been reported (Teeri et al., 1989; Daniell et al., 1990), there are few reports of successful stable organelle transformation in higher plants (Cornelissen et al., 1987). One of the greatest problems is to avoid gene expression derived from gene copies being integrated in the nuclear genome rather than that of the organelles. Therefore, the recombinant DNA must only be biologically active in the target organelles. The usual way to discriminate between different sources of transient gene expression is to isolate the organelles and to measure expression of the transferred foreign genes.

3. APPLIED TRANSIENT GENE EXPRESSION ASSAYS

3.1. Optimizing DNA-Delivery Conditions

Transient assays are commonly used for the analysis of gene expression and rapid monitoring of gene transfer. A prerequisite for obtaining maximal gene expression is determination of the optimal conditions for gene transfer. Once established, these parameters may also be utilized for establishing gene transfer systems.

3.1.1. The Fate of Transferred Molecules

How much of the applied recombinant DNA reaches the nucleus and, of that, which amount becomes stably integrated into the plant genome? The answer is still open. It is likely that due to the membranous barriers only a small portion of the introduced DNA will enter the nucleus in a functional state, and once there, the majority remains extrachromosomal where it can be attacked efficiently by nucleases (Werr and Lörz, 1986). These factors will influence the strength of transient gene expression and make comparisons and quantitative evaluations very difficult, as specific gene uptake conditions and target cells differ in this respect. In part, this can be overcome by using the same experimental conditions and comparing results obtained with the same batch of protoplasts or cells (Pröls et al., 1988; Töpfer et al., 1988b). Alternatively, a gene construction that includes a constitutively expressed reporter gene as an "internal control" may be used. For translational studies direct mRNA transfer can circumvent problems concerning DNases and nuclear membrane properties (Callis et al., 1987b; Gallie et al., 1989).

What happens to foreign DNA transferred into plant cells? Kinetic studies have demonstrated the transient nature of gene expression (Fromm et al., 1985; Werr and Lörz, 1986; Hauptmann et al., 1987; Rosenberg et al., 1988; Pröls et al., 1988, 1989; Jones et al., 1989; Kartha et al., 1989). DNA introduced into the nucleus is lost with time and only a minor part becomes stably integrated into the genome. The fate of foreign DNA transfected into animal cells has been monitored using fluorescent dyes (Loyter et al., 1982) as well as injection into frog oocytes. One idea that arose from these studies is that the DNA is partly assembled into chromatin where it is more resistant to attack by DNases than naked DNA (Gurdon and Melton, 1981; Ryoji and Worcel, 1985). Using plant cells, Wirtz *et al.* (1987) showed that frequently recombination occurs between foreign DNA copies, which may lead to rearrangements, concatenation, tandem molecules, etc. (Czernilofsky *et al.*, 1986; Riggs and Bates, 1986). Under specific conditions this may also influence transient expression studies, and this has to be considered in designing experiments.

3.1.2. Polyethyleneglycol-Mediated Gene Transfer

Polyethyleneglycol (PEG) was first used to promote protoplast fusion. As PEG affects membranes, it was successfully tested for its ability to promote uptake of DNA molecules by plant protoplasts. This method was first applied to tobacco protoplasts, being the most established system. PEG-mediated gene transfer has now been used successfully with a large number of plant systems, as it is the cheapest and easiest method and has also been shown to be one of the most efficient and reproducible (Negrutiu *et al.*, 1987; Pröls *et al.*, 1988, 1989; Töpfer *et al.*, 1988b). This is of great importance; for example, when plant promoter analysis studies are performed, it is essential to compare many individual gene transfer experiments, which must be easily performed and as reproducible as possible. Recently, Maas and Werr (1989) studied mechanisms and optimized conditions for PEG-mediated DNA transfection using maize and rice protoplasts and transient gene expression assays.

3.1.3. Electroporation

The electroporation method is based on the use of electrical pulses differing in length, strength, field decay, and number of repeats. These physical differences may result in different DNA uptake mechanisms. Nevertheless, all approaches will lead to transient gene expression, but with different intensities and survival ratios of the target cells. Electroporation was first successfully applied in the uptake of DNA using animal cells (Neumann *et al.*, 1982; Potter *et al.*, 1984). The technique was then used for the uptake of DNA into tobacco protoplasts. This was facilitated by the simultaneous development of plant expression vectors suitable for transient expression studies (Bevan *et al.*, 1983; Herrera-Estrella *et al.*, 1983, 1988). The electroporation efficiency has been improved dramatically using transient gene expression assays to optimize parameters involved in gene transfer (Fromm *et al.*, 1985, 1987; Ou-lee *et al.*, 1986; Cutler and Saleem, 1987; Lindsey and Jones, 1987; Bates *et al.*, 1988; Bekkaoui *et al.*, 1988; Seguin and Lalonde, 1988; Shigekawa and Dower, 1988; Taylor and Larkin, 1988; Vasil *et al.*, 1988; Jones *et al.*, 1989; Tyagi *et al.*, 1989; Joersbo and Brunstedt, 1990), and electroporation is now widely used to obtain stably transformed plants. Additionally, a combination of PEG and electroporation leads to a more effective gene transfer than PEG or electroporation alone (Shillito et al., 1985; Boston et al., 1987).

3.1.4.   The Biolistic Process

In principle, the PEG and electroporation methods are applicable for gene transfer into protoplasts of a wide range of plant cells. However, regeneration of protoplasts remains difficult for a variety of crops. It would be simpler and faster to regenerate plants directly from intact cells and organs instead of protoplasts. Since the establishment of the particle acceleration system (biolistics) by Klein et al. (1987, 1988a,b,c), Sanford et al. (1987), McCabe et al. (1988), and Morikawa et al. (1989), many technical improvements have been made, e.g., the development of a new airgun apparatus (Oard et al., 1990), overcoming the problem of "gunpowder," which can contaminate the target and necessitates continual cleaning of the whole apparatus.

This gene transfer technique is in a fast-growing phase (see also Section 2.3) and is undergoing further improvements; in particular, survival of the target cell after particle bombardment will dictate whether this technique will surpass all other gene transfer methods for routine transformation studies.

3.1.5.   Gene Uptake by Seed-Derived Embryos

An interesting alternative to the other methods for transient gene expression studies cited above has been developed by Töpfer et al. (1989, 1990). It was found that dry embryos can take up foreign DNA applied during imbibition and express the genes in a transient fashion. The main reason for developing such an alternative was to find a method that is simple, very cheap, can in principle be applied to every plant and every genotype, and can be performed routinely every day, as a stock of dry seed embryos can be easily made. This technique needs only a very short, hormone-free tissue culture step and avoids possible somaclonal variation. The system was first used to study transient expression of various reporter genes in legumes and cereal embryos (Töpfer et al., 1989, 1990).

A method that shares some of the advantages discussed above has recently been published by Ahokas (1989). In this study exogenous DNA was transferred electrophoretically to germinating seeds. DNA uptake was also monitored using transient gene expression assays.

3.2.   Comparison of Reporter Genes

The availability of several reporter genes that have sensitive, convenient, reliable enzymatic assays has greatly increased the utility of transient assays. The genes coding for the bacterial chloramphenicol acetyltransferase (CAT) (Gorman *et al.*, 1982; Fromm *et al.*, 1985; Ballas *et al.*, 1987; Nishiguchi *et al.*, 1987b, 1988; Pröls *et al.*, 1988; Rosenberg *et al.*, 1988; Töpfer *et al.*, 1988b), the *E. coli* β-glucuronidase gene (*GUS*) (Jefferson *et al.*, 1986, 1987; Jefferson, 1987), the bacterial luciferase gene (Koncz *et al.*, 1987, 1990; Olsson *et al.*, 1988, 1989), and firefly luciferase Fromm *et al.*, 1985; Ow *et al.*, 1986, 1987; Riggs and Chrispeels, 1987; De Wet *et al.*, 1987; Ballas *et al.*, 1987; Gupta *et al.*, 1988; Nguyen *et al.*, 1988; Brasier *et al.*, 1989; Maxwell and Maxwell, 1988; Coker *et al.*, 1989; Planckaert and Walbot, 1989) have all been successfully used as reporter genes for transient expression studies.

The luciferase enzyme assay has an advantage in that there is very little endogenous luciferase activity in plants, whereas endogenous enzymes present in some plant species or specific tissues are able to acetylate chloramphenicol (Charest *et al.*, 1989) and function similarly to glucuronidase (Bekkaoui *et al.*, 1988; Hu *et al.*, 1990). Luciferase (Ow *et al.*, 1986; Legocki *et al.*, 1986; Shaw *et al.*, 1987; O'Kane *et al.*, 1988; Schauer, 1988) has been widely used as a visual marker gene in cells and tissues.

Use of the bacterial neomycin phosphotransferase II (*NPTII*) gene as a reporter gene provides a sensitive means of monitoring transient gene expression, albeit less convenient, as detection requires an *in situ* gel assay (Reiss *et al.*, 1984a,b). Several efforts have been made to establish a more convenient and quantitative *NPTII* assay (McDonnell *et al.*, 1987; Platt and Young, 1987; Cabanes-Bastos *et al.*, 1989; Weide *et al.*, 1989; Staebell *et al.*, 1990; Roy and Sahasradbudke, 1990). *NPTII* also has the advantage of being a selectable marker gene in many important crops, but it is less efficient in monocots. Therefore, "G418" (Potrykus *et al.*, 1985; Okada *et al.*, 1986a,b; Hauptmann *et al.*, 1988b) as well as hygromycin resistance (Waldron *et al.*, 1985; Van den Elzen *et al.*, 1985) have been favored in preference to kanamycin resistance in the selection of transgenic monocot cells.

The *E. coli* β-glucuronidase has a sensitive enzymatic assay and can be detected by an *in situ* histochemical assay that allows the visualization of a single *GUS*-expressing cell (Jefferson *et al.*, 1987). This has proven useful for optimizing DNA delivery conditions and detecting the recipients of DNA transfer; e.g., the type and distribution of cells receiving foreign genes by microprojectile bombardment was determined successfully using expression of *GUS* (see Section 3.1.4).

Very recently Ludwig *et al.* (1990) used microprojectiles to transfer into maize aleurone cells a vector containing the transcription unit of a gene responsible for a specific pigmentation pattern fused to a constitutive promoter. The chimeric gene induces cell autonomous pigmentation in tissues that are not normally pigmented by this gene. This provides a novel reporter gene for expression studies in maize, simply by counting spots. This marker can be quantified and may also be useful as a visible marker for selecting stably transformed cell lines.

3.3.   DNA Topology

A critical step in PEG-mediated gene transfer is the change in the structure of the applied DNA. This effect has often been neglected in planning DNA transfer experiments. Maas and Werr (1989) have shown that the precipitation of extracellular DNA by various PEG/salt mixtures is very sensitive to the experimental conditions used for transformation and can strongly influence transient gene expression due to altered DNA uptake efficiencies.

In prokaryotes the toplogical state of the DNA is often a crucial parameter in processes involving replication, transcription, and recombination (for review, see Wang, 1985). In eukaryotes, studies using animal cells suggest that DNA topology is also important for proper gene expression (Smith, 1981; Harland et al., 1983; North, 1985; Weintraub, 1985; Pina et al., 1990). Weintraub et al. (1986) found that supercoiled DNAs can be over 100-fold more effective as templates in animal cells than linear DNAs. This difference cannot be explained by trivial effects, such as variations in DNA uptake or assembly into chromatin or preferential degradation of linear molecules. By contrast, Ballas et al. (1988) obtained 10-fold higher gene expression using linear DNA than that observed in protoplasts transfected by the supercoiled template.

3.4.   The Physiological State of the Target Cells

Kartha et al. (1989) claimed that the target cells have to be in an active state of cell division in order for the introduced gene to be expressed, as mature zygotic and somatic embryos failed to reveal any gene expression. Okada et al. (1986b) found that gene expression is higher if genes are transferred into cells during the M phase of the cell cycle compared to the other cell cycle phases. They claimed that the absence of the nuclear membrane in mitotic cells favors delivery of DNA to the nucleus once introduced into the cytoplasm. By contrast, Junker et al. (1987) demonstrated that transient gene expression appears also in nondividing cells, so that the mitotic cell stage has only an enhancement, if any, effect. Cell-stage-specific expression can also be influenced by the promoter fused to the reporter gene (Nagata et al., 1987).

By injection of DNA directly into nuclei at specific cell cycle stages of animal cells, Wong and Capecchi (1985) found that at all stages of the cell cycle there is efficient expression of genes introduced on a supercoiled plasmid. In plant systems this type of experiment can be done efficiently only by aid of chemically induced synchronization. Aphidicolin alone (Okada et al., 1986b; Nagata et al., 1987; Sala et al., 1986) or in combination with a herbicide (2, 6-dichlorobenzonitrile) preventing cell wall formation (Meyer et al., 1985; Kartzke et al., 1990) has been used to obtain highly synchronized cells. At the chromosomal level aphidicolin increases the amount of spontaneous sister chromatid exchanges in Chinese hamster cells. Aphidicolin also induces moderate chromatid abberrations; in other words, aphidicolin is a mutagenic substance (Smith and Paterson, 1983). It is not yet clear whether such effects will influence transient expression assays, but if cell-cycle-dependent integration of genes is to be studied, this effect has to be considered. As Köhler et al. (1989, 1990) and Hilson et al. (1990) found that damaging of DNA and the consequent switching on of the DNA repair system are strongly involved in incorporation of exogenous DNA introduced into the plant genome.

3.5. Promoter Function Studies

Transient gene expression is extremely useful in studying gene structure and function relationships for a number of reasons. The most important is the speed with which results on gene expression can be obtained. The transfected cells are typically ready for analysis of gene expression 24–48 hr after DNA transfer. A second important aspect is the lack of host chromosomal flanking sequences that may influence gene expression. Quite often there is variation in gene expression in stable transformants depending on the site of insertion. Such variation can be greater than the variation between the gene structures being compared (Odell et al., 1985; Velten et al., 1984). Additionally, gene expression can be negatively influenced or even silenced after integration into the genome (Al-Shawi et al., 1990; position effect). Transient gene expression assays are free of interaction with host flanking sequences, allowing a much simpler comparison of normal and altered gene structures.

The rapidity of transient assays makes them particularly suitable to select and evaluate promoters prior to their deployment in stable transformation systems (Fromm et al., 1985; Werr and Lörz, 1986; Ebert et al., 1987; Boston et al., 1987; Sanders et al., 1987; Wang et al., 1988; Hauptmann et al., 1988a; Planckaert and Walbot, 1989; Jones et al., 1989; Bekkaoui et al., 1990; Sanger et al., 1990). However, promoter strength is strongly influenced by the cell lines used (Hobbs et al., 1990) as well as the stage of the cell cycle (Nagata et al., 1987).

Ebert et al. (1987) identified essential upstream elements in the nopaline synthase promoter using transient assays. This was achieved via deletion experiments as well as by the duplication of distinct DNA regions. They also found that the transient analysis gave results quite different from those from stable expression analysis. DNA sequences lying upstream of the octopine synthase promoter having enhancer functions were found by Ellis et al. (1987b) via transient gene expression. Enhancer-like sequences from Nicotiana plumbaginifolia also were characterized by transient assays (Horth et al., 1987).

3.6. Expression Improvement by Use of Introns

In plant cells, the first examples of chimeric gene expression utilized genes that lack introns (Bevan et al., 1983; Fraley et al., 1983; Herrera-Estrella et al.,

1983, 1988; Reynaerts *et al.*, 1988). In one case where the effect of introns was examined in plant cells, the gene for the seed storage protein phaseolin was found to be expressed at similar levels with or without its introns (Chee *et al.*, 1986). Consequently, it was assumed that introns were not important for plant gene expression.

However, in monocots it has recently been shown that the expression of reporter genes can be dramatically increased by introducing an intron between the promoter and the structural gene. The *AdhI* intron of maize has been used in such studies (Callis *et al.*, 1987a; Klein *et al.*, 1988a; Kartha *et al.*, 1989; Planckaert and Walbot, 1989), but also more recently the first intron of the shrunken-1 (*Sh1*) locus of maize (Vasil *et al.*, 1989). It was found that the *Sh1* intron gave gene expression approximately 10 times higher than any other plant intron used to date (Oard *et al.*, 1989; McElroy *et al.*, 1990). Additionally, the minimal length of pre-mRNA introns has been determined by transient gene expression assays (Goddall and Filipowicz, 1990).

3.7. Regulation of Gene Expression

For investigation of the mechanisms regulating expression of genes in different cell types, transient gene expression assays have many advantages. They are not time consuming, and position effects, which may occur in stably transformed cells, do not appear.

Ha and An (1989) studied the regulatory elements controlling temporal and organ-specific expression of the nopaline synthase gene by analyzing deletion mutants of the promoter. Do we have conclusive evidence that transient gene expression generally correlates with the gene expression in stable transformants? Jacobsen and Beach (1985) have shown that expression of the α-amylase gene increases after gibberellic acid treatment but is suppressed by abscisic acid in barley aleurone protoplasts. This response is the same as in intact aleurone, indicating it should be possible to study hormone-regulated gene expression transiently in aleurone protoplasts.

Aleurone cells of maize have also been used as target cells by Ludwig *et al.* (1990) using microprojectiles. Klein *et al.* (1989) showed transient expression of the Bronze I gene in maize aleurone, demonstrating further that genes introduced into intact tissues can be regulated as in the whole plant. These examples show that transient assays allow tissue-specific gene regulation and that this technique is useful particularly in such plants as cereals currently lacking stable transformation assays (Schwall and Feix, 1988; Klein *et al.*, 1989). Transient gene expression assays have been successfully used to study gene expression in pollen (Twell *et al.*, 1989) as well as expression of photosynthesis-related gene fusions (Harkins *et al.*, 1990).

One of the first examples that regulated genes can be examined by transient assays was the anaerobic regulation of the alcohol dehydrogenase (*Adh1*) promoter in a chimeric construct. Howard *et al.* (1987) and Ellis *et al.* (1987a) found *cis*-acting sequences responsible for the anaerobic regulation of the *Adh1* gene. Heat (Callis *et al.*, 1988) and UV-inducible (Dangl *et al.*, 1987; Lipphardt *et al.*, 1988) promoters have also been studied and specific *cis*-regulatory elements have been found and analyzed. Additionally, transient assays have been developed to study phytochrome (Bruce *et al.*, 1989), gibberellin (Huttly and Baulcombe, 1989), and abscisic acid (Marcotte *et al.*, 1988) regulated genes in wheat and rice.

3.8. Silencing Gene Expression

Antisense constructs have recently been widely used to inhibit the expression of specifically targeted genes in eukaryotic organisms (Van der Krol *et al.*, 1988; Rothstein and Lagrimini, 1989). Ecker and Davis (1986) used electroporation-mediated transient assays to demonstrate that an antisense *CAT* expression plasmid inhibits the expression of a *CAT* expression plasmid *in vivo*, indicating that antisense RNA inhibition of gene expression can occur in plants as it does in animal cells (Izant and Weintraub, 1985). As plant lines containing an antisense gene can be crossed without losing their gene silencing capacity (Cheon *et al.*, 1990), this approach has great potential in future transient expression studies.

3.9. Transient Assays in Virus Research

Transient gene expression assays have been performed mainly with DNA to follow transcription and translation in plant cells as a prerequisite for obtaining stable transformed cells. However, this review would not be complete without a brief discussion of the possibility of using transient assays for virus research.

Fusion of *E. coli* spheroplasts carrying a plasmid containing a complete cauliflower mosaic virus (CaMV) genome resulted in transient expression of the CaMV capsid protein as detected by immunofluorescence (Tanaka *et al.*, 1984). Similarly, Walden and Howell (1983) and Lebeurier *et al.* (1982) performed transient DNA recombination assays on CaMV genomes using subsequent viral propagation in plant cells as a detection system.

A common way to test functions of cloned viral RNA is to transfect plant protoplasts with viral RNA and to study the resulting products (Hibi *et al.*, 1986; Okada *et al.*, 1986a; Watts *et al.*, 1987; Nishiguchi *et al.*, 1986, 1987a; Saunders *et al.*, 1989). In recent years some progress has been made in the development of RNA plant viruses as autonomously replicating vectors (Ahlquist *et al.*, 1987). Out of the large group of RNA viruses, brome mosaic virus (BMV) and tobacco mosaic virus (TMV) have been successfully engineered to replicate and express a foreign gene in protoplasts and in whole plants, generally by replacing the coat protein cistron by the bacterial *CAT* gene (French *et al.*, 1986; Takamatsu *et al.*, 1987). Furthermore, viral DNA delivery into plant cells and tissues has been studied with great success (Hanley-Bowdoin *et al.*, 1988; Töpfer *et al.*, 1989, 1990; Laufs *et al.*, 1990; Creissen *et al.*, 1990).

These selected cases clearly demonstrate that transient gene expression assays are also very helpful for virus research particularly when plant viruses have been engineered as expression vectors (Gronenborn and Matzeit, 1989).

4. CONCLUDING REMARKS

Only a small portion of the DNA introduced into the cell by gene transfer methods becomes stably integrated into the chromosome of the plant cell. The introduced DNA is lost with time and cell division. Fortunately, this transient DNA is expressed in the cell and forms the basis of extremely useful transient assays. Transient assays are commonly used for the analysis of gene expression and rapid monitoring of gene transfer. The establishment of gene transfer conditions leading to stable transformation systems can also be monitored via transient gene expression. The analysis of gene expression by transient assays has been used to define promoter strength, inducible promoter regions, and the stimulation of gene expression with introns and with different 3′ ends. Additionally, the expression of various coding regions and the stability of their protein products can be rapidly evaluated. Transient assays of gene expression are a rapid and valuable tool to evaluate selectable marker genes prior to their deployment in stable transformation assays.

All these important effects have been found within the last few years, demonstrating that such an assay has become fertile field and showing that transient gene assays are of great importance for the future of molecular biology of higher plants.

5. REFERENCES

Abdullah, R., Cocking, E. C., and Thompson, J. A., 1986, Efficient plant regeneration from rice protoplasts through somatic embryogenesis, *Bio/Technology* **4**:1087–1090.

Ahlquist, P., French, R., and Bujarski, J. J., 1987, Molecular studies of brome mosaic virus using infectious transcripts from cloned cDNA, *Adv. Virus Res.* **32**:188–197.

Ahokas, H., 1989, Transfection of germinating barley seeds electrophoretically with exogenous DNA, *Theor. Appl. Genet.* **77**:469–472.

Al-Shawi, R., Kinnaird, J., Burke, J., and Bishop, J. O., 1990, Expression of a foreign gene in a line of transgenic mice is modulated by a chromosomal position effect, *Mol. Cell. Biol.* **10**:1192–1198.

Ballas, N., Zakai, N., and Loyter, A., 1987, Transient expression of the plasmid pCaMVCAT in plant protoplasts following transformation with PEG, *Exp. Cell. Res.* **170**:228–234.

Ballas, N., Zakai, N., Friedberg, D., and Loyter, A., 1988, Linear forms of plasmid DNA are superior to supercoiled structures as active templates for gene expression in plant protoplasts, *Plant Mol. Biol.* **11**:517–527.

Bates, G. W., Piastuch, W., Riggs, C. D., and Rabussay, D., 1988, Electroporation for DNA delivery to plant protoplasts, *Plant Cell Tissue Organ Cult.* **12**:213–218.

Bekkaoui, F., Pilon, M., Laine, E., Raju, D. S. S., Crosby, W. L., and Dunstan, D. I., 1988, Transient gene expression in electroporated *Picea glauca* protoplasts, *Plant Cell Rep.* **7**:481–484.

Bekkaoui, F., Datla, R. S. S., Pilon, M., Tautorus, T. E., Crosby, W. L., and Dunstan, D. I., 1990, The effects of promoter on transient expression in conifer cell lines, *Theor. Appl. Genet.* **79**:353–359.

Bevan, M. W., Flavell, R. B., and Chilton, M. D., 1983, A chimaeric antibiotic resistance gene as a selectable marker for plant cell transformations, *Nature* **304**:184–187.

Blowers, A. D., Bogorad, L., Shark, K., and Sanford, J. C., 1989, Studies on *Chlamydomonas* chloroplast transformation: Foreign DNA can be stably maintained in the chromosome, *Plant Cell* **1**:123–132.

Boston, R. S., Becwar, M. R., Ryan, R. D., Goldsbrough, P. B., Larkins, B. A., and Hodges, T. K., 1987, Expression from heterologous promoters in electroporated carrot protoplasts, *Plant Physiol.* **83**:742–746.

Boynton, J. E., Gillham, N. W., Harris, E. H., Hoster, J. P., Johnson, A. M., Jones, A. R., Randolph-Anderson, B. L., Robertson, D., Klein, T. M., and Shark, K. B., 1988, Chloroplast transformation in *Chlamydomonas* with high velocity microprojectiles, *Science* **240**:1534–1538.

Brasier, A. R., Tate, J. E., and Habener, J. F., 1989, Optimized use of the firefly luciferase assay as a reporter gene in mammalian cell lines, *Bio/Techniques* **7**:1116–1122.

Bruce, W. B., Christensen, A. H., Klein, T., Fromm, M., and Quail, P. H., 1989, Photoregulation of a phytochrome gene promoter from oat transferred into rice by particle bombardment, *Proc. Natl. Acad. Sci. USA* **86**:9692–9696.

Cabanes-Bastos, E., Day, A. G., and Lichtenstein, C. P., 1989, A sensitive and simple assay for neomycin phosphotransferase II activity in transgenic tissue, *Gene* **77**:169–176.

Cai, Q-G., Kuo, C-S, Quian, Y-Q., Jiong, R-X., and Zhou, Y-L., 1987, Plant regeneration from protoplasts of corn (*Zea mays*), *Acta Bot. Sin.* **29**:453–458.

Callis, J., Fromm, M., and Walbot, V., 1987a, Introns increase gene expression in cultured maize cells, *Genes Dev.* **1**:1183–1200.

Callis, J., Fromm, M., and Walbot, V., 1987b, Expression of mRNA electroporated into plant and animal cells, *Nucleic Acids Res.* **15**:5823–5831.

Callis, J., Fromm, M., and Walbot, V., 1988, Heat inducible expression of a chimeric maize *hsp70CAT* gene in maize protoplasts, *Plant Physiol.* **88**:965–968.

Cao, J., Wang, Y-C., Klein, T. M., Sanford, J., and Wu, R., 1990, Transformation of rice and maize using the biolistic process, in *Plant Gene Transfer* (C. J. Lamb and R. N. Beachy, eds.), pp. 21–33, Wiley-Liss, New York.

Charest, P. J., Iyer, V. N., and Miki, B. L., 1989, Factors affecting the use of chloramphenicol acetyltransferase as a marker for *Brassica* genetic transformation, *Plant Cell Rep.* **7**:628–631.

Chee, P. P., Klassy, C., and Slightom, J., 1986, Expression of a bean storage protein "phaseolin minigene" in foreign plant tissues, *Gene* **41**:47–57.

Cheon, C-I., Delauney, A. J., and Verma, D. P. S., 1990, Maintenance of a plant line containing an antisense gene and silencing of the target gene following sexual crosses, *Plant Sci.* **66**:231–236.

Christou, P., McCabe, D., and Swain, W. F., 1988, Stable transformation of soybean callus by DNA-coated gold particles, *Plant Physiol.* **87:**671–674.

Coker, G. T., Vinnedge, L., and O'Malley, K. L., 1989, 8-Br-cAMP inhibits the transient expression of firefly luciferase, *FEBS Lett.* **249:**183–185.

Cornelissen, M. J., De Block, M., Van Montagu, M., Leemans, J., Schreier, P. H., and Schell, J., 1987, Plastid transformation: a progress report, in *Plant DNA Infectious Agents* (Th. Hohn and J. Schell, eds.), pp. 311–320, Springer Verlag, Vienna, New York.

Creissen, G., Smith, C., Francis, R., Reynolds, H., and Mullineaux, P., 1990, *Agrobacterium*- and microprojectile mediated viral DNA delivery into barley microspore-derived cultures, *Plant Cell Rep.* **8:**680–683.

Cutler, A. J., and Saleem, M., 1987, Permeabilizing soybean protoplasts to macromolecules using electroporation and hypotonic shock, *Plant Physiol.* **83:**24–28.

Czernilofsky, A. P., Hain, R., Baker, B., and Wirtz, U., 1986, Studies on the structure and functional organization of foreign DNA integrated into the genome of *Nicotiana tabacum*, *DNA* **5:**473–482.

Dangl, J., Hauffe, K., Lipphardt, S., Hahlbrock, K., and Scheel, D., 1987, Parsley protoplasts retain differential responsiveness to U.V. light and fungal elicitor, *EMBO J* **4:**2731–2737.

Daniell, H., and McFadden, B. A., 1987, Uptake and expression of bacterial and cyanobacterial genes by isolated cucumber etioplasts, *Proc. Natl. Acad. Sci. USA* **84:**6349–6353.

Daniell, H., Vivekananda, J., Nielsen, B. L., Ye, G. N., Tewari, K. K., and Sanford, J. C., 1990, Transient foreign gene expression in chloroplasts of cultured tobacco cells after biolistic delivery of chloroplast vectors, *Proc. Natl. Acad. Sci. USA* **87:**88–92.

De Cleene, M., 1985, The susceptibility of monocotyledons to *Agrobacterium tumefaciens*, *Phytopathol. Z.* **113:**81–89.

Deshayes, A., Herrera-Estrella, L., and Caboche, M., 1985, Liposome-mediated transformation of tobacco mesophyll protoplasts by an *Escherichia coli* plasmid, *EMBO J.* **4:**2731–2737.

De Wet, J. R., Wood, J. V., De Luca, M., Helinski, D. R., and Subramani, S., 1987, The firefly luciferase gene: Structure and expression in mammalian cells, *Mol. Cell Biol.* **7:**725–737.

Ebert, P. R., Ha, S. B., and An, G., 1987, Identification of an essential upstream element in the nopaline synthase promoter by stable and transient assays, *Proc. Natl. Acad. Sci. USA* **84:**5745–5749.

Ecker, J., and Davis, R., 1986, Inhibition of gene expression in plant cells by expression of antisense RNA, *Proc. Natl. Acad. Sci. USA* **83:**5372–5376.

Ellis, J. G., Llewellyn, D. J., Dennis, E. J., and Peacock, W. J., 1987a, Maize *Adh-1* promoter sequences control anaeobic regulation: addition of upstream promoter elements from constitutive genes is necessary for expression in tobacco, *EMBO J.* **6:**11–16.

Ellis, J. G., Llewellyn, D. J., Walker, J. C., Dennis, E. S., and Peacock, W. J., 1987b, The *ocs* element: a 16 base pair palindrome essential for activity of the octopine synthase enhancer, *EMBO J.* **6:**3203–3208.

Evans, D. A., 1989, Somaclonal variation—Genetic basis and breeding applications, *Trends in Genetics* **5:**46–50.

Fox, T. D., Sanford, J. C., and McMullin, T. W., 1988, Plasmids can stably transform yeast mitochondria lacking endogenous *mt* DNA, *Proc. Natl. Acad. Sci. USA* **85:**7288–7292.

Fraley, R. T., Rogers, S. G., Horsch, R. B., Sanders, P. R., Flick, J. S., Adams, S. P., Bittner, M. L., Brans, L. A., Fink, C. L., Fry, J. S., Galuppi, G. R., Goldberg, S. B., Hoffman, N. L., and Woo, S. C., 1983, Expression of bacterial genes in plant cells, *Proc. Natl. Acad. Sci. USA* **80:**4803–4807.

Fraley, R. T., Horsch, R. B., Matzke, A., Chilton, M. D., and Sanders, P. R., 1984, *In vitro* transformation of *Petunia* cells by an improved method of co-cultivation with *A. tumefaciens*, *Plant Mol. Biol.* **3:**371–378.

French, R., Janda, M., and Ahlquist, P., 1986, Bacterial gene inserted in an engineered RNA virus: efficient expression in monocotyledonous plant cells, *Science* **231**:1294–1297.

Fromm, M. E., Taylor, L. P., and Walbot, V., 1985, Expression of genes transferred into monocot and dicot plant cells by electroporation, *Proc. Natl. Acad. Sci. USA* **82**:5824–5828.

Fromm, M., Callis, J., Taylor, L. P., and Walbot, V., 1987, Electroporation of DNA and RNA into plant protoplasts, *Methods Enzymol.* **153**:351–366.

Gallie, D. R., Lucas, W. J., and Walbot, V., 1989, Visualizing mRNA expression in plant protoplasts: Factors influencing efficient mRNA uptake and translation, *Plant Cell* **1**:301–311.

Goddall, G. J., and Filipowicz, W., 1990, The minimum functional length of pre-mRNA introns in monocots and dicots, *Plant Mol. Biol.* **14**:727–733.

Gorman, C., Moffat, L. F., and Howard, B. H., 1982, Recombinant genomes which express chloramphenicol acetyltransferase in mammalian cells, *Mol. Cell. Biol.* **2**:1044–1051.

Gronenborn, B., and Matzeit, V., 1989, Plant gene vectors and genetic transformation: Plant viruses as vectors, *Cell Cult. Somatic Cell Genet. Plants* **6**:69–100.

Gupta, P. K., Dandekar, A. M., and Durzan, D. J., 1988, Somatic proembryo formation and transient expression of a luciferase gene in douglas fir and loblolly pine protoplasts, *Plant Sci.* **58**:85–92.

Gurdon, J. B., and Melton, D. A., 1981, Gene transfer in amphibian eggs and oocytes, *Annu. Rev. Genet.* **15**:189–218.

Ha, S-B., and An, G., 1989, Cis-acting regulatory elements controlling temporal and organ specific activity of nopaline synthase promoter, *Nucleic Acids Res.* **17**:215–223.

Hain, R., Stabel, P., Czernilofsky, A. P., Steinbiss, H. H., Herrera-Estrella, L., and Schell, J., 1985, Uptake, integration and genetic transmission of a selectable chimaeric gene by plant protoplasts, *Mol. Gen. Genet.* **199**:161–168.

Hanley-Bowdoin, L., Elmer, J. S., and Rogers, S. G., 1988, Transient expression of heterologous RNAs using tomato golden mosaic virus, *Nucleic Acids Res.* **16**:10511–10528.

Harkins, K. R., Jefferson, R. A., Kavanagh, T. A., Bevan, M. W., and Galbraith, D. W., 1990, Expression of photosynthesis-related gene fusions is restricted by cell type in transgenic plants and in transfected protoplasts, *Proc. Natl. Acad. Sci. USA* **87**:816–820.

Harland, R. M., Weintraub, H., and McKnight, S. L., 1983, Transcription of DNA injected into *Xenopus* oocytes is influenced by template topology, *Nature* **302**:38–43.

Hauptmann, R. M., Ozias-Akins, P., Vasil, V., Tabaeizadeh, Z., Rogers, S. G., Horsch, R. B., Vasil, I. K., and Fraley, R. T., 1987, Transient gene expression of electroporated DNA in monocotyledonous and dicotyledonous species, *Plant Cell Rep.* **6**:265–270.

Hauptmann, R. M., Ashraf, M., Vasil, V., Hannah, L. C., Vasil, I. K., and Ferl, R., 1988a, Promoter strength comparisons of maize shrunken 1 and alcohol dehydrogenase 1 and 2 promoters in mono- and dicotyledonous species, *Plant Physiol.* **88**:1063–1066.

Hauptmann, R. M., Vasil, V., Ozias-Akins, P., Tabaeizadeh, Z., Rogers, S. G., Fraley, R. T., Horsch, R. B., and Vasil, I. K., 1988b, Evaluation of selectable markers for obtaining stable transformants in the Gramineae, *Plant Physiol.* **86**:602–606.

Herrera-Estrella, L., Depicker, A., Van Montagu, M., and Schell, J., 1983, Expression of chimaeric genes transferred into plant cells using a Ti-plasmid-derived vector, *Nature* **303**:209–213.

Herrera-Estrella, L., Teeri, T. H., and Simpson, J., 1988, Use of reporter genes to study gene expression in plant cells, *Plant Mol. Biol. Manual* **B1**:1–22.

Hibi, T., Kano, H., Sugiura, M., Kazami, T., and Kimura, S., 1986, High efficiency electrotransfection of tobacco mesophyll protoplasts with tobacco mosaic virus RNA, *J. Gen. Virol.* **67**:2037–2042.

Hilson, P., Dewulf, J., Delporte, F., Installe, P., Jacquemin, J-M., Jacobs, M., and Negrutiu, J., 1990, Yeast RAS2 affects cell viability, mitotic division and transient gene expression in *Nicotiana* species, *Plant Mol. Biol.* **14**:669–685.

Hobbs, S. L. A., Jackson, J. A., Baliski, D. S., Delong, C. M. O., and Mahon, J. D., 1990, Genotype- and promoter-induced variability in transient β-glucuronidase expression in pea protoplasts, *Plant Cell Rep.* **9**:17–20.

Horth, M., Negrutiu, I., Burny, A., Van Montagu, M., and Herrera-Estrella, L., 1987, Cloning of a *Nicotiana plumbaginifolia* protoplast-specific enhancer-like sequence, *EMBO J.* **6**:2525–2530.

Howard, E. A., Walker, J. C., Dennis, E. S., and Peacock, W. J., 1987, Regulated expression of an alcohol dehydrogenase 1 chimeric gene introduced into maize protoplasts, *Planta* **170**:535–540.

Hu, C-Y., Chee, P. P., Chesney, R. H., Zhou, J. H., Miller, P. D., and O'Brien, W. T., 1990, Intrinsic *GUS*-like activities in seed plants, *Plant Cell Rep.* **9**:1–5.

Huttly, A. K., and Baulcombe, D. C., 1989, A wheat (alpha)-Amy 2 promoter is regulated by gibberellin in transformed oat aleurone protoplasts, *EMBO J.* **8**:1907–1913.

Izant, J. G., and Weintraub, H., 1985, Constitutive and conditional suppression of exogenous and endogenous genes by anti-sense RNA, *Science* **229**:345–352.

Jacobsen, J. V., and Beach, L. R., 1985, Control of transcription of alpha-amylase and rRNA genes in barley aleurone protoplasts by gibberellin and abscisic acid, *Nature* **316**:275–277.

Janssen, B-J., and Gardner, R. C., 1989, Localized transient expression of GUS in leaf discs following cocultivation with *Agrobacterium*, *Plant Mol. Biol.* **14**:61–72.

Jefferson, R. A., 1987, Assaying chimeric genes in plants: The *GUS* gene fusion system, *Plant Mol. Biol. Rep.* **5**:387–405.

Jefferson, R. A., Burgess, S. M., and Hirsch, D., 1986, β-Glucuronidase from *Escherichia coli* as a gene fusion marker, *Proc. Natl. Acad. Sci. USA* **83**:8447–8451.

Jefferson, R. A., Kavanagh, T. A., and Bevan, M. W., 1987, Gus fusions: β-Glucuronidase as a sensitive and versatile gene fusion marker in higher plants, *EMBO J.* **6**:3901–3907.

Joersbo, M., and Brunstedt, J., 1990, Direct gene transfer to plant protoplasts by electroporation by alternating, rectangular and exponentially decaying pulses, *Plant Cell Rep.* **8**:701–705.

Johnston, S. A., Anziano, P. Q., Shark, K. B., Sanford, J. C., and Butow, R. A., 1988, Mitochondria transformation of yeast by bombardment with microprojectiles, *Science* **240**:1538–1541.

Jones, H., Ooms, G., and Jones, M. G. K., 1989, Transient gene expression in electroporated *Solanum* protoplasts, *Plant Mol. Biol.* **13**:503–511.

Junker, B., Zimny, J., Lührs, R., and Lörz, H., 1987, Transient expression of chimaeric genes in dividing and non dividing cereal protoplasts after PEG induced DNA uptake, *Plant Cell Rep.* **6**:329–332.

Kartha, K. K., Chibbar, R. N., Georges, F., Leung, N., Caswell, K., Kendall, E., and Qureshi, J., 1989, Transient expression of chloramphenicol acetyltransferase (*CAT*) gene in barley cell cultures and immature embryos, *Plant Cell Rep.* **8**:429–432.

Kartzke, S., Saedler, H., and Meyer, P., 1990, Molecular analysis of transgenic plants derived from transformations of protoplasts at various stages of the cell cycle, *Plant Sci.* **67**:63–72.

Kay, R., Chan, A., Daly, M., and McPherson, J., 1987, Duplication of CaMV 35S promoter sequences creates a strong enhancer for plant genes, *Science* **236**:1299–1302.

Klein, T. M., Wolf, E. D., Wu, R., and Sanford, J. C., 1987, High-velocity microprojectiles for delivering nucleic acids into living cells, *Nature* **327**:70–73.

Klein, T. M., Fromm, M., Weissinger, A., Tomes, D., Schaaf, S., Sletten, M., and Sanford, J. C., 1988a, Transfer of foreign genes into intact maize cells with high-velocity microprojectiles, *Proc. Natl. Acad. Sci. USA* **85**:4305–4309.

Klein, T. M., Gradziel, T., Fromm, M. E., and Sanford, J. C., 1988b, Factors influencing gene delivery into *Zea mays* cells by high-velocity microprojectiles, *Bio/Technology* **6**:559–563.

Klein, T. M., Harper, E. C., Svab, Z., Sanford, J. C., and Fromm, M. E., 1988c, Stable genetic transformation of intact *Nicotiana* cells by the particle bombardment process, *Proc. Natl. Acad. Sci. USA* **85**:8502–8505.

Klein, T. M., Roth, B. A., and Fromm, M. E., 1989, Regulation of anthocyanin biosynthetic genes introduced into intact maize tissues by microprojectiles, *Proc. Natl. Acad. Sci. USA* **86:**6681–6685.

Kobayashi, S., and Uchimiya, H., 1989, Expression and integration of a foreign gene in orange (*Citrus sinensis Osb.*) protoplasts by direct gene transfer, *Jpn. J. Genet.* **64:**91–97.

Köhler, F., Cardon, G., Pöhlman, M., Gill, R., and Schieder, O., 1989, Enhancement of transformation rates in higher plants by low dose irradiation: Are DNA repair systems involved in the incorporation of exogenous DNA into the plant genome? *Plant Mol. Biol.* **12:**189–199.

Köhler, F., Benediktsson, I., Cardon, G., Andreo, C. S., and Schieder, O., 1990, Effect of various irradiation treatments of plant protoplasts on the transformation rates after direct gene transfer, *Theor. Appl. Genet.* **79:**679–685.

Koncz, C., Olsson, O., Langridge, W. H. R., Schell, J., and Szalay, A. A., 1987, Expression and assembly of functional bacterial luciferase in plants, *Proc. Natl. Acad. Sci. USA* **84:**131–135.

Koncz, C., Langridge, W. H. R., Olsson, O., Schell, J., and Szalay, A. A., 1990, Bacterial and firefly luciferase genes in transgenic plants: Advantages and disadvantages of a reporter gene, *Dev. Genet.* **11:**224–232.

Krens, F. H., Molendijk, L., Wullems, G. J., and Schilperoort, R. A., 1982, *In vitro* transformation of plant protoplasts with Ti-plasmid DNA, *Nature* **296:**72–74.

Krüger-Lebus, S., and Potrykus, I., 1987, Direct gene transfer to *Petunia hybrida* without electroporation, *Plant Mol. Biol. Rep.* **5:**289–294.

Kyozuka, J., Hayashi, Y., and Shimamoto, K., 1987, High frequency plant regeneration from rice protoplasts by novel nurse culture methods, *Mol. Gen. Genet.* **206:**408–413.

Kyozuka, J., Otoo, E., and Shimamoto, K., 1988, Plant regeneration from protoplasts of indica rice: Genotypic differences in culture response, *Theor, Appl. Genet.* **76:**887–890.

Laufs, J., Wirtz, U., Kamann, M., Matzeit, V., Schaefer, S., Schell, J., Czernilofsky, A. P., Baker, B., and Gronenborn, B., 1990, WDV-Ac/Ds: Expression and excision of transposable elements introduced into various cereals by a viral replicon, *Proc. Natl. Acad. Sci. USA* **87:**7752–7756.

Lazzeri, P. A., and Lörz, H., 1990, Regenerable suspension and protoplast cultures of barley and stable transformation via DNA uptake into protoplasts, in *Genetic Engineering of Crop Plants* (G. W. Lycett and D. Grierson, eds.), pp. 231–238, Butterworths, London.

Lebeurier, G., Hirth, L., Hohn, B., and Hohn, T., 1982, *In vivo* recombination of cauliflower mosaic virus DNA, *Proc. Natl. Acad. Sci. USA* **79:**2932–2936.

Lee, B., Murdoch, K., Topping, J., Kreis, M., and Jones, G. K., 1989, Transient gene expression in aleurone protoplasts isolated from developing caryopses of barley and wheat, *Plant Mol. Biol.* **13:**21–29.

Lee, M., and Phillips, R. L., 1988, The chromosomal basis of somaclonal variation, *Annu. Rev. Plant. Physiol.* **39:**413–437.

Legocki, R., Legocki, M., Baldwin, T. O., and Szalay, A. A., 1986, Bioluminescence in soybean root nodules: Demonstration of a general approach to assay gene expression *in vivo* by using bacterial luciferase, *Proc. Natl. Acad. Sci. USA* **83:**9080–9084.

Lindsey, K., and Jones, M. G. K., 1987, Transient gene expression in electroporated protoplasts and intact cells of sugar beet, *Plant Mol. Biol.* **10:**43–52.

Lipphardt, S., Brettschneider, R., Kreuzaler, F., Schell, J., and Dangl, J. L., 1988, UV-inducible transient expression in parsley protoplasts identifies regulatory *cis*-elements of a chimeric *Antirrhinum majus* chalcone synthase gene, *EMBO J.* **7:**4027–4033.

Lörz, H., Göbel, E., and Brown, P., 1988, Advances in tissue culture and progress towards genetic transformation of cereals, *Plant Breeding* **100:**1–25.

Loyter, A., Scangos, G. A., and Ruddle, F. H., 1982, Mechanisms of DNA uptake by mammalian cells: Fate of exogenously added DNA monitored by the use of fluorescent dyes, *Proc. Natl. Acad. Sci. USA* **79:**422–426.

Ludwig, S. R., Bowen, B., Beach, L., and Wessler, S. R., 1990, A regulatory gene as a novel visible marker for maize transformation, *Science* **247**:449–450.

Maas, C., and Werr, W., 1989, Mechanisms and optimized conditions for PEG mediated DNA transfection into plant protoplasts, *Plant Cell Rep.* **8**:148–151.

Marcotte, W. R., Bayley, C. C., and Quatrano, R. S., 1988, Regulation of a wheat promoter by abscisic acid in rice protoplasts, *Nature* **335**:454–457.

Maxwell, I. H., and Maxwell, F., 1988, Electroporation of mammalian cells with a firefly luciferase expression plasmid: kinetics of transient expression differ markedly among cell types, *DNA* **7**:557–562.

McCabe, D. E., Swain, W. F., Martinell, B. J., and Christou, P., 1988, Stable transformation of soybean (*Glycine max*) by particle acceleration, *Bio/Technology* **6**:923–926.

McDonnell, R. E., Clark, R. D., Smith, W. A., and Hinchee, M. A., 1987, A simplified method for the detection of neomycin phosphotransferase II activity in transformed plant tissue, *Plant Mol. Biol. Rep.* **5**:380–386.

McElroy, D., Zhang, W., Cao, J., and Wu, R., 1990, Isolation of an efficient actin promoter for use in rice transformation, *Plant Cell* **2**:163–171.

Mendel, R. R., Müller, B., Schulze, J., Kolesnikov, V., and Zelenin, A., 1989, Delivery of foreign genes to intact barley cells by high-velocity microprojectiles, *Theor. Appl. Genet.* **78**:31–34.

Meyer, P., Walgenbach, E., Bussmann, K., Hombrecher, G., and Saedler, H., 1985, Synchronized tobacco protoplasts are efficiently transformed by DNA, *Mol. Gen. Genet.* **201**:513–518.

Morikawa, H., Iida, A., and Yamada, Y., 1989, Transient expression of foreign genes in plant cells and tissues obtained by a simple biolistic device (particle gun), *Appl. Microbiol. Biotechnol.* **31**:320–322.

Nagata, T., Okada, K., Kawazu, T., and Takebe, I., 1987, Cauliflower mosaic virus 35S promoter directs S phase specific expression in plant cells, *Mol. Gen. Genet.* **207**:242–244.

Negrutiu, I., Shillito, R., Potrykus, I., Biasini, G., and Sala, F., 1987, Hybrid genes in the analysis of transformation conditions I. Setting up a simple method for direct gene transfer in plant protoplasts, *Plant Mol. Biol.* **8**:363–373.

Neumann, E., Schaefer-Ridder, M., Wang, Y., and Hofschneider, P. H., 1982, Gene transfer into mouse lyoma cells by electroporation in high electric fields, *EMBO J.* **1**:841–845.

Nguyen, V. T., Morange, M., and Bensaude, O., 1988, Firefly luciferase luminescence assays using scintillation counters for quantitation in transfected mammalian cells, *Analyt. Biochem.* **171**:404–408.

Nishiguchi, M., Langridge, W. H. R., Szalay, A. A., and Zaitlin, M., 1986, Electroporation-mediated infection of tobacco leaf mesophyll protoplasts with tobacco mosaic virus RNA and cucumber mosaic virus RNA, *Plant Cell Rep.* **5**:57–60.

Nishiguchi, M., Sato, T., and Motoyoshi, F., 1987a, An improved method for electroporation into plant protoplasts: Infection of tobacco protoplasts by tobacco mosaic virus particles, *Plant Cell Rep.* **6**:90–93.

Nishiguchi, M., Sato, T., and Motoyoshi, F., 1987b, Factors influencing the transfer of the chloramphenicol acetyltransferase gene by electroporation and its transient expression in tobacco mesophyll protoplasts, *Bull. Natl. Inst. Agrobiol.* **3**:105–114.

Nishigughi, M., Kohno, M., and Motoyoshi, F., 1988, Electroporation-mediated gene transfer into melon mesophyll protoplasts: transient expression of the chloramphenicol acetyltransferase (*CAT*) gene, *Bull. Natl. Inst. Agrobiol.* **4**:177–187.

North, G., 1985, Eukaryotic topoisomerases come into the limelight, *Nature* **316**:394–396.

Oard, J. H., Paige, D., and Dvorak, J., 1989, Chimeric gene expression using maize intron in cultured cells of breadwheat, *Plant Cell Rep.* **8**:156–160.

Oard, J. H., Paige, D. F., Simmonds, J. A., and Gradziel, T. M., 1990, Transient gene expression in maize, rice, and wheat cells using an airgun apparatus, *Plant Physiol.* **92**:334–339.

Odell, J. T., Nagy, F., and Chua, N-H., 1985, Identification of DNA sequences required for the activity of the cauliflower mosaic virus 35S promoter, *Nature* **313**:810–812.

Okada, K., Nagata, T., and Takebe, I., 1986a, Introduction of functional RNA into plant protoplasts by electroporation, *Plant Cell Physiol.* **27**:619–626.

Okada, K., Takebe, I., and Nagata, T., 1986b, Expression and integration of genes introduced into highly synchronized plant protoplasts, *Mol. Gen. Genet.* **205**:398–403.

O'Kane, D. J., Lingle, W. L., Wampler, J. E., Legocki, M., and Szalay, A. A., 1988, Visualization of bioluminescence as a marker of gene expression in rhizobium-infected soybean root nodules, *Plant Mol. Biol.* **10**:387–399.

Olsson, O., Koncz, C., and Szalay, A. A., 1988, The use of the luxA gene of the bacterial luciferase operon as a reporter gene, *Mol. Gen. Genet.* **215**:1–9.

Olsson, O., Escher, A., Sandberg, G., Schell, J., Koncz, C., and Szalay, A. A., 1989, Engineering of monomeric bacterial luciferases by fusion of LuxA and LuxB genes in *Vibrio harveyi, Gene* **81**:335–347.

Ou-Lee, T-M., Turgeon, R., and Wu, R., 1986, Expression of a foreign gene linked to either a plant-virus or a *Drosophila* promoter, after electroporation of protoplasts of rice, wheat, and sorghum, *Proc. Natl. Acad. Sci. USA* **83**:6815–6819.

Ow, D. W., Wood, K. V., DeLuca, M., deWet, J. R., Helsinki, D. R., and Howell, S. H., 1986, Transient and stable expression of the firefly luciferase gene in plant cells and transgenic plants, *Science* **234**:856–859.

Ow, D. W., Jacobs, J., and Howell, S. H., 1987, Functional regions of the cauliflower mosaic virus 35S RNA promoter determined by the use of the firefly luciferase gene as a reporter of promoter activity, *Proc. Natl. Acad. Sci. USA* **84**:4870–4874.

Pietrzak, M., Shillito, R. D., Hohn, Th., and Potrykus, I., 1986, Expression in plants of two bacterial antibiotic resistance genes after protoplast transformation with a new plant expression vector, *Nucleic Acid Res.* **14**:5857–5868.

Pina, B., Hache, R. J. G., Arnemann, J., Chalepakis, G., Slater, E. P., and Beato, M., 1990, Hormonal induction of transfected genes depends on DNA topology, *Mol. Cell. Biol.* **10**:625–633.

Planckaert, F., and Walbot, V., 1989, Transient gene expression after electroporation of protoplasts derived from embryogenic maize callus, *Plant Cell Rep.* **8**:144–147.

Platt, S. G., and Young, N. S., 1987, Dot assay for neomycin phosphotransferase II activity in crude cell extracts, *Anal. Biochem.* **162**:529–535.

Potrykus, I., Saul, M. W., Petruska, J., Paszkowski, J., and Shillito, R. D., 1985, Direct gene transfer to cells of a graminaceous monocot, *Mol. Gen. Genet.* **199**:183–188.

Potter, H., Weir, L., and Leder, P., 1984, Enhancer-dependent expression of human K immunoglobulin genes introduced into mouse pre-B lymphocytes by electroporation, *Proc. Natl. Acad. Sci. USA* **81**:7161–7165.

Pröls, M., Töpfer, R., Schell, J., and Steinbiss, H-H., 1988, Transient gene expression in tobacco protoplasts. I. Time course of *CAT* appearance, *Plant Cell Rep.* **7**:221–224.

Pröls, M., Schell, J., and Steinbiss, H-H., 1989, Critical evaluation of electromediated gene transfer and transient expression in plant cells, in *Electroporation and Electrofusion in Cell Biology* (E. Neumann, A. E. Sowers, and C. A. Jordan, eds.), pp. 367–375, Plenum Press, New York.

Reiss, B., Sprengel, R., and Schaller, H., 1984a, Protein fusions with the kanamycin resistance gene from transposon Tn5, *EMBO J.* **3**:3317–3322.

Reiss, B., Sprengel, R., Will, H., and Schaller, H., 1984b, A new and sensitive method for qualitative and quantitative analysis of neomycinphosphotransferase in crude cell extracts, *Gene* **30**:211–218.

Reynaerts, A., DeBlock, M., Hernalsteens, J-P., and Van Montagu, M., 1988, Selectable and screenable markers, *Plant Mol. Biol. Manual* **A9**:1–16.

Rhodes, C. A., Lowe, K. S., and Ruby, K. L., 1988a, Plant regeneration from protoplasts isolated from embryogenic maize cell cultures, *Bio/Technology* **6**:56–60.

Rhodes, C. A., Pierce, D. A., Mettler, I. J., Mascarenhas, D., and Detmer, J. J., 1988b, Genetically transformed maize plants from protoplasts, *Science* **240**:204–207.

Riggs, C. D., and Bates, G. W., 1986, Stable transformation of tobacco by electroporation: Evidence for plasmid concatenation, *Proc. Natl. Acad. Sci. USA* **83**:5602–5606.

Riggs, C. D., and Chrispeels, M. J., 1987, Luciferase reporter cassettes for plant gene expression studies, *Nucleic Acids Res.* **15**:8115.

Rogers, S. G., Klee, H. J., Horsch, R. B., and Fraley, R. T., 1987, Improved vectors for plant transformation: Expression cassette vectors and new selectable markers, *Methods Enzymol.* **113**:253–277.

Rosenberg, N., Gad, A. E., Altman, A., Navot, N., and Czosnek, H., 1988, Liposome-mediated introduction of the chloramphenicol acetyltransferase (*CAT*) gene and its expression in tobacco protoplasts, *Plant Mol. Biol.* **10**:185–191.

Rothstein, S. J., and Lagrimini, L. M., 1989, Silencing gene expression in plants, 1989, *Oxford Surv. Plant Mol. Cell Biol.* **6**:221–246.

Roy, P., and Sahasradbudke, N., 1990, A sensitive and simple paper chromatographic procedure for detecting neomycin phosphotransferase II(NPT II) gene expression, *Plant Mol. Biol.* **14**:873–876.

Ryoji, M., and Worcel, A., 1985, Structure of the two distinct types of minichromosomes that are assembled on DNA injected in *Xenopus* oocytes, *Cell* **40**:923–932.

Sala, F., Galli, M. G., Pedrali-Noy, G., and Spadari, S., 1986, Synchronization of plant cells in culture and in meristems by aphidicolin, *Methods Enzymol.* **118**:87–96.

Sanders, P. R., Winter, J. A., Barnason, A. R., Rogers, S. G., and Fraley, R. T., 1987, Comparison of cauliflower mosaic virus 35S and nopaline synthase promoters in transgenic plants, *Nucleic Acid Res.* **15**:1543–1558.

Sanford, J. C., 1988, The biolistic process, *Trends Biotechnol.* **6**:299–302.

Sanford, J. C., Klein, T. M., Wolf, E. D., and Allen, N., 1987, Delivery of substances into cells and tissues using a particle bombardment process, *Particulate Sci. Technol.* **5**:27–37.

Sanger, M., Daubert, S., and Goodman, R. M., 1990, Characteristics of a strong promoter from figwort mosaic virus: Comparison with the analogous 35S promoter from cauliflower mosaic virus and the regulated mannopine synthase promoter, *Plant Mol. Biol.* **14**:433–443.

Saunders, J. A., Rhodes Smith, C., and Kaper, J. M., 1989, Effects of electroporation pulse wave on the incorporation of viral RNA into tobacco protoplasts, *Bio/Techniques* **7**:1124–1131.

Schauer, A. T., 1988, Visualizing gene expression with luciferase fusions, *Tibtech* **6**:2327.

Schwall, M., and Feix, G., 1988, Zein promoter activity in transiently transformed protoplasts from maize, *Plant Sci.* **56**:161–166.

Scowcroft, W. R., and Larkin, P. J., 1988, Somaclonal variation, in *Application of Plant Cell and Tissue Culture*, CIBA Foundation Symposium 137, pp. 21–35, Wiley, Chichester.

Seguin, A., and Lalonde, M., 1988, Gene transfer by electroporation in Betulaceae protoplasts: *Alnus incana*, *Plant Cell Rep.* **7**:367–370.

Shaw, J. J., Rogowsky, P., Close, T. J., and Kado, C. I., 1987, Working with bacterial bioluminescence, *Plant Mol. Biol. Rep.* **5**:225–236.

Shigekawa, K., and Dower, W. J., 1988, Electroporation of eucaryotes and procaryotes: a general approach to the introduction of macromolecules into cells, *BioTechniques* **6**:742–751.

Shillito, R. D., Saul, M. W., Paszkowski, S. J., Muller, M., and Potrykus, I., 1985, High efficiency direct gene transfer to plants, *Bio/Technology* **3**:1099–1103.

Shimamoto, K., Terada, R., Izawa, T., and Fujimoto, H., 1988, Fertile transgenic rice plants regenerated from transformed protoplasts, *Nature* **338**:274–276.

Smith, G. R., 1981, DNA supercoiling: Another level for regulating gene expression, *Cell* **24**:599–600.

Smith, P. J., and Paterson, M. C., 1983, Effect of aphidicolin on de novo DNA synthesis, DNA repair and cytotoxicity in gamma-irradiated human fibroblasts, Biochim. Biophys. Acta 739:17–26.

Staebell, M., Tomes, D., Weissinger, A., Maddock, S., Marsh, W., Huffman, G., Bauer, R., Ross, M., and Howard, J., 1990, A quantitative assay for neomycin phosphotransferase activity in plants, Analyt. Biochem. 185:319–323.

Steinbiss, H-H., and Davidson, A., 1989, Genetic manipulation of plants: from tools to agronomical applications, Sci. Prog. (Oxf.) 73:147–168.

Takamatsu, N., Ishikawa, M., Meshi, T., and Okada, Y., 1987, Expression of bacterial chloramphenicol acetyltransferase gene in tobacco plants mediated by TMV-RNA, EMBO J. 6:307–311.

Tanaka, N., Ikesami, M., Hohn, T., Matsui, C., and Watanabe, I., 1984, E. coli spheroplast-mediated transfer of cloned cauliflower mosaic virus DNA into plant protoplasts, Mol. Gen. Genet. 195:378–380.

Tautorus, T. E., Bekkaoui, F., Pilon, M., Datla, R. S. S., Crosby, W. L., Fowke, L. C., and Dunstan, D. I., 1989, Factors affecting transient gene expression in electroporated black spruce (Picea mariana) and jack pine (Pinus banksiana) protoplasts, Theor. Appl. Genet. 78:531–536.

Taylor, B. H., and Larkin, P. J., 1988, Analysis of electroporation efficiency in plant protoplasts, Aust. J. Biotechnol. 1:52–57.

Teeri, T. H., Patel, G. K., Aspegren, K., and Kauppinen, V., 1989, Chloroplast targeting of neomycin phosphotransferase II with a pea transit peptide in electroporated barley mesophyll protoplasts, Plant Cell Rep. 8:187–190.

Terada, R., Kyozuka, J., Nishibayashi, S., and Shimamoto, K., 1987, Plantlet regeneration from somatic hybrids of rice (Oryza sativa L.) and barnyard grass (Echinochloa oryzicola Vasing), Mol. Gen. Genet. 210:39–43.

Töpfer, R., Matzeit, V., Gronenborn, B., Schell, J., and H-H. Steinbiss, 1987, A set of plant expression vectors for transcriptional and translational fusions, Nucleic Acid Res. 14:5890.

Töpfer, R., Schell, J., and Steinbiss, H-H., 1988a, Versatile cloning vectors for transient gene expression and direct gene transfer, Nucleic Acids Res. 16:8725.

Töpfer, R., Pröls, M., Schell, J., and Steinbiss, H-H., 1988b, Transient gene expression in tobacco protoplasts: II. Comparison of the reporter gene system for CAT, NPTII, and GUS, Plant Cell Rep. 7:225–228.

Töpfer, R., Gronenborn, B., Schell, J., and Steinbiss, H-H., 1989, Uptake and transient expression of chimeric genes in seed-derived embryos, Plant Cell 1:133–139.

Töpfer, R., Gronenborn, B., Schäfer, S., Schell, J., and Steinbiss, H-H., 1990, Expression of engineered wheat dwarf virus in seed-derived embryos, Physiol. Plant 79:158–162.

Toriyama, K., Arimoto, Y., Uchimiya, H., and Hinata, K., 1988, Transgenic rice plants after direct gene transfer into protoplasts, Bio/Technology 6:1072–1074.

Twell, D., Klein, T. M., Fromm, M. E., and McCormick, S., 1989, Transient expression of chimeric genes delivered into pollen by microprojectile bombardment, Plant Physiol. 91:1270–1274.

Tyagi, S., Spörlein, B., Tyagi, A. K., Herrmann, R. G., and Koop, H. U., 1989, PEG- and electroporation-induced transformation in Nicotiana tabacum: Influence of genotype on frequencies, Theor. Appl. Genet. 78:287–292.

Vancanneyt, G., Schmidt, R., O'Connor-Sanchez, Willmitzer, L., and Rocha-Sosa, M., 1990, Construction of an intron containing marker gene: Splicing of the intron in transgenic plants and its use in monitoring early events in Agrobacterium-mediated plant transformation, Mol. Gen. Genet. 220:245–250.

Van den Elzen, P. J. M., Townsend, J., Lee, K. Y., and Bedbrook, J. R., 1985, A chimeric hygromycin resistance gene as a selectable marker in plant cells, Plant Mol. Biol. 5:299–302.

Van der Krol, A. R., Mol, J. N. M., and Stuitje, A. R., 1988, Regulation of eukaryotic gene expression by complementary RNA or DNA sequences: An overview, Bio/Techniques 6:958–976.

Vasil, I. K., 1988, Progress in regenerating and genetic manipulation of cereal crops, Bio/Technology **6**:397–402.

Vasil, V., Hauptmann, R. M., Morrish, F. M., and Vasil, I. K., 1988, Comparative analysis of free DNA delivery and expression into protoplasts of *Panicum maximum* Jacq. (guinea grass) by electroporation and polyethylene glycol, *Plant Cell Rep.* **7**:499–503.

Vasil, V., Clancy, M., Ferl, R. J., Vasil, I. K., and Hannah, L. C., 1989, Increased gene expression by the first intron of maize shrunken-1 locus in grass species, *Plant Physiol.* **91**:1575–1579.

Vasil, V., Redway, F., and Vasil, I. K., 1990, Regeneration of plants from embryogenic suspension culture protoplasts of wheat (*Triticum aestivum* L.), *Bio/Technology* **8**:429–434.

Velten, J., Velten, L., Hain, R., and Schell, J., 1984, Isolation of a dual promoter fragment from Ti plasmid of *Agrobacterium tumefaciens, EMBO J.* **3**:2723–2730.

Walden, R. M., and Howell, S. H., 1983, Uncut recombination plasmids bearing nested cauliflower mosaic virus genomes infect plants by intragenomic recombinations, *Plant Mol. Biol.* **2**:27–31.

Waldron, C., Murphy, E. B., Roberts, J. L., Gustafson, G. D., Armour, S. L., and Malcom, S. K., 1985, Resistance to hygromycin B, *Plant Mol. Biol.* **5**:103–108.

Wang, J., 1985, Topoisomerase, *Annu. Rev. Biochem.* **54**:685–699.

Wang, Y-C., Klein, T. M., Fromm, M., Cao, J., Sanford, J. C., and Wu, R., 1988, Transient expression of foreign genes in rice, wheat and soybean cells following particle bombardment, *Plant Mol. Biol.* **11**:433–439.

Watts, J. W., King, J. M., and Stacey, N. J., 1987, Inoculation of protoplasts with viruses by electroporation, *Virology* **157**:40–46.

Wei, Z-M., and Xu, Z-H., 1990, Regeneration of fertile plants from embryogenic suspension culture protoplasts of *Sorghum vulgare, Plant Cell Rep.* **9**:51–53.

Weide, R., Koorneef, M., and Zabel, P., 1989, A simple, nondestructive spraying assay for the detection of an active kanamycin resistance gene in transgenic tomato plants, *Theor. Appl. Genet.* **78**:169–172.

Weintraub, H., 1985, Assembly and propagation of repressed and derepressed chromosomal states, *Cell* **42**:705–711.

Weintraub, H., Chang, P. F., and Conrad, K., 1986, Expression of transfected DNA depends on DNA topology, *Cell* **46**:115–122.

Werr, W., and Lörz, W., 1986, Transient gene expression in a Gramineae cell line, *Mol. Gen. Genet.* **202**:471–475.

Wilson, S. M., Thorpe, T. A., and Moloney, M. M., 1989, PEG-mediated expression of *GUS* and *CAT* genes in protoplasts from embryogenic suspension cultures of *Picea glauca, Plant Cell Rep.* **7**:704–707.

Wirtz, U., Schell, J., and Czernilowsky, A. P., 1987, Recombination of selectable marker DNA in *Nicotiana tabacum, DNA* **6**:245–253.

Wong, E. A., and Capecchi, M. R., 1985, Effect of cell cycle position on transformation by microinjection, *Somatic Cell Mol. Genet.* **11**:43–51.

Yamada, Y., Zhi-Qi, Y., and Ding-Tai, T., 1986, Plant regeneration from protoplast-derived callus of rice (*Oryza sativa* L.), *Plant Cell Rep.* **5**:85–88.

Yang, Z. Q., Shikanai, T., Moriu, K., and Yamada, Y., 1989, Plant regeneration from cytoplasmic hybrids of rice (*Oryza sativa* L.), *Theor. Appl. Genet.* **77**:305–310.

Zhang, W., and Wu, R., 1988, Efficient regeneration of transgenic plants from rice protoplasts and correctly regulated expression of the foreign gene in the plants, *Theor, Appl. Genet.* **76**:835–840.

Zhang, H. M., Yang, H., Rech, E. L., Golds, T. J., Davis, A. S., and Mulligan, B. J., 1988, Transgenic rice plants produced by electroporation-mediated plasmid uptake into protoplasts, *Plant Cell Rep.* **7**:379–384.

Chapter 8

Restriction Fragment Length Polymorphism in Plants and Its Implications

Gary Kochert

1. INTRODUCTION

Botstein *et al.* (1980) were the first to outline convincingly how cloned pieces of eukaryotic DNA could be used as genetic markers. They pointed out that organisms differing in base sequence would be expected to yield fragments of different sizes when their DNA was digested with a suitable restriction enzyme. These DNA fragments could then be fractionated by size on an agarose gel and detected by hybridization with a cloned probe homologous to a portion of the fragment. Such differences in the size of restriction fragments were termed restriction fragment length polymorphisms (RFLPs), and they would be expected to arise from nucleotide changes in the DNA or by insertions, deletions, or other forms of rearrangement.

RFLPs would behave as codominant markers, and linkage maps (RFLP maps) could be constructed using conventional methods. Further, an essentially unlimited number of such markers should be available in nearly every organism, removing one of the principal limitations of conventional map construction: a

Gary Kochert   Department of Botany, University of Georgia, Athens, Georgia 30602.

Subcellular Biochemistry, Volume 17: Plant Genetic Engineering, edited by B. B. Biswas and J. R. Harris. Plenum Press, New York, 1991.

shortage of easily scorable markers. Since the human genome was thought to be about 33 Morgans in length, about 150 RFLP markers would suffice for a genetic map with 20 cM resolution. RFLP markers would not have to represent genes or parts of genes; any fragment of DNA should serve as an RFLP marker provided it was not a highly repetitive sequence, which would give RFLP patterns too complex to interpret. Botstein *et al.* (1980) also explained how it would be possible to map the gene for any trait by using an RFLP map. With a 20-cM map, any trait would lie within 10 map units of an RFLP marker and could be mapped relative to the flanking markers. It was also pointed out that the number of fragments examined is the crucial parameter in discovering RFLPs caused by single base changes, but the length of the fragment examined is more important in discovering RFLPs caused by rearrangement events.

In subsequent years RFLP analysis has been applied to many different organisms and the usefulness of RFLP analysis has been repeatedly confirmed. RFLP maps have been constructed for several animals and plants (Bernatzky and Tanksley, 1986; Bonierbale *et al.*, 1988; Gebhardt *et al.*, 1989b; Chang *et al.*, 1988; Chao *et al.*, 1989; Donis-Keller *et al.*, 1987; Landry *et al.*, 1987; Mc-Couch *et al.*, 1988), and they are being applied to a variety of biological problems. Important human disease genes have been cloned with the aid of linked RFLP markers (Rommens *et al.*, 1989), and this technology is beginning to be applied to plants (Somerville, 1989). Genetic maps of phenotypic markers are quite rare; for only a few organisms has it been possible to apply the large resources of time and money needed to construct a detailed map. RFLP analysis has provided a quantum leap in ability to construct genetic maps. A relatively small research effort can produce a detailed map, and in plants this can usually be accomplished with only one cross and a small mapping population. Thus it becomes possible to undertake projects such as comparative mapping of a cultivated species and its wild relatives, which would have been impractical before. Soller and Beckmann (1983) and Beckmann and Soller (1983, 1986) were among the first to analyze the possibilities of RFLP analysis in plants. They pointed out possibilities for varietal identification, surveys of genetic polymorphism, marker-assisted introgression, and identification and mapping of quantitative traits. Advantages of RFLP markers for genetic analysis included absence of pleiotropic effects on economic traits, multiple allelic forms, and lack of dominance (codominance).

Now we are coming to another frontier in genetic mapping. Instead of simply looking for relatively large differences in the size of restriction fragments by conventional RFLP analysis, new methods with enhanced sensitivity are being implemented (Beckmann, 1988; Orita *et al.*, 1989a,b; Cotton, 1989). These new methods promise to simplify techniques, make them less labor intensive, less expensive, and thus accessible to a larger group of scientists.

In addition to their use in genetic maps, unmapped RFLP markers can also be used in several ways. Surveys of genetic polymorphism or application of RFLP data to systematic or evolutionary studies can be carried out without knowledge of the chromosomal position of the markers, although this sometimes weakens the inferences that can be made. Identification of individual cultivars or plants (DNA fingerprinting) can also be done with unmapped markers.

In this review, I will attempt to summarize current knowledge about using molecular markers in plant research, particularly in plant breeding, and to outline the areas where progress can be expected in the immediate future. I will emphasize RFLP analysis of plant nuclear DNA. Several previous reviews have been published and the reader is referred to these for alternative points of view (Evola *et al.*, 1986; Helentjaris *et al.*, 1985; Landry and Michelmore, 1987; Pethe *et al.*, 1989; Tanksley, 1989; Tanksley *et al.*, 1988a). Organelle DNA, particularly chloroplast DNA, has been extensively used in taxonomic and phylogenetic studies of plants, and recent reports may be consulted for an entry into this literature (Chase and Palmer, 1989; Schilling and Jansen, 1989).

2. METHODS

2.1. Probe Libraries

For an RFLP mapping project several hundred to several thousand clones of chromosomal DNA are needed as hybridization probes. The clones need to represent low copy-number nuclear sequences, so some method must be used to screen out repeated sequences and organelle sequences. Both random genomic libraries and cDNA libraries have been used as sources of RFLP markers. Bacterial plasmids have been the most common vector, since libraries in plasmids are easily constructed and complete libraries are not needed. The size of the cloned insert has usually been small (1–2 kb). Because the interspersion pattern of repeated sequences is usually not known in detail in the plant to be studied, small clones are used to make sure that most clones do not contain repeats.

2.1.1. Random Genomic Libraries

Random genomic libraries in plasmids are very simple to construct and have been widely used in RFLP mapping. Plant DNA is highly methylated at the cytosine of CpG and CpNpG residues (Belanger 'and Hepburn, 1990; Brown, 1989), and repeated sequences tend to be more highly methylated than low copy sequences. Therefore, it is possible to enrich a library for low copy sequences by selecting a methylation-sensitive restriction enzyme, such as *Pst*I for its construction (McCouch *et al.*, 1988; Tanksley *et al.*, 1988a). Smaller fragments (typically 1–2 kb) are selected by cutting a band from an agarose gel or by sucrose gradient fractionation. Many repeated sequences are left as partially digested larger fragments which are not cloned. Usually, it is not worth the effort to try to isolate nuclear DNA, and most researchers have used total DNA extractions for library construction.

Most of the clones in a random genomic library will represent intergenic sequences, since coding regions make up a relatively small portion of the genome of plants. If the library has been constructed from total DNA, organelle sequences will also be present. These are usually screened from the library, along with repeated sequences that have survived the initial selection. Organelle sequences can be detected by probing the library with labeled chloroplast or mitochondrial DNA; nuclear repeated sequences are detected by probing dot blots or by colony hybridization with labeled total DNA as probe (Landry and Michelmore, 1985). Only repeated sequences are present in sufficient concentration to produce a signal. Members of the library that do not give a signal with either organelle DNA or total DNA are selected as potential low-copy-number mapping probes.

2.1.2. cDNA Libraries

Since cDNA libraries are constructed from RNA transcripts, they represent coding regions. Most clones will represent low copy sequences, but some repeated sequences will also be present since a fraction of these is transcribed in plants. Coding sequences would be expected to be more conserved between different taxa than intergenic sequences which have no known function. One would thus expect random genomic libraries to yield more polymorphism between closely related organisms than cDNA libraries. However, this may not always be the case. In lettuce, Landry *et al.* (1987a) showed that probes from a cDNA library actually detected more polymorphisms than random genomic probes. One has to realize, however, that the restriction fragments detected by cDNA probes may also contain flanking sequences that may not be as conserved as the coding regions themselves. Since cDNAs are likely to be more conserved between taxa than random genomic clones, cDNA libraries would normally be used for comparative RFLP analysis, which will necessarily involve heterologous probes.

2.1.3. Heterologous Probes

In addition to the anonymous probes from libraries such as those described above, any cloned sequence can be used as an RFLP probe, provided only that sufficient homology exists with the test organism's DNA to give an unambiguous hybridization signal. This makes it possible to rather rapidly map known genes, since many such probes are available.

2.2. Detection of RFLPs

In the most widely used procedure for detection of RFLPs, total DNA from the plants of interest is digested with restriction enzymes and separated by electrophoresis in agarose gels. The DNA is then transferred from the gel to a nylon filter by Southern blotting (Southern, 1975), and RFLPs are detected by hybridization with cloned probes followed by autoradiography. Disadvantages of this method include its high cost, relatively low resolution, and intensive labor requirement.

2.2.1. Selection of Restriction Enzymes

Restriction enzymes to be used in RFLP analysis should be selected on the basis of cost and likelihood of detecting polymorphisms. The restriction enzymes that are affordable enough to use in an RFLP project are those with six base pair recognition sites (six-cutters) and those with four base pair restriction sites (four-cutters). Fewer, larger DNA fragments would be expected to be produced by six-cutters (assuming randomness, a six-cutter site should occur about every 4 kb, and a four-cutter site about every 256 bp).

RFLP differences in plants may be caused either by base substitution or by rearrangements such as insertions or deletions. If RFLPs are caused by base substitution, it would be best to survey as many restriction sites as possible, since RFLPs will be detected only in the restriction sites for the enzyme being used. Nucleotide changes between the restriction sites for the enzyme would not cause a change in the length of the fragment and would be undetected. In this case, four-cutters should be preferred. In a 4-kb sequence (again assuming random sequence), a six-cutter would survey 12 nucleotides, while a four-cutter would survey 52 nucleotides (from the 13 sites that would be expected to occur in a 4-kb sequence). If RFLPs, however, are caused by insertions, deletions, and other rearrangement events, then enzymes that produce larger fragments would be preferred, since there is more chance of encountering such an event in a larger fragment.

Available evidence from plants (Apuya et al., 1988; McCouch et al., 1988) suggests that RFLPs are more often caused by rearrangements. In rice (McCouch et al., 1988), for example, a positive correlation was found between the length of fragments detected in digests of a given restriction enzyme and the probability of the fragment being polymorphic between rice cultivars. It was also found that probes that detect polymorphisms with one enzyme usually also detect poly-morphisms with other enzymes. This result would be expected if the poly-morphism is being caused by an insertion–deletion event. Therefore, a reason-able first strategy is to use the least expensive six-cutter restriction enzymes for initial polymorphism surveys.

2.2.2. Selection of Plant Material

The degree of RFLP polymorphism found between any two plant cultivars or taxa is variable. Most RFLP polymorphism surveys have been done with crop plants preparatory to the construction of an RFLP map. Maize inbred lines exhibit a tremendous amount of RFLP polymorphism; virtually any two inbred lines can be distinguished with one or two probe/enzyme combinations (Evola *et al.*, 1986). Tomato (Bernatzky and Tanksley, 1986) and peanut (Kochert, unpublished), however, have very low levels of polymorphism between cultivars. This could reflect the history of domestication of these plants, perhaps reflective of a genetic bottleneck. Alternatively, genome evolution may vary between plant groups, perhaps because of differing levels of transposon activity. In cases where low polymorphism is found between cultivars, a cross with a land race or a related wild species will have to be used to generate a mapping population.

2.2.3. Labeling of Probes

Usually probes are labelled with ^{32}P either by nick-translation (Rigby *et al.*, 1977) or by using random primers (Feinberg and Vogelstein, 1984). Whole plasmids may be labelled and, in plants with a small genome, will usually give an adequate hybridization signal. Insert DNA can also be isolated if a better signal-to-noise ratio is required. This has usually been done by extracting the insert from agarose gels after digestion with the appropriate restriction enzyme to free the insert. However, a newer method uses the polymerase chain reaction (PCR) with universal primers homologous to areas of the vector just outside the insert. One set of primers can be used for all the members of the library, and a large amount of insert can be obtained very easily (Tanksley, personal communication). The small amount of recombinant vector DNA needed means that one can isolate inserts from single bacterial colonies or single phage plaques.

Nonradioactive methods for labeling probes are also rapidly being developed, and some are commercially available as kits. In areas where radioactive isotopes are difficult to obtain, these methods will be more suitable. However, what is really needed is new methods for detecting RFLPS without the necessity of doing Southern blots and hybridization (see below).

2.3. Linkage Analysis

Most plant RFLP maps have been constructed with plants that are normally self-pollinating or can be self-fertilized to produce inbred lines. Such plants are normally homozygous at all loci, and genetic analysis is relatively straightforward. In such inbred plant species, three types of mapping populations are suitable for construction of RFLP maps: backcross, F_2, and recombinant inbreds

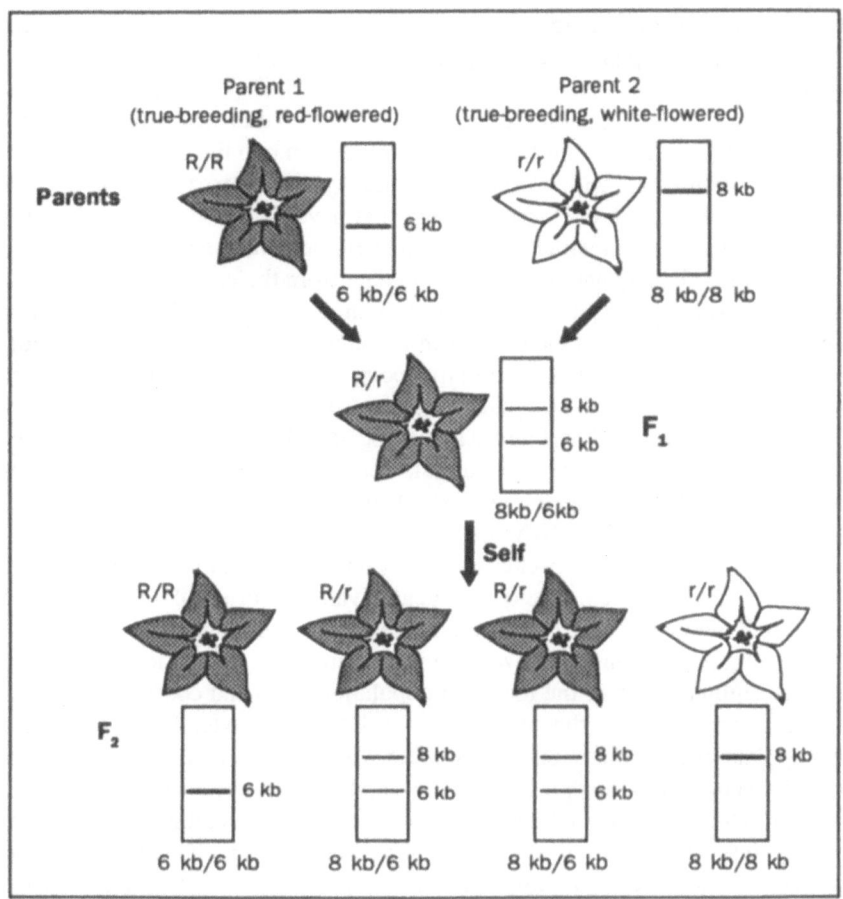

FIGURE 1. Inheritance of an RFLP marker compared with inheritance of a conventional single gene marker controlling flower color. The conventional gene shows dominance, with colored flowers being dominant over white flowers. The plant is assumed to be a diploid, selfing species, which is homozygous at all loci. A segregation ratio of 1:2:1 is shown for the RFLP marker in the F_2.

(RI). Backcross populations are formed when an F_1 hybrid is crossed to one of its parents. Since the F_1 is heterozygous at all loci for which the parents were polymorphic, backcross progeny should be homozygotes and heterozygotes in a 1:1 ratio for any given single copy RFLP marker. If the F_1 is self-pollinated rather than crossed to one of the parents, an F_2 population results. Such a population will contain homozygotes similar to parent 1, heterozygotes, and homozygotes similar to parent 2 in a ratio of 1:2:1 (Figure 1). For populations containing equivalent numbers of individuals, F_2 populations contain more information because recombination events can be detected in both of the parents (Allard, 1956). Most plant RFLP maps have been constructed using about 50 F_2 progeny.

There is a measure of ambiguity, however, in F_2 data (Lander *et al.*, 1987). When one is trying to measure recombination between two linked probes, one of the classes of progeny will be doubly heterozygous (heterozygous with both probes). This class contains those progeny that have two recombinant chromosomes as well as those in which neither chromosome is recombinant. Therefore, one cannot directly calculate recombination data from the progeny arrays, and a statistical approach must be used. Several methods are available. One can use two-point analysis based on the maximum-likelihood equations of Allard (1956), which can be solved by a simple iterative computer program. Most researchers now use either the Linkage computer program (Suiter *et al.*, 1983) or the Mapmaker program developed by Lander and associates (Lander *et al.*, 1987). Linkage is available in a DOS version suitable for IBM personal computers; Mapmaker is supplied as C source code for use on a VAX computer or a personal computer that can run the UNIX operating system.

One problem with both backcross and F_2 populations is that in annual plants they are a limited resource. One of the great advantages of RFLP markers is that an unlimited number of markers can be mapped using the progeny from one cross (one mapping population). However, this approach works only as long as one can keep the original plants of the mapping population alive and continue to extract DNA from them. When the plants and the DNA are gone, the accumulated mapping data can no longer be used in adding new probes to a developing map. One must start a new mapping population and transfer the map. Obviously, a perennial plant has a great advantage in this respect, since cuttings can be made and the plants maintained indefinitely. If it is desirable to continue to add probes to the map after an F_2 population is no longer available, one can reconstitute the F_2 if selfed seeds (F_3 seed) from the F_2 plants are available. Several F_3 plants from a given F_2 can be combined for DNA extraction, and the combined DNA will be equivalent to the F_2 from which the plants are descended and can be used in place of the F_2 DNA.

Recombinant inbreds offer an attraction alternative for RFLP mapping, but they have been little used because they are time consuming to develop. Basically, one starts with an F_2 mapping population. One seed from each of the F_2 plants is grown and the resultant plant is selfed. One seed is then removed from the F_3 plant, grown to maturity, and selfed. Such a pattern of single seed descent is then continued for six to eight rounds. The resultant plants are then homozygous at nearly all loci, and each chromosome is a mosaic of segments from each of the original parents that were crossed to produce the original F_2 population. When analyzed with RFLP probes, each progeny RI plant will be a homozygote for either one parent or the other. Probes that were tightly linked in the parents will tend to be on common parental chromosome segments, and linkage can be calculated by scoring successive probes on each of the RI lines (Burr *et al.*, 1988). Since RI lines are completely homozygous, they are true-breeding and exactly reproduce themselves from seed. Thus data accumulated in one set of RI plants is not lost as long as one has seed from the plants, and seed from RI lines can be distributed to other researchers who can use the accumulated data as a basis for mapping other probes.

In plants that are not self-fertile or show severe inbreeding depression, a different approach may be needed. Obviously, it will not be possible to develop RI lines. Further, such outcrossing plants are likely to be heterozygous at many loci, and mapping in the F_1 is a practical approach. Phase of the parental alleles (coupling or repulsion) may have to be deduced on a locus-by-locus basis by observing the apparent numbers of recombinants and nonrecombinants.

2.3.1. Use of Aneuploid Lines

If aneuploid lines are available in the plant of interest, they can be of great value in the construction of RFLP maps (Helentjaris *et al.*, 1986; Weber and Helentjaris, 1989; Young *et al.*, 1987). These would include trisomics, monosomics, and substitution or addition lines. Probes can be rapidly assigned to chromosomes by probing these lines, and in this way linkage groups derived from recombination analysis can be reconciled with cytological chromosomes. These assignments make use of the fact that the hybridization signal will be higher in a trisomic plant, because it contains three copies of the locus, than in a diploid plant, which has only two copies (monosomics would have half the signal strength of a diploid plant). Young *et al.* (1987) have demonstrated how it is possible, by using several probes at once, to make comparisons of hybridization signal strength in a given gel lane rather than between lanes. This avoids DNA quantification problems that arise if one tries to compare different lanes containing trisomic and disomic plants.

2.3.2. Mapping Conventional Genes

Conventional genes can be mapped relative to RFLP markers. Indeed, this is one of the principal reasons for making an RFLP map. It is necessary to have a population segregating for the trait in question which is also polymorphic for the mapped RFLP markers. Plants of the mapping population are then scored for the phenotypic trait and for segregation of the previously mapped RFLP markers. Cosegregation of the phenotypic trait and the RFLP markers would indicate linkage, and commonly used computer programs will accommodate phenotypic data.

2.4. Newer Approaches

2.4.1. PCR-Based Methods

The polymerase chain reaction (White *et al.*, 1989) can be used to amplify a selected part (usually up to 2–3 kb in size) of the genome of a plant. To accomplish this it is necessary to have oligonucleotide primers (20-mers are commonly used) that are complementary to opposite strands at the ends of the sequence of interest. The sequence information necessary for synthesis of such primers can be obtained from published sequences, if the region of interest has been sequenced in the organism being studied. One can also use a sequence published from another organism if conserved regions of the sequence can be identified. Alternatively, one can sequence the ends of any cloned insert relatively simply by using universal sequencing primers, which are homologous to the vector. Potential oligonucleotide PCR primers can be identified from the sequence thus obtained.

When primers are obtained, they are then used to amplify the segment of interest from the plants to be compared. Polymorphisms of the PCR products (POPPs?) can then be determined in several ways. If the region between the primers contains an insertion or deletion in one of the plants, this could be detected by running the products in an agarose gel. Base sequence differences in the amplified fragment can be located by digesting the PCR product with a variety of four-cutters and looking for RFLPs after electrophoresis on high-percentage agarose gels.

The PCR methods described above have the advantage of being able to directly visualize RFLP differences. They eliminate the requirement for radioactive isotopes, Southern blots, nucleic acid hybridization, and autoradiography. This makes the process easier, more rapid, less expensive, safer, and possible to do in areas of the world where radioactive isotopes are difficult to obtain.

Various attempts are being made to eliminate the need for detailed sequence information for PCR analysis. Using random primers is one possible approach. On the basis of the base composition of a plant, one can calculate the size of random oligonucleotide primer, which should occur frequently enough so that both ends of a suitable-sized genomic sequence would be primed. Different short random primers would then be used to amplify various parts of the genome. Preliminary experiments in soybean (Tingey, personal communication) have shown that a high percentage of random 10-mers will work in this way and produce one to several PCR product fragments. The amplified products can then be checked for polymorphisms as outlined above for regular PCR products. If a change in the sequence homologous to one of the primers occurs, that fragment is not amplified, presumably because the primer will not hybridize to the genomic sequence if even one base pair is mismatched. Thus scoring is on a plus–minus basis similar to a dominant–recessive phenotypic marker.

2.4.2. High-Resolution Gels

If more resolution is needed, denaturing gels (Kreitman and Aquade, 1986) or denaturing gradient gels (Uitterlinden *et al.*, 1989; Fischer and Lerman, 1983) can be used. Both these methods have the ability to detect base changes that are not in the restriction site of the enzyme being used. Gebhardt *et al.* (1989a,b) have demonstrated the use of denaturing gels in analysis of potato RFLPs. The level of polymorphism was expected to be low in cultivated potato, but these researchers were able to show that abundant polymorphism could be detected through use of denaturing gels.

2.4.3. Oligonucleotide Probes

Conventional RFLP analysis detects only base sequence changes that occur in the recognition site of the enzyme being used or rather large-scale insertion–deletion events. Further, the methodology is costly and time-consuming. Beckmann (1988) has outlined the potential advantages of using oligonucleotide probes as genetic markers. It can be calculated that oligonucleotides in the range of 18 base pairs should occur only once in a plant genome and should be usable as allele-specific probes. Hybridization conditions can be adjusted such that even a single base pair change in the sequence homologous to the oligonucleotide probe will cause it to fail to hybridize. Screening with such allele-specific oligonucleotides could be done with a relatively simple dot-blot protocol and would score the absence or presence of an allele.

3. USES OF RFLPs AND RFLP MAPS

3.1. DNA Fingerprinting

The pattern of restriction fragments detected by one or more cloned probes can often be used as a taxonomic character (the DNA fingerprint) to identify a cultivar, a clone, or even an individual plant. Two basic approaches to DNA fingerprinting have been used. One way is to use single-copy or low-copy-number probes. Each probe yields a relatively small amount of information, so information from several probes must be combined to yield fingerprints with high resolving power. However, the allelic relationships of the individual fragments detected can be inferred, particularly if they have been mapped. This simplifies the assumptions that must be made in the analysis. Wang and Tanksley (1989) have used this approach to fingerprint cultivars of cultivated rice. They analyzed 10 single-copy RFLP markers on 70 rice cultivars selected to represent the total genetic diversity present in cultivated rice. Fifty-eight of the 70 rice cultivars could be uniquely determined by combinations of the probes used.

A contrasting approach is to use probes that detect sequences which occur several to many times in the genome. One probe thus yields more information, but at the cost of a loss of knowledge about allelic relationships. Particularly useful in this respect are hypervariable probes, which are reported to be more polymorphic than conventional probes (Jeffreys *et al.*, 1985a; Tyler-Smith and Taylor, 1988). These probes have been extensively used in human DNA fingerprinting, and they are admissible as evidence in the courts of several countries (Dodd, 1985; Gill *et al.*, 1985; Jeffreys *et al.*, 1985b; Moody, 1989). Some of these probes are more variable than simple sequences because they detect loci made up of tandemly repeated sequences, [also called minisatellite loci or variable number of tandem repeat (VNTR) loci]. These loci occur at several places in the genome, and each locus can have a variable number of the repeat units. If genomic DNA is digested with a restriction enzyme that does not cut in the repeat unit, fragments of differing sizes related to the number of repeat units present will be produced. These fragments are detected by using the repeat, or a part of the repeat, as the probe. The number of repeat units in each locus can change over relatively short evolutionary time by mechanisms that are as yet unclear and may or may not involve unequal crossing-over (Wahls *et al.*, 1990; Wolff *et al.*, 1989; Jarman and Wells, 1989). The result is that individual humans can be distinguished by the DNA fingerprints produced with probes of this sort (Jeffreys *et al.*, 1985c).

Several attempts have been made in plants to find this sort of hypervariable sequence or to use the hypervariable sequences found in other organisms. Dallas (1988) was able to use a human minisatellite probe to fingerprint rice cultivars. Since rice cultivars are produced by inbreeding, each cultivar more or less corresponds to an individual. Dallas was able to unambiguously fingerprint each of the nine cultivars he studied. He was also able to show that segregation of the hypervariable fragments could be demonstrated in crosses. The human probes are not easy to use in plants, however; hybridization conditions are critical and repeatable results are difficult to obtain (Dallas, personal communication). Rogstad *et al.* (1988b) showed that a human minisatellite probe could distinguish individual plants. Another hypervariable probe that has been widely used is a sequence from the bacteriophage M13, which detects hypervariable sequences in humans and other animals (Vassart *et al.*, 1987). Several studies (Nybom *et al.*, 1990; Rogstad *et al.* 1988a; Zimmerman *et al.*, 1989) have shown that this probe may have some utility in fingerprinting plants, both angiosperms and gymnosperms. Gebhardt *et al.* (1989b) was able to use a combination of denaturing gels and highly variable homologous probes to distinguish 20 potato varieties. Only two probes and one or two enzymes were needed, and the patterns produced were found to be highly stable and repeatable. In these cases, however, the molecular basis of the apparent hypervariability has not been determined, but a minisatellite array has been reported from wheat (Martienssen and Baulcombe, 1989), so the phenomenon may be widespread in animals and plants.

3.2. Systematics

3.2.1. Use of RFLP Data as a Phenetic Character

Chloroplast RFLP data has been widely used in plant systematics, and it is now regarded as a basic tool with general applicability. Analysis of nuclear DNA RFLPs for systematic use is much more difficult, however, and has been sparingly used so far. The problem is one of complexity. Chloroplast genomes are small and highly conserved, and complete libraries of cloned probes are widely available. The complete chloroplast DNA sequence is available for three species (Hiratsuka et al., 1989; Ohyama et al., 1986; Shinozaki et al., 1986). Since chloroplast DNA evolves mostly by base substitution (Palmer and Stein, 1986), RFLP data are directly usable as an indicator of change at the DNA sequence level. Whole chloroplast genomes can thus be compared from one plant to another without much fear of artifact.

With nuclear DNA these sorts of analyses are much more problematic. RFLP data cannot be correlated directly with sequence change, or genetic distance measures based on sequence change, because the reason for the polymorphism observed is often not known. Most plant nuclear RFLPs apparently are the result of rearrangements (McCouch et al., 1988; Apuya et al., 1988), and these are not handled well by existing analytic methods. If the fragments being scored have not been mapped, one cannot even be sure about allelic relationships. Thus a basic phenetic approach based on band sharing must be used. Despite the uncertainties involved in such an approach, it appears to work well in some cases. Song et al. (1988a,b) analyzed several species and cultivars of the genus *Brassica*. A phylogeny, based on band-sharing and parsimony analysis, produced a tree that was highly consistent with information from other sorts of taxonomic characters. A phylogeny of rice cultivars, based on band-sharing, was also produced by Wang and Tanksley (1989) and appears to be consistent with known taxonomic relationships.

3.2.2. Comparative RFLP Mapping

Where RFLP maps are available, valuable data can be gathered on taxonomic relationships and chromosome evolution by comparative RFLP mapping. This approach has been rather extensively used in mammalian systems, where conservation of linkage units has been demonstrated in animals as diverse as humans and mouse (Buchberg et al., 1989). In mammals, some groups exhibit rather stable chromosome organization, and different members of the group have large conserved linkage blocks. Other groups have undergone extensive chromosome rearrangements (O'Brien *et al.*, 1988).

Comparative RFLP mapping has just begun in plants. Tanksley's laboratory has transferred their tomato (*Lycopersicon esculentum*) RFLP map to two related species of the Solanaceae: potato (*Solanum tuberosum*) and pepper (*Capsicum* sp.). All three organisms have a haploid chromosome number of 12, but pepper has a much larger nuclear genome than tomato. The chromosome organization of tomato and potato was shown to be very similar: only four paracentric inversions differed between the two species (Bonierbale *et al.*, 1988). Pepper and tomato, however, exhibited extensive rearrangement (Tanksley *et al.*, 1988b). The exact sequence of events could not be reconstructed, but at least 32 breakages of the tomato chromosomes would be necessary to rearrange the markers into the arrangement seen in pepper. Based on this information, it might be possible to substitute a potato chromosome for a tomato chromosome, because they contain the same complement of genes. This would not work with pepper, however, because the gene complements of the individual chromosomes are not the same.

Since it is now possible to make genetic maps using RFLPs in plants where this would not be practical using conventional markers, comparative RFLP mapping will be able to address many interesting questions. For example, genome analysis using cytological and chromosome pairing data has been carried out on several crop plants and different genomes have often been defined. Rice has A, B, C, D, E, F, and some unassigned genome types among its cultivated and related wild species (Chang, 1984). Comparative RFLP mapping of these organisms should yield basic information on the degree of relationship of these genome and the types of changes that have led to their apparent divergence. Such information might be directly applicable to wide-cross breeding programs also.

3.3. Plant Breeding

RFLP maps give the plant breeder the ability to follow all the chromosome segments of two parent plants through a cross and into the progeny. This can be done with a fairly high level of resolution, depending on the density of the RFLP map. Such a detailed level of chromosome segregation analysis cannot be done with conventional markers, where only a few markers can be followed in any one cross. This ability to follow chromosome segments through a cross leads to a wide variety of uses in plant breeding, and more are continually being devised.

3.3.1. Tagging Genes

In plant-breeding protocols, progeny from crosses often have to be screened for a trait of interest controlled by a major gene. For a conventional gene this screening may be fairly straightforward, or it may be time-consuming and expensive. Dominant genes affecting some visible trait that is expressed at all stages of the life-cycle would be the ideal situation. Unfortunately, most agronomically important genes are more difficult to score. The gene may be recessive, in which case progeny testing is often necessary. The gene may be expressed only at certain stages of the life-cycle, such as the waxy gene in maize or rice, which can only be scored in pollen or endosperm tissue. It may involve interactions with another organism, as do insect-resistance or disease-resistance genes. In such cases it is necessary to screen by inoculating the plant with the test organisms if these can be obtained and successfully reared. In other cases, expression of the gene may be affected by the environment or by interactions with other genes, making scoring difficult. In such cases, it may be easier to score for a closely linked RFLP marker than to score for the trait itself. When a conventional gene is found to be closely linked to an RFLP marker, the gene is said to be "tagged" with the RFLP marker or markers. Segregation of the trait can be followed by following segregation of the linked RFLP marker.

The idea of scoring for one gene, which is difficult to score, by scoring a closely linked gene is not new to plant breeding. Sax (1923) demonstrated the utility of such an approach long ago. However, fortuitously linked phenotypic markers such as Sax used are very rare. RFLPs, however, provide an unlimited number of markers. If a high-density RFLP map is available, any conventional gene will be flanked by closely linked RFLP markers and can be tagged. Several agronomically important genes have already been tagged in various plants (Tanksley, 1989).

Another case where gene tagging might be useful is when it is desirable to combine (or "pyramid") two or more genes that have the same phenotype into a single cultivar. It would frequently be desirable, for example, to incorporate multiple genes for insect or disease resistance into a cultivar. This should make the resistance more durable. However, it is difficult to score for the presence of more than one gene conferring the same phenotype by conventional methods. Progeny testing at each generation would be necessary. If, however, each gene is tagged with an RFLP marker, segregation of all the genes can be followed simultaneously and plants containing multiple genes for resistance can be selected from progeny arrays.

3.3.2. Introgression

In cases where a gene for a valuable agronomic trait is being transferred from another cultivar, a land race, or a related wild species, RFLP analysis can be used to follow the course of the introgression, to quantitatively document the degree of introgression, and to dramatically speed up backcross programs designed to achieve introgression. In backcross progeny, for example, RFLP analysis can be used to locate all the chromosome segments derived from each parent and to estimate their size. Young and Tanksley (1989) used RFLP analysis to analyze tomato cultivars that had been produced by backcross programs designed to incorporate a virus resistance gene from a related wild species. The cultivars had been subjected to a varying number of backcross-selection cycles designed to keep the virus resistance, but to get rid of extra chromosome segments linked to the desired trait but potentially containing genes for undesirable agronomic traits. Since the breeding programs that produced these cultivars had to rely on phenotypic analysis, there was no way to directly determine the amount of undesired chromosome segments from the wild species remaining in the cultivars. When the cultivars were examined with RFLPs, however, direct measurement of introgressed chromosome segments was possible. Contrary to expectations, some cultivars retained large chromosome segments from the wild species, despite as many as 20 backcrosses being used in their generation. One cultivar contained an entire chromosome arm (57 cM) from the wild species.

By using RFLP analysis in a backcross breeding program, it should be possible to make the process much more efficient. If it is desired, for example, to introduce a single gene from a wild species, the cross should be made and the F_1 should be backcrossed to the recurrent parent as usual. The progeny of the first backcross generation should then be subjected to RFLP analysis. Progeny that exhibit a crossover close to the desired marker on one side should be selected. These should be again crossed to the recurrent parent to produce the second backcross generation. These progeny should be selected for a crossover close to and on the other side of the desired marker. In this fashion, lines containing the desired gene on a small segment of introgressed chromosome can be produced in only two backcross generations with small populations. With conventional methods, as many as 100 generations would be required (Tanksley, 1989; Young and Tanksley, 1989).

3.3.3. Analysis of Quantitative Traits

Many important agronomic traits are controlled by one gene or a small number of genes. However, most of the plant characteristics important in agriculture are controlled by several genes. These include such characters as yield, tolerance to environmental stress, and resistance to many diseases or insects. Each of the several genes contributes a small positive or negative effect to the character in question. Since all the genes contribute to the same phenotype, only the collective effect of the genes can be measured, often in a quantitative fashion. Such polygenic traits are also often called quantitative traits and the genes controlling them are called quantitative trait loci (QTL). It is difficult to apply Mendelian genetic methods to the analysis of QTL, since the effects of the individual genes cannot be discerned from their collective phenotype. To make progress in the improvement of plant for quantitative traits, it has been necessary to resort to statistical methods. Using these biometric methods it is very difficult to determine the number of QTL controlling any trait or the chromosomal location of these genes.

RFLP analysis is bringing about a revolution in the analysis of QTL. With RFLPs every chromosomal segment of both parents can be followed through a cross and correlated with the quantitative trait being studied. A correlation between the presence of a given chromosome segment and the trait is evidence that a QTL is located on the segment. It then becomes possible to identify individual QTL and to apply Mendelian genetic analysis to the study of quantitative traits (Lander and Botstein, 1989; Beckmann and Soller, 1988). In the most complete study to date, interval analysis was used to identify the chromosomal location of several QTL controlling fruit weight, pH, and soluble solids content in tomato fruit (Paterson *et al.*, 1988, 1990). These QTL are now being isolated into isogenic lines so that a diallel cross can be used to study the effect of various combinations of the individual QTL (Tanksley, personal communication).

3.4. Cloning Genes

A large number of genes have been cloned in plants and other organisms. In most cases these have been genes that produce a large amount of product, such as histone genes or actin genes, or genes that are highly expressed under certain conditions, such as heat shock genes. Most of the methods used to clone genes depend on obtaining the product of the gene as a direct or indirect probe to isolate the gene from a library. The problem with cloning agronomically important genes is that we seldom know the gene product. It is also likely that the product of many important genes is present only in small quantities so that any cloning strategy based on isolating the gene product would be difficult to apply.

Two basic methods are being used to clone genes for which little or nothing is known about the gene product. One of these involves insertional mutagenesis by transposon tagging (Hake *et al.*, 1989; Gerats *et al.*, 1989; Pan and Peterson, 1989) or by the introduction of T-DNA by transformation with *Agrobacterium* (Feldmann *et al.*, 1989). The other method is cloning by starting from a linked marker, sometimes called reverse genetics or chromosome walking. Basically the technique involves establishing close linkage by conventional genetic analysis between the gene and flanking RFLP markers. Then, starting from the flanking markers and proceeding toward the gene by cloning adjacent, overlapping segments (chromosome walking) or by "jumping" down the chromosome to nonadjacent segments, clones containing the desired gene are eventually obtained.

Cloning genes by walking from linked markers is a technique that is easy to describe but difficult to implement. One difficulty is the relatively large amount of DNA that separates markers that are quite closely linked in conventional genetic terms. Markers that are 1 cM apart would be separated by several hundred thousand to more than a million base pairs of DNA in various plant species. This makes the cloning of all the intervening segments with the current generation of vectors (which can handle inserts of 20–40 kb) a formidable undertaking.

Several recent developments are combining to make the eventual cloning of genes from linked RFLP markers a much more feasible enterprise. Reconciliation of RFLP maps with physical maps is currently underway in some plant model systems. In *Arabidopsis*, which has a very small genome, the entire genome will soon be available as a set of overlapping, ordered cosmid clones and yeast artificial chromosomes (YACs) (Somerville, 1989). If close linkage can be established between a gene and flanking RFLP markers, it should be relatively easy to search through the clones that are known to be between the RFLP markers for the desired gene. Identification of the gene from the set of clones that are thought to contain it is still a problem, however, if the product of the gene is not known. Several methods have been suggested to address this problem. Complementation of a mutant by transformation of candidate clones is feasible in plants where transformation is fairly efficient. Using models derived from human research, it may also be possible to screen tissues known to be expressing the gene with candidate clones (Rommens *et al.*, 1989) or to look for CpG islands known to be associated with genes (Antequera and Bird, 1988).

In tomato, a YAC library is also being constructed. (Tanksley, personal communication). YACs can contain fairly large inserts of genomic DNA, so there is a much greater possibility that a clone containing a flanking marker might also contain the gene of interest. Walking through an individual YAC insert is much more straightforward than trying to walk in total genomic DNA isolations.

Any chromosome walking method depends on having RFLP markers that are closely linked to the gene of interest. The most desirable scenario would be several markers that are very closely linked and are known to flank the gene on both sides. Knowledge of linkage can only come from genetic studies of recombination between the markers and the gene, and plant systems are advantageous for such genetic studies since controlled crosses can be made and large mapping populations can be easily produced. Certain types of plant genetic stocks, such as near-isogenic lines, may also be valuable in locating closely linked RFLP markers. Near-isogenic lines are produced by repeatedly backcrossing an F_1 to one of the parents and selecting each backcross generation for a characteristic controlled by a single gene. This ultimately produces a breeding line that has only a small segment of one chromosome from the nonrecurrent parent; the rest of that chromosome and all the other chromosomes are derived from the recurrent parent. Young *et al.* (1988) demonstrated that a probe library can be rapidly screened for markers tightly linked to a gene if near-isogenic lines for that gene are available. Several candidate probes are used to simultaneously probe the two parents and the near-isogenic line. Those probes which detect the restriction fragments common to the nonrecurrent parent and the near-isogenic line will be in the introgressed chromosome segment and, therefore, closely linked to the gene of interest.

The technique of pulsed-field electrophoresis is also being applied to genetic mapping project in plants. In tomato it has been possible to fractionate restriction fragments as large as several hundred kb (Ganal and Tanksley, 1989). The ability to analyze such large fragments will also make it possible to begin to reconcile genetic maps with physical maps and to physically map complex loci, such as gene families or tandem repeats.

4. SUMMARY

In the last decade RFLP analysis has evolved from an idea that seemed promising to a well-established tool that has led to fundamental advances in several fields. Construction of genetic maps has now become feasible in many organisms where it would previously have been impractical. Since genetic maps are of general utility for many sorts of biological research, they cannot fail to have a significant impact in the immediate future. As genetic maps become reconciled with physical maps in several plants, it will become possible to clone virtually any gene. For a plant breeder this will have the effect of broadening the gene pool available for plant improvement to include virtually all organisms, including animals and microorganisms.

Much remains to be done, however. We need basic studies of the biochemistry, physiology, and genetics of plants and the insects and pathogens infesting them to be able to identify target genes for cloning. We need basic studies of transformation and gene expression to be able to have introduced genes expressed in transformed plants in the proper amounts and in the desired tissues. It must also be kept in mind that the best of our present technologies only suffice to clone and transform single genes. We will have to make another large jump in capabilities to be able to transfer QTL between plants. Since the most important agronomic traits are controlled by QTL, this effort will have to be undertaken.

However, the future looks promising for plant breeding and RFLP analysis. The molecular genetic revolution now has every indication of being transferrable to practical problems such as plant breeding, and the first steps in this transferral have already occurred through the medium of RFLP analysis.

5. REFERENCES

Allard, R. W., 1956, Formulas and tables to facilitate the calculation of recombination values in heredity, *Hilgardia* **24**:235–278.

Antequera, F., and Bird, A. P., 1988, Unmethylated CpG islands associated with genes in higher plant DNA, *EMBO J.* **7:**2295–2299.

Apuya, N. R., Frazier, B. L., Keim, P., Roth, E. J., and Lark, K. G., 1988, Restriction fragment length polymorphisms as genetic markers in soybean, *Glycine max* (L.) merrill, *Theor. Appl. Genet.* **75:**889–991.

Beckmann, J. S., 1988, Oligonucleotide polymorphisms: A new tool for genomic genetics, *Bio/Technology* **6:**1061–1064.

Beckmann, J. S., and Soller, M., 1983, Restriction fragment length polymorphisms in genetic improvement: methodologies, mapping and costs, *Theor. Appl. Genet.* **67:**35–43.

Beckmann, J. S., and Soller, M., 1986, Restriction fragment length polymorphisms and genetic improvement of agricultural species, *Euphytica* **35:**111–124.

Beckmann, J. S., and Soller, M., 1988, Detection of linkage between marker loci and loci affecting quantitative traits in crosses between segregating populations, *Theor. Appl. Genet.* **76:**228–236.

Belanger, F. C., and Hepburn, A. G., 1990, The evolution of CpNpG methylation in plants, *J. Mol. Evol.* **30:**26–35.

Bernatzky, R., and Tanksley, S. D., 1986, Toward a saturated linkage map in tomato based on isozymes and random cDNA sequences, *Genetics* **112:**887–898.

Bonierbale, M. W., Plaisted, R. L., and Tanksley, S. D., 1988, RFLP maps based on a common set of clones reveal modes of chromosomal evolution in potato and tomato, *Genetics* **120:**1095–1103.

Botstein, D., White, R. L., Skolnick, M., and Davis, R. W., 1980, Construction of a genetic linkage map in man using restriction fragment length polymorphisms, *Am. J. Hum. Genet.* **32:**314–331.

Brown, P. TH., 1989, DNA methylation in plants and its role in tissue culture, *Genome* **31:**717–729.

Buchberg, A. M., Brownell, E., Nagata, S., Jenkins, N. A., and Copeland, N. G., 1989, A comprehensive genetic map of murine chromosome 11 reveals extensive linkage conservation between mouse and human, *Genetics* **122:**153–161.

Burr, B., Burr, F. A., Thompson, K. H., Albertson, M. C., and Stuber, C. W., 1988, Gene mapping with recombinant inbreds in maize, *Genetics* **118:**519–526.

Chang, C., Bowman, J. L., DeJohn, A. W., Lander, E. S., and Meyerowitz, E. M., 1988, Restriction fragment length polymorphism linkage map for *Arabidopsis thaliana*, *Proc. Natl. Acad. Sci. USA* **85:**6856–6860.

Chang, T. T., 1984, Conservation of rice genetic resources: Luxury or necessity, *Science* **224:**251–256.

Chao, S., Sharp, P. J., Worland, A. J., Warham, E. J., Koebner, R. M. D., and Gale, M. D., 1989, RFLP-based genetic maps of wheat homoeologous group-7 chromosomes, *Theor. Appl. Genet.* **78:**495–504.

Chase, M. W., and Palmer, J. D., 1989, Chloroplast DNA systematics of Lilioid monocots—Resources, feasibility, and example from the Orchidaceae, *Am. J. Bot.* **76:**1720–1730.

Cotton, R. G. H., 1989, Detection of single base changes in nucleic acids, *Biochem. J.* **263:**1–10.

Dallas, J. F., 1988, Detection of DNA "fingerprints" of cultivated rice by hybridization with a human minisatellite DNA probe, *Proc. Natl. Acad. Sci. USA* **85:**6831–6835.

Dodd, B. E., 1985, DNA fingerprinting in matters of family and crime, *Nature* **318:**506–507.

Donis-Keller, H., Green, P., Helms, C., Cartinhour, S., Weiffenbach, B., Stephens, K., Keith, T. P., Bowden, D. W., Smith, D. R., Lander, E. S., Botstein, D., Akots, G., Rediker, K. S., Gravius, T., Brown, V. A., Rising, M. B., Parker, C., Powers, J. A., Watt, D. E., Kauffman, E. R., Bricker, A., Phipps, P., Muller-Kahle, H., Fulton, T. R., Ng, S., Schumm, J. W., Braman, J. C., Knowlton, R. G., Barker, D. F., Crooks, S. M., Lincoln, S. E., Daly, M. J., and Abrahamson, J., 1987, A genetic linkage map of the human genome, *Cell* **51:**319–337.

Evola, S. V., Burr, F. A., and Burr, B., 1986, The suitability of restriction fragment length polymorphisms as genetic markers in maize, *Theor. Appl. Genet.* **71:**765–771.

Feinberg, A. P., and Vogelstein, B., 1984, A technique for radiolabeling DNA restriction fragments to a high specific activity, *Anal. Biochem.* **132:**6–13.

Feldmann, K. A., Marks, M. D., Christianson, M. L., and Quatrano, R. S., 1989, A dwarf mutant of *Arabidopsis thaliana* generated by T-DNA insertion mutagenesis, *Science* **243:**1351–1354.

Fischer, S. G., and Lerman, L. S., 1983, DNA fragments differing by single base-pair substitutions are separated in denaturing gradient gels: Correspondence with melting theory, *Proc. Natl. Acad. Sci. USA* **80:**1579–1583.

Ganal, M. W., and Tanksley, S. D., 1989, Analysis of tomato DNA by pulsed field gel electrophoresis, *Plant Mol. Biol. Rep.* **7:**17–28.

Gebhardt, C., Blomendahl, C., Schachtschabel, U., Debener, T., Salamini, F., and Ritter, E., 1989a, Identification of 2n breeding lines and 4n varieties of potato (*Solanum tuberosum*, ssp. *tuberosum*) with RFLP-fingerprints, *Theor. Appl. Genet.* **78:**16–22.

Gebhardt, C., Ritter, E., Debener, T., Schachtschable, U., Walkemeir, B., Uhrig, H., and Salamini, F., 1989b, RFLP analysis and linkage mapping in *Solanum tuberosum*, *Theor. Appl. Genet.* **78:**65–75.

Gerats, A. G. M., Beld, M., Huits, H., and Prescott, A., 1989, Gene tagging in *Petunia hybrida* using homologous and heterologous transposable elements, *Dev. Genet.* **10:**561–568.

Gill, P., Jeffreys, A. J., and Werrett, D. J., 1985, Forensic applications of DNA "fingerprints," *Nature* **318:**577–579.

Hake, S., Vollbrecht, E., and Freeling, M., 1989, Cloning *Knotted,* the dominant morphological mutant in maize using *Ds2* as a transposon tag, *EMBO J.* **8:**15–22.

Helentjaris, T., King, G., Slocum, M., Sidenstrang, C., and Wegmen, S., 1985, Restriction fragment polymorphisms as probes for plant diversity and their development as tools for applied plant breeding, *Plant Mol. Biol.* **5:**109–118.

Helentjaris, T., Weber, D. F., and Wright, S., 1986, Use of monosomics to map cloned DNA fragments in maize, *Proc. Natl. Acad. Sci. USA* **83:**6035–6039.

Hiratsuka, J., Shimada, H., Whittier, R., Ishibashi, T., Sakamoto, M., Mori, M., Kondo, C., Honji, Y., Sun, C-R., Meng, B-Y., Li, Y-Q., Kanno, A., Nishizawa, Y., Hirai, A., Shinozaki, K., and Sugiura, M., 1989, The complete sequence of the rice (*Oryza sativa*) chloroplast genome: Intermolecular recombination between distinct tRNA genes accounts for a major plastid DNA inversion during the evolution of the cereals, *Mol. Gen. Genet.* **217:**185–194.

Jarman, A. P., and Wells, R. A., 1989, Hypervariable minisatellites—Recombinators or innocent bystanders? *Trends in Genetics* **5:**367–371.

Jeffreys, A. J., Wilson, V., and Thein, S. L., 1985a, Hypervariable "minisatellite" regions in human DNA, *Nature* **314:**67–73.

Jeffreys, A. J., Brookfield, J. F., and Semeonoff, R., 1985b, Positive identification of an immigration test-case using human DNA fingerprints, *Nature* **317:**818–822.

Jeffreys, A. J., Wilson, V., and Thein, S. L. 1985c, Individual-specific "fingerprints" of human DNA, *Nature* **316:**76–79.

Kreitman, M., and Aquade, M., 1986, Genetic uniformity in two populations of *Drosophila melanogaster* as revealed by filter hybridization by four-nucleotide recognizing enzyme digests, *Proc. Natl. Acad. Sci. USA* **83:**3562–3566.

Lander, E. S., and Botstein, D., 1989, Mapping Mendelian factors underlying quantitative traits using RFLP linkage maps, *Genetics,* **121:**185–199.

Lander, E. S., Green, P., Abrahamson, J., Barlow, A., Daly, M. J., Lincoln, S. E., and Newburg, L., 1987, MAPMAKER: An interactive computer package for constructing primary genetic linkage maps of experimental and natural populations, *Genomics* **1:**174–181.

Landry, B. S., and Michelmore, R. W., 1985, Selection of probes for restriction fragment length analysis from plant genomic clones, *Plant Mol. Biol. Rep.* **3:**174–179.

Landry, B. S., and Michelmore, R. W., 1987, Methods and applications of restriction fragment length polymorphism analysis to plants, in: *Tailoring Genes for Crop Improvement* (T Kosuge, and A. Hollaender, eds.), pp. 25–44, Plenum Press, New York.

Landry, B. S., Kesseli, R. V., Farrara, B., and Michelmore, R. W., 1987, A genetic map of lettuce (*Lactuca sativa* L.) with restriction fragment length polymorphism, isozyme, disease resistance and morphological markers, *Genetics* **116:**331–337.

Martienssen, R. A., and Baulcombe, D. C., 1989, An unusual wheat insertion sequence (WIS1) lies upstream of an α-amylase gene in hexaploid wheat, and carries a "minisatellite" array, *Mol. Gen. Genet.* **217:**401–410.

McCouch, S. R., Kochert, G., Yu, Z. H., Wang, Z. Y., Khush, G. S., Coffman, W. R., and Tanksley, S. D., 1988, Molecular mapping of rice chromosomes, *Theor. Appl. Genet.* **76:**148–149.

Moody, M. D., 1989, DNA analysis in forensic science, *BioScience* **39:**31–36.

Nybom, H., Rogstad, S. H., and Schaal, B. A., 1990, Genetic variation detected by use of the M13 DNA fingerprint Probe in *Malus, Prunus,* and *Rubus* (Rosaceae), *Theor. Appl. Genet.* **79:**153–156.

O'Brien, S. J., Seuánez, H. N., and Womack, J. E., 1988, Mammalian genome organization: An evolutionary view, *Annu. Rev. Genet.* **22:**323–351.

Ohyama, K., Fukuzawa, H., Kohchi, T., Shirai, H., Sano, T., Sano, S., Umesono, K., Shiki, Y., Takeuchi, M., Chang, Z., Aota, S. I., Inokuchi, H., and Ozeki, H., 1986, Chloroplast gene organization deduced from complete sequence of liverwort *Marchantia polymorpha* chloroplast DNA, *Nature* **322:**572–574.

Orita, M., Iwahana, H., Kanazawa, H., Hayashi, K., and Sekiya, T., 1989a, Detection of polymorphisms of human DNA by gel electrophoresis as single-strand conformation polymorphisms, *Proc. Natl. Acad. Sci. USA* **86:**2766–2770.

Orita, M., Suzuki, Y., Sekiya, T., and Hayashi, K., 1989b, Rapid and sensitive detections of point mutations and DNA polymorphisms using the polymerase chain reaction, *Genomics* **5:**874–879.

Palmer, J. D., and Stein, D. B., 1986, Conservation of chloroplast genome structure among vascular plants, *Curr. Genet.* **10:**823–833.

Pan, Y. B., and Peterson, P. A., 1989, Tagging of a maize gene involved in kernel development by an activated Uq transposable element, *Mol. Gen. Genet.* **219:**324–327.

Paterson, A. H., Lander, E. S., Hewitt, J. D., Peterson, S., Lincoln, S. E., and Tanksley, S. D., 1988, Resolution of quantitative traits into Mendelian factors by using a complete linkage map of restriction fragment length polymorphisms, *Nature* **335:**721–726.

Paterson, A. H., Deverna, J. W., Lanini, B., and Tanksley, S. D., 1990, Fine mapping of quantitative trait loci using selected overlapping recombinant chromosomes, in an interspecies cross of tomato, *Genetics* **124:**735–742.

Pethe, V., Lagu, M., Chitnis, P. K., Gupta, V., and Ranjekar, P. K., 1989, Restriction fragment length polymorphism—A recent approach in plant breeding. I. *J. Biochem. Biophys.* **26:**285–288.

Rigby, P., Dieckmann, M., Rhodes, C., and Berg, P., 1977, Labelling deoxyribonucleic acid to high specific activity in vitro by nick-translation with DNA polymerase I, *J. Mol. Biol.* **113:**237–251.

Rogstad, S. H., Patton, J. C. II, and Schaal, B. A., 1988a, M13 repeat probe detects DNA minisatellite-like sequences in gymnosperms and angiosperms, *Proc. Natl. Acad. Sci. USA* **85:**9176–9178.

Rogstad, S. H., Patton, J. C., and Schaal, B. A., 1988b, A human minisatellite probe reveals RFLPs among individuals of two angiosperms, *Nucleic Acids Res.* **16:**11378.

Rommens, J. M., Iannuzzi, M. C., Kerem, B.-S., Drumm, M. L., Melmer, G., Dean, M., Rozmahel, R., Cole, J. L., Kennedy, D., Hidada, N., Zsiga, M., Buchwald, M., Riordan, J.

R., Tsui, L-C., and Collins, F. S., 1989, Identification of the cystic fibrosis gene: Chromosome walking and jumping, *Science* **245**:1059–1065.

Sax, K., 1923, The association of size differences with seed-coat pattern and pigmentation in *Phaseolus vulgaris*, *Genetics* **8**:552–560.

Schilling, E. E., and Jansen, R. K., 1989, Restriction fragment analysis of chloroplast DNA and the systematics of *Viguiera* and related genera (Asteraceae, Heliantheae), *Am. J. Bot.* **76**:1769–1778.

Shinozaki, K., Ohme, M., Tanaka, M., Wakasugi, T., Hayashida, N., Matsubayashi, T., Zaita, N., Chunwongse, J., Obokata, J., Yamaguchi-Shinozaki, K., Ohto, C., Torazawa, K., Meng, B. Y., Sugita, M., Deno, H., Kamogashira, T., Yamada, K., Kusuda, J., Takaiwa, F., Kato, A., Tohdoh, N., Shimada, H., and Sugiura, M., 1986, The complete nucleotide sequence of the tobacco chloroplast genome: its gene organization and expression, *EMBO J.* **5**:2043–2049.

Soller, M., and Beckmann, J. S., 1983, Genetic polymorphism in varietal identification and genetic improvement, *Theor. Appl. Genet.* **67**:25–33.

Somerville, C., 1989, *Arabidopsis* Blooms, *Plant Cell* **1**:1131–1135.

Song, K. M., Osborn, T. C., and Williams, P. H., 1988a, *Brassica* taxonomy based on nuclear restriction fragment length polymorphisms (RFLPs). 2. Preliminary analysis of subspecies within *B. rapa* (syn. campestris) and *B. oleracea*, *Theor. Appl. Genet.* **76**:593–600.

Song, K. M., Osborn, T. C., and Williams, P. H., 1988b, *Brassica* taxonomy based on nuclear restriction fragment length polymorphisms (RFLPs), *Theor. Appl. Genet.* **75**:784–794.

Southern, E. M., 1975, Detection of specific sequences among DNA fragments separated by gel electrophoresis, *J. Mol. Biol.* **98**:503–517.

Suiter, K. A., Wendel, J. F., and Case, J. S., 1983, Linkage-1: A PASCAL computer program for the detection and analysis of genetic linkage, *J. Hered.* **74**:203–204.

Tanksley, S. D., 1989, RFLP mapping in plant breeding: New tools for an old science, *Bio/Technology* **7**:257–263.

Tanksley, S. D., Miller, J., Paterson, A., and Bernatzky, R., 1988a, Molecular Mapping of plant chromosomes, in *Chromosome Structure and Function* (J. P. Gustafson and R. Appels, eds.), pp. 157–173, Plenum Press, New York.

Tanksley, S. D., Bernatzky, R., Lapitan, N. L., and Prince, J. P., 1988b, Conservation of gene repertoire but not gene order in pepper and tomato, *Proc. Natl. Acad. Sci. USA* **85**:6419–6423.

Tyler-Smith, C., and Taylor, L., 1988, Structure of a hypervariable tandemly repeated DNA sequence on the short arm of the human Y chromosome, *J. Mol. Biol.* **203**:837–848.

Uitterlinden, A. G., Slagboom, P. E., Knook, D. L., and Vijg, J., 1989, Two-dimensional DNA fingerprinting of human individuals, *Proc. Natl. Acad. Sci. USA* **86**:2742–2746.

Vassart, G., Georges, M., Monsieur, R., Brocas, H., Lequarre, A. S., and Christophe, D., 1987, A sequence in M13 detects hypervariable minisatellites in human and animal DNA, *Science* **235**:683–684.

Wahls, W. P., Wallace, L. J., and Moore, P. D., 1990, Hypervariable minisatellite DNA is a hotspot for homologous recombination in human cells, *Cell* **60**:95–103.

Wang, Z. Y., and Tanksley, S. D., 1989, Restriction fragment length polymorphism in *Oryza sativa* L, *Genome* **32**:1113–1118.

Weber, D., and Helentjaris, T., 1989, Mapping RFLP loci in maize using B-A translocations, *Genetics* **121**:583–590.

White, T. J., Arnheim, N., and Erlich, H. A., 1989, The polymerase chain reaction, *Trends in Genetics* **5**:185–189.

Wolff, R. K., Plaetke, R., Jeffreys, A. J., and White, R., 1989, Unequal crossing over between homologous chromosomes is not the major mechanism involved in the generation of new alleles at VNTR loci, *Genomics* **5**:382–384.

Young, N. D., and Tanksley, S. D., 1989, RFLP analysis of the size of chromosomal segments retained around the Tm-2 locus of tomato during backcross breeding, *Theor. Appl. Genet.* **77:**353–359.

Young, N. D., Miller, J. C., and Tanksley, S. D., 1987, Rapid chromosomal assignment of multiple genomic clones in tomato using primary trisomics, *Nucleic Acids Res.* **15:**9339–9348.

Young, N. D., Zamir, D., Ganal, M. W., and Tanksley, S. D., 1988, Use of isogenic lines and simultaneous probing to identify DNA markers tightly linked to the *TM-2a* gene in tomato, *Genetics* **120:**579–585.

Zimmerman, P. A., Langunnasch, N., and Cullis, C. A., 1989, Polymorphic regions in plant genomes detected by an M13 probe, *Genome* **32:**824–828.

Chapter 9

Fundamentals of Light-Regulated Gene Expression in Plants

Richard J. Mural

1. INTRODUCTION

Plants must respond to light in order to live. By affecting gene expression both qualitatively and quantitatively, light is a major regulator of both plant development and metabolism. When a seedling emerges into the light, it undergoes a number of light-induced changes. Chloroplasts become organized, chlorophyll and the proteins required for assembling the photosynthetic apparatus are made, and the enzymes needed for carbon fixation are synthesized, preparing the plant for its autotrophic existence. Mature plants deal with diurnal fluctuations of light and adjust their metabolism accordingly. Gene expression is regulated in plants by light at many levels. The level of a gene product may be controlled by regulation of the level of transcription of its gene or, alternatively, by regulation of translation of its mRNA into protein. Turnover of the mRNA or the gene product is another point at which gene expression can be regulated. Finally, if the product of a gene is an enzyme, its activity level can respond to an environmental signal, thereby modulating the effective level of gene expression. As will be discussed in this review, light can regulate the expression of a given gene at one

Richard J. Mural   Biology Division, Oak Ridge National Laboratory, Oak Ridge, Tennessee 37831-8077.

Subcellular Biochemistry, Volume 17: Plant Genetic Engineering, edited by B. B. Biswas and J. R. Harris. Plenum Press, New York, 1991.

or more of these levels. Light also plays a critical role in plant development and a number of genes respond to light as both a developmental signal and a metabolic regulator.

This review will address the regulation of plant gene expression by light and will concentrate on illustrating recent experiments that have provided insight into our understanding of this regulation at a molecular level. As such, it is not meant to be an exhaustive review of the literature, but rather a guide to the key concepts that have emerged regarding this important aspect of plant physiology. A number of other recent reviews are relevant to this topic (Tobin and Silverthorne, 1985; Hoober, 1987; Kuhlemeier et al., 1987b).

2. LIGHT SENSORS

In order to respond to an environmental signal an organism must first be able to sense a change in the relevant signal. Plants have evolved several photoreceptor systems that provide information on the quantity, the wavelength (blue/UV-A, UV-B, or red/far red light), and the direction of the source of light (see Shropshire and Mohr, 1983; Kendrick and Kronenberg, 1986). Of these systems, phytochrome, the red/far red sensor, is the best characterized (see Furuya, 1987). The blue and near UV sensors are less studied and not well understood. A detailed discussion of plant photoreceptor systems is beyond the scope of this review; however, some features are relevant and will be discussed. The volumes edited by Shropshire and Mohr (1983), Kendrick and Kronenberg (1986), and Furuya (1987) will provide the reader with a more extensive discussion of plant photoreceptors.

2.1. Phytochrome

Phytochrome, as a regulator of plant gene expression, was first recognized by its effects on plant growth and development. Only recently have some of the molecular events that underlie these effects been clarified (for reviews, see Silverthorne and Tobin, 1987; Nagy et al., 1988; Cuozzo et al., 1988). Phytochrome exists in two interconvertible forms. Photoconversion of the P_r (red absorbing phytochrome) to P_{fr} (far-red absorbing phytochrome) by red light acts as a signal that affects gene expression in a number of ways. P_{fr} is converted back to P_r upon exposure to far red light, canceling the signal and completing the regulatory cycle.

Depending on the plant species, phytochromes consist of a dimer of 120–127-kD polypeptide subunits which each have a covalently attached linear tetrapyrrole chromophore. In *Arabidopsis thaliana* phytochrome is encoded by a small gene family with four or five members (Sharrock and Quail, 1989). Clones of phytochrome genes from a number of different plant species have been obtained and characterized. cDNA clones have been isolated from oat (Hershey *et al.*, 1985), zucchini (Sharrock *et al.*, 1986), pea (Sato, 1988), rice (Kay *et al.*, 1989), corn (Christensen and Quail, 1989), and *A. thaliana* (Sharrock and Quail, 1989). Genomic clones have been isolated from oat (Hershey *et al.*, 1985, 1987), rice (Kay *et al.*, 1989), and *A. thaliana* (Sharrock and Quail, 1989).

Virtually all plant tissues that have been examined contain phytochrome-including roots (Pratt, 1986). In dark-grown (etiolated) plant tissue the P_r form of phytochrome is the most abundant. Exposure to light converts the P_r form to the P_{fr} form, signaling the molecular events that are phytochrome regulated. There is also a dramatic decrease in phytochrome levels since P_{fr} is degraded much more rapidly than P_r (Quail *et al.*, 1973). In a number of plants exposure of etiolated tissues to light also causes a decrease in phytochrome mRNA levels, which is the result of a phytochrome-dependent reduction in transcription of the phytochrome gene(s) (Otto *et al.*, 1984; Colbert *et al.*, 1985; Quail *et al.*, 1987; Lissemore *et al.*, 1987; Lissemore and Quail, 1988; Sato, 1988; Kay *et al.*, 1989; Sharrock and Quail, 1989). As is the case with all phytochrome-mediated regulatory responses, the mechanism of the autoregulation of phytochrome synthesis is not known. The phytochrome-mediated regulation of phytochrome gene transcription has been studied most extensively in oats (Lissemore and Quail, 1988) and in rice (Kay *et al.*, 1989). Kay *et al.* (1989) found sequences in the 5′ upstream region of the rice phytochrome gene that are homologous to the core binding site of GT-1. GT-1 is a nuclear protein that has been shown to bind to light-responsive elements (LRE) of the pea *rbcS-3A* gene (Green *et al.*, 1987, 1988). The pea *rbcs-3A* gene is light regulated by phytochrome, and its regulation and the role of GT-1 will be discussed in detail below. It was also shown that one of these regions, −242/−220, was bound by a protein found in nuclear extracts from etiolated rice leaf nuclei and that this binding was competed by DNA containing a pea GT-1 binding site. The −242/−220 sequence contains two GT-1 binding cores (GGTTAA and GGTAAT; Green *et al.*, 1988). Furthermore, they found a similar region, 20 identities in a sequence 23 nucleotides long, in the 5′ upstream region of the oat *phy3* gene (Hershey *et al.*, 1987). The finding of known light-responsive elements in the 5′ region of the rice phytochrome gene and the conservation of these sequences in the 5′ upstream region of the oat phytochrome gene should facilitate the study of the phytochrome response.

There have been several reports of low-abundance phytochromes found in green tissues that are immunologically and physically distinct from the major etiolated-tissue form of phytochrome (Abe *et al.*, 1985; Shimazaki and Pratt, 1985; Tokuhisa *et al.*, 1985). Of the three phytochrome cDNA clones isolated from *A. thaliana* by Sharrock and Quail (1989), one, *phyA*, appears, based on its deduced amino acid sequence, to represent the abundant form of phytochrome normally found in etiolated tissues. In etiolated tissue the mRNA for *phyA* is abundant and its level decreases following exposure to white light. In contrast, the other two phytochrome genes, *phyB* and *phyC,* are more closely related to one another, at the protein sequence level, than they are to *phyA* or any other known phytochrome. The level of mRNA for these genes is low in etiolated tissues and the level of mRNA is not light regulated. Whether *phyB* and *phyC* represent genes for the low-abundance green tissue form of phytochrome remains to be demonstrated.

2.2. Other Sensors

Compared with phytochrome, very little is known about the plant receptors that perceive light from the blue and near UV portions of the spectrum. In fully green tissue there is evidence that the expression of the *rbcS* genes requires the action of both phytochrome and a blue light receptor (Fluhr and Chua, 1986), and it has also been demonstrated that anthocyanin synthesis requires two photoreceptors (Oelmuller and Mohr, 1985). Determining the role of photoreceptors other than phytochrome in plant development and gene regulation awaits further study.

3. LIGHT REGULATION OF TRANSCRIPTION

Light can regulate gene expression at the transcriptional level both qualitatively and quantitatively. The effect of light on transcription in plants is easily observed during plant development. Many experiments that have contributed to an understanding of the light regulation of transcription are based on comparing transcription of particular genes in dark-grown plant tissue to the levels found following illumination. If plants are allowed to develop in the dark, development is inhibited and these plants cannot perform photosynthesis because they lack chlorophyll as well as many of the polypeptides required for photosynthesis. Illumination leads to "greening," a process in which chloroplasts become organized, chlorophyll is synthesized, and proteins needed for photosynthesis accumulate (Buchanan, 1980; Ellis, 1981). Part of this response includes the appearance of new mRNA species and increases in the level of others.

3.1. Light-Induced Regulation of Genes Involved
in Chloroplast Development

Since many of the proteins found in the chloroplast are encoded in the nuclear DNA, chloroplast development involves the coordinate expression of plastid genes and a set of nuclear genes whose products are involved in assembling the photosynthetic apparatus. The accumulation of plastid-encoded gene products can be regulated at the level of transcription, RNA processing, or RNA stability (Mullet and Klein, 1987; Klein and Mullet, 1987, 1990; Gamble *et al.,* 1988). The state of leaf development also influences the response of various genes to light. Leaf development in the dark varies widely in different plant species. In pea and spinach, leaf development is very restriction in darkness, less restricted in maize, and essentially light independent in barley (for references see Klein and Mullet, 1990).

Nelson *et al.* (1984) examined the light-dependent accumulation of mRNA and protein for the large (LSU) and small (SSU) subunits of ribulose-1, 5-bisphosphate carboxylase/oxygenase (Rubisco), chlorophyll a/b binding protein (LHCP or Cab), and phosphoenolpyruvate carboxylase (PEPcase) during maize leaf development. The gene for the large subunit of Rubisco is located in the chloroplast DNA and the remainder of these are nuclear genes. Nine days after germination, the levels of the various mRNAs in dark-grown seedlings was very low compared to light-grown seedlings. For LHCP(Cab) and PEPcase the amount of mRNA reached less than 15–30% of the light-grown controls. The SSU of Rubisco was about 0.5% of the control and the LSU of mRNA reached about 13% of control levels. The various proteins accumulate to a much higher level. The protein for the LSU of Rubisco accumulated to about 50% and SSU to about 30% of the illuminated control. Since the PEPcase and mRNA was not detectable in 9-day-old dark-grown seedlings, it was surprising that the protein reached a level of about 20% of the control, suggesting that there is a low level of transcription for this gene in the dark and that, once synthesized, the protein is very stable. The LHCP(Cab) polypeptide, which is not detectable in 9-day dark-grown seedlings, is the exception, indicating that LHCP synthesis is absolutely light dependent. Under normal conditions LHCP mRNA and protein accumulate in parallel. Illumination of seedlings after 7 days of growth in the dark causes a rapid increase in the mRNAs for all these genes followed by a parallel increase in the various proteins. One unanticipated result of this study involved the accumulation of mRNA and protein during normal development. For Rubisco LSU and PEPcase the accumulation of the mRNAs precedes the accumulation of the respective proteins by 12–24 hr. For the SSU of Rubisco the mRNA reaches a peak 2–3 days before the SSU polypeptide appears. The reason for this early accumulation of SSU mRNA is not known. This study illustrates the range of responses that can be seen among various light-regulated genes. The mRNAs for the LSU and SSU of Rubisco of maize show a three- to sixfold increase on illumination while the mRNA levels for LHCP(Cap) and PEPcase are induced 200-fold. Protein synthesis is also differentially affected by light since, in the dark, no LHCP(Cab) protein is detected even though the amount of LHCP mRNA is similar to the level of PEPcase mRNA and PEPcase accumulates to about 20% of the light-grown control.

For the Rubisco subunits, LHCP(Cab), and PEPcase it has been demonstrated by a number of groups that the light-dependent accumulation of mRNA is primarily due to an increase in transcription. This was demonstrated for the Rubisco SSU and the LHCP(Cab) of pea by Gallagher and Ellis (1982) and in *Lemna* by Silverthorne and Tobin (1984). Both studies used the ability to label mRNA being transcribed in isolated nuclei. Hybridization of this labeled RNA to cloned probes of SSU and LHCP genes provides a means of quantitating gene-specific transcripts. For both these genes there is more transcription in nuclei isolated from light-grown plants than in nuclei from dark-grown plants. Silverthorne and Tobin (1984) also demonstrated that this increase in transcription is phytochrome dependent.

Light regulation at the level of mRNA accumulation has been shown for a number of other nuclear genes that encode proteins destined for the chloroplast. Malate dehydrogenase (NADP) from sorghum (Cretin *et al.*, 1988), chloroplast glutamine synthetase from pea (Edwards and Coruzzi, 1989), chloroplast fructose-1,6-bisphosphatase from wheat (Raines *et al.*, 1988), and phosphoribulokinase from wheat (Raines *et al.*, 1989) all show light-dependent accumulation of mRNA. Since none of these studies provide nuclear run-on data, it is not clear whether this accumulation is due to increased transcription or differences in mRNA turnover.

The studies by Raines *et al.* (1988, 1989) on fructose-1,6-bisphosphatase (FBPase) and phosphoribulokinase (PRK) illustrate a combination of light-regulated and developmental-regulated expression. Both enzymes are critical to carbon fixation. In the chloroplast FBPase takes part in the regeneration of ribulose bisphosphate, the primary CO_2 acceptor in the Calvin cycle, and PRK is the enzyme that actually synthesizes ribulose bisphosphate by the ATP-dependent phosphorylation of ribulose monophosphate. PRK mRNA is not detectable and the mRNA for FBPase is barely detectable in etiolated tissue. Within 24 hr of illumination of etiolated wheat seedlings mRNA for both genes has accumulated to high levels. Wheat provides a convenient model for studying developmental expression of these genes. In a wheat leaf there is a developmental gradient, beginning with the youngest cells in the zone of division at the base of the leaf to the oldest tissue at the tip of the leaf (Boffey *et al.*, 1979; Dean and Leech, 1982). Examining the steady-state levels of mRNA for FBPase and PRK in 2-cm sections taken along the length of a 5-day-old light-grown wheat seedling, one finds that both messages accumulate from a very low level at the base of the leaf to a maximum near the midpoint of the leaf and then decrease until they are barely detectable at the tip. Hence the state of tissue development also affects the accumulation of transcripts for these genes. A similar pattern has been reported for the wheat *cab* gene (Lamppa *et al.*, 1985).

As illustrated for the LSU of Rubisco, plastid gene expression can also be light-regulated at the level of transcription. Klein and Mullet (1990) examined the effect of light on the transcription of the *psbA* (a 32-kDa reaction center polypeptide of photosystem II), *rbcL,* and the chloroplast 16S rRNA genes of barley and maize. They found that the overall rate of plastid transcription, as measured by UTP incorporation into isolated plastids, was maximal in 4.5-day-old dark-grown barley seedlings and decreased to a minimum (about 10% of the 4.5-day-old maximum) after 8 days of growth in the dark. Transcription increased only 5–10% upon illumination of the 4.5-day-old dark-grown seedlings, but illumination of the 8-day-old dark-grown seedlings showed about a fourfold increase in overall plastid transcription. Examining the RNA made 4 hr after illumination of the 8-day-old dark-grown seedling showed that *psbA* transcripts increased 13-fold and the *rbcL* and 16S rRNA transcripts were enhanced 3.7- and 4-fold, respectively. Thus the *psbA* transcript is stimulated more than either the *rbcL* or the 16S rRNA. The increase in the transcription of the *rbcL* and the 16S rRNA genes reflects the light-induced increase of overall plastid transcription, while the *psbA* gene shows differential stimulation. Klein and Mullet found a similar pattern of gene expression in maize. There may also be differences in the stabilities of these various RNAs that would explain, in part, their differential accumulation. In spinach, Deng and Gruissem (1987) have shown that light affects the overall level of plastid transcription, but they found no differential stimulation of the various transcripts, including the one for *psbA*.

3.2. The *rbcS* Gene as a Model for the Mechanism of Light Regulation of Transcription

In higher plants, ribulose-1,5-bisphosphate carboxylase/oxygenase is a heterohexadecamer made up of eight LSU, which are encoded by the chloroplast genome, and eight SSU, which are the product of nuclear genes. This enzyme is responsible for net CO_2 fixation in the biosphere and as such is critical to life on earth. It is also the most abundant protein in plants, accounting for up to 50% of their soluble leaf protein. The SSU of Rubisco is encoded by a small multigene family (*rbcS* genes). These genes have been characterized in a number of different plants, including soybean (Berry-Lowe *et al.,* 1982), wheat (Broglie *et al.,* 1983), *Lemna* (Wimpee *et al.,* 1983), pea (Coruzzi *et al.,* 1984), tobacco (Mazur and Chui, 1985), petunia (Dean *et al.,* 1987a), and tomato (Sugita *et al.,* 1987). Accumulation of SSU mRNA is light dependent and is regulated at the level of transcription (Gallagher and Ellis, 1982; Silverthorne and Tobin, 1984). Silverthorne and Tobin (1984) also showed that this regulation is phytochrome dependent. As will be discussed below, there is also developmental and tissue-specific regulation of *rbcS* transcription. The light-regulated expression of the *rbcS* genes has provided a model system for studying the mechanisms of the light regulation of transcription in plants (Broglie *et al.,* 1984; Morelli *et al.,* 1985).

A number of features are conserved in the regulation of transcription in eukaryotes (for reviews, see Dynan and Tjian, 1985; Maniatis *et al.,* 1987). The promoters of most eukaryotic genes contain an element known as the "TATA box," located 25–30 base pairs (bp) from the start of transcription, which plays a role in determining the transcription start site. Further upstream (5') from the transcription start site, sequences related to "CCAAT" or "GGGCGG" are found. The role of such sequences is variable, but they often serve as binding sites for promoter-specific transcriptional factors. Finally, the activity of many promoters is modulated by an element known as an enchancer, which is required for optimal gene expression. Enhancers are generally located within 1000 bp of the start site of transcription but can be either upstream or downstream. Some are tissue specific and others modulate gene expression in response to extracellular signals such as hormones.

How do the promoters for the various *rbcS* genes compare to the canonical eukaryotic promoter and what elements are responsible for the light-regulated and tissue-specific transcription of the *rbcS* genes? Several groups have dissected the *rbcS* promoters from a number of different plant species, and their structure and regulation is discussed below.

3.2.1. Localization of the Minimal Light-Responsive *rbcS* Promoter

Two groups have defined the minimal light-responsive promoters for two *rbcS* genes from pea. The pea *rbcS-E9* gene was transferred into petunia cells using a Ti-plasmid vector from *Agrobacterium tumefaciens* (Fraley *et al.*, 1983), and it was shown that this gene was accurately transcribed from its own promoter (Broglie *et al.*, 1984). Morelli *et al.* (1985) used a derivative of the *rbcS-E9* gene (*rbcS-E9 5' del-1052*), which contains 1052 bp of the 5' upstream region of the *rbcS-E9* gene and is transcribed in a light-dependent manner in petunia calli, as starting point for a deletion analysis of the *rbcS-E9* promoter. The first 100 bp upstream of the start site for transcription contains both a "TATA box" and a "CAAT box." A 5' deletion that removes all the DNA to within 35 bp of the transcription start site still retains light-dependent transcription of the *rbcS-E9* gene. The TATA box is retained within these first 35 bp, but the CAAT box, which is located at −85, is not. This demonstrates that the CAAT box is not required for the transcription of the pea *rbcS-E9* gene in petunia calli. Using the 5' upstream region from another pea SSU gene (*rbcS ss3.6*) fused to a bacterial chloramphenicol acetyltransferase (*CAT*) gene, Timko *et al.* (1985) did a similar analysis in transgenic tobacco calli. They found light regulation of the *CAT* reporter gene in deletions retaining 90 bp proximal to the transcription start site. From these studies we can conclude that the minimal light-responsive promoter for the pea *rbcS* genes is a region within 35 bp of the transcription start site containing the TATA box. The region containing the *rbcS-E9* TATA box (−35 to −24) is also conserved in the sequences of six other *rbcS* promoters from several species (Morelli *et al.*, 1985).

3.2.2. Enhancer-like Elements in the *rbcS* Promoter

In the studies of Morelli *et al.* (1985) and Timko *et al.* (1985) it was found that although the light responsiveness of the *rbcS* gene was retained in promoters where all but the proximal 35 bp were deleted, these constructs show only about 15% of the normal amount of transcription. Timko *et al.* showed that, independent of orientation, the region between −90 and −973 of the *rbcS ss3.6* gene was able to support the full level of *CAT* gene expression. This suggests that this region contains an enhancer-like element. They found, however, that moving this sequence to the 3' side of the transcription start site did not support full *CAT* expression. Interestingly, they also found that the −90 to −973 region could confer light responsiveness to a normally light-insensitive promoter. Fusing the *rbcS ss3.6* upstream region to a normally light-insensitive nopaline synthase (*nos*) promoter driving the *CAT* gene created a promoter that shows a 13- to 16-fold induction by light. This induction was independent of the orientation of the −90 to −973 upstream *rbcS ss3.6* fragment.

This enhancer-like sequence was further defined by Fluhr *et al.* (1986) to a region between −50 and −330. Analysis of the sequence of the 280 bp in this region reveals a number of interesting features (Coruzzi *et al.*, 1984; Fluhr *et al.*, 1986; Kuhlemeir *et al.*, 1987a). Of particular interest is the region between −169 and −112, which contains three short sequences, referred to as box I, II, and III, which are conserved in the sequences of all known *rbcS* genes (Fluhr *et al.*, 1986). Box II resembles the SV40 core enhancer GT motif (Coruzzi *et al.*, 1984) and, in the reverse orientation, box III shows strong homology to the adenovirus 5 E1A −200 and the human interferon-β enhancers (Kuhlemeier *et al.*, 1987a). Kuhlemeier *et al.* (1987a) found that deletion of the upstream region of the pea *rbcS-3A* down to −166 had no effect on the level of expression or light regulation of their test gene in transgeneic tobacco. Deletion to −149 shows a dramatic decrease in the level of gene expression. That the low level of transcription still observed in the −149 construct is still light dependent is consistent with earlier observations placing a light-responsive element in the region of the TATA box. To test for enhancer activity they synthesized a 58-bp fragment corresponding to the sequence −169 to −112 and studied its effect on the expression of a *CAT* reporter gene in a construction where the TATA box and the transcriptional start site are supplied by a fragment from the cauliflower mosaic virus (CaMV) 35S gene (Odell *et al.*, 1985). Expression of the *CAT* gene in transgeneic tobacco plants was no greater than that observed with the 35S-*CAT* vector lacking the 58-bp insert. Thus the region containing box I, II, and III (−169 to −112) does not act as an enhancer. An enhancement of transcription of the 35S-*CAT* gene was observed only when the region spanning −330 to −112 was used. To test for other possible modulating effects of the −166 to −112 region, the synthetic 58-bp fragment was placed between the CaMV 35S promoter and its normal enhancer and this construct was used to drive transcription of a *CAT* reporter gene. The 35S-*CAT* gene used in this experiment is constitutively expressed and not regulated by light (Fluhr and Chua, 1986). Insertion of the 58-bp fragment made the expression of the 35S-*CAT* gene light dependent by turning off transcription in the dark! Boxes II and III are implicated in this down-regulation. An 87-bp sequence containing box III causes down-regulation of transcription in the dark. Multiple copies of box II can also function as dark-dependent down-regulators. Deletion of the region from −169 to −112 from the normal *rbcS-3A* upstream region has no effect on transcription of the *rbcS-3A* gene, implying that redundant light regulatory elements exist in this promoter.

The role of the redundancy of the light regulatory elements and the richness of the regulatory repertoir of the *rbcS* gene is illustrated in another study by Kuhlemeir *et al.* (1988). Although it was previously demonstrated that the region between −169 and −112 of the *rbcS-3A* gene was, by itself, not a positive regulatory element, the observation that a deletion to −149 is very much impaired in its transcriptional ability compared to a deletion to −166 remained to be explained. To address this question, Kuhlemeier *et al.* modified box I, II, and III by site-directed mutagenesis. In constructs that had been truncated to −175, substitution of box I had no effect on expression, but substitution of box II and box III abolished *rbcS-3A* transcription. Since box II has homology to the SV40 core enhancer motif, GTGGWWWG (where W = A or T), this core motif was modified by changing the GG dinucleotide to CC by site-directed mutagenesis. This transversion rendered the −175 construct transcriptionally inactive. As was the case with the deletion of the −169 to −112 region from the intact upstream region, none of these substitutions had any effect on transcription when introduced into the full-length *rbcS-3A* promoter. These studies show that in addition to the negative regulatory elements that had been found in the −169 to R−112 region, positive regulatory elements are also present in this region. The redundancy of regulatory elements within the *rbcs-3A* promoter is influenced by the stage of development of the plant tissue in which the gene is expressed. While the −166 deletion is fully active in mature leaves of transgenic tobacco, its transcription is very reduced in very young leaves and seedlings. Thus the enhancer properties of the upstream region (−410 to −171) may be redundant in fully mature leaves but it is necessary for full expression at earlier stages in development.

Similar promoter structure is seen in the *rbcS-3A* gene from tomato (Ueda *et al.*, 1989).

3.2.3. Nuclear Proteins that Bind to the *rbcS* Promoter

Using gel retardation assays and DNase footprinting experiments, Green *et al.* (1987) identified a protein factor in pea nuclear extracts that interacts specifically with regulatory sequences upstream of the pea *rbcS-3A* gene. This nuclear protein, designated GT-1, binds sepecifically to the light-responsive elements box II and box III. In addition, GT-1 binds to two DNA sequences found in the interval −172 to −291, which show homology to box II and box III. These sequences referred to as boxes II* and III* may be responsible for the enhancer properties of the −171 to −410 region of the *rbcs-3A* upstream region. Nuclear extracts prepared from either light-grown or dark-grown pea leaves contain GT-1 and the synthesis of GT-1 does not appear to be light regulated. Since boxes II and III are involved in both up- and down-regulation, the role GT-1 in *rbcS* transcription remains to be elucidated.

GT-1 is also found in nuclear extracts from rice, and binding sites for GT-1 are found in the genes for rice and oat phytochrome genes (Kay *et al.*, 1989).

Another DNA binding protein factor has been identified in nuclear extracts from tomato and *A. thaliana* (Giuliano *et al.*, 1988). This factor (GBF) binds to the upstream sequences of *rbcS* genes and recognizes a short conserved sequence (G box) that has the consensus TCTTACACGTGGCAYY (where Y = pyrimidine). This factor is found in both light-grown and dark-adapted tomato leaf extracts. GBF levels are low in extracts prepared from roots, suggesting some degree of tissue specificity. Understanding of the precise function of GBF in the regulation of *rbcS* gene expression awaits further experiments.

3.3.   Interaction of Light and Developmental Stage in *rbcS* Regulation

A number of investigators (for examples, see Simpson *et al.*, 1986; Kuhlemeier *et al.*, 1987a; Aoyagi *et al.*, 1988) have found that the expression of light-regulated genes whose products are involved in photosynthesis appears to be limited to chloroplast-containing cells. This has led to the speculation that the light regulation and organ specificity of these genes are linked phenomena. Some of the constructs discussed above, for instance those using the −410 to −50 *rbcS-3A* enchancer (Fluhr *et al.*, 1986), have both light regulation and tissue specificity. Kuhlemeier *et al.* (1989) found that expression from the region around the TATA box of pea *rbcS* genes, which as we have seen contains a light-responsive element, is light-regulated in transgenic plants, when associated with a weak enhancer, but that it is also expressed in roots. Thus the light-responsive and cell-type-specific components of *rbcS* regulation are separable phenomena. Poulsen and Chua (1988) show a similar uncoupling in an analysis of the *rbcS-8B* gene of *Nicotiana plumbaginifolia* though, in this case, tissue specificity is maintained but light regulation is lost.

Not all the various *rbcS* genes in the small multigene family of *rbcS* genes are expressed to the same extent (Dean *et al.*, 1985; Sugita and Gruissem, 1987). The five genes, *rbcS-1, -2, -3A, -3B,* and *-3C*, of tomato show differential expression (Sugita and Gruissem, 1987). All five genes are highly expressed in leaves, and the transcripts of two of these genes, *rbcS-3B* and *rbcS-3C*, account for about 60% of all leaf transcripts. Transcripts of these two genes are not detectable in dark-grown seedlings or in immature fruit. The *rbcS-1* and *rbcS-2* genes are expressed during fruit development, and the *rbcS-1*, *rbcS-2*, and *rbcS-3A* genes are all transcribed in etiolated tissue. Following light exposure, these transcripts accumulate rapidly but the *rbcS-3B* and *rbcS-3C* transcripts, which become the most abundant transcripts in mature leaves, accumulate slowly. This complex pattern of expression may reflect several regulatory mechanisms.

The eight *rbcS* genes of petunia vary over 100-fold in their level of expression in leaves (Dean *et al.*, 1985, 1987b). *SSU301* and *SSU611*, the most highly expressed of the petunia *rbcS* genes, share nucleotide sequences in their 5' upstream regions that are not present in the other petunia *rbcS* genes (Dean *et al.*, 1985). To investigate the basis for the observed difference in petunia *rbcS* gene expression, Dean *et al.* (1989a,b) constructed chimeric genes between *SSU301*, the most highly expressed petunia *rbcS* gene, and *SSU911*, the least expressed of the petunia *rbcS* genes, and examined their expression in transgenic tobacco plants. When introduced into transgenic tobacco plants, the *SSU301* and *SSU911* genes are expressed to the same relative levels as they are in petunia, demonstrating that these genes contain the determinants of their levels of expression. The chimeric genes were constructed by exchanging the 5' regions of the two genes. The *SSU301/SSU911* chimera has the 5' upstream region from the *SSU301* gene and the coding sequence and 3' untranslated region of the *SSU911* gene. The *SSU911/SSU30* chimera is the reciprocal construction. The most surprising result of these studies was that placing the *SSU301* 3' region with the 5' region of the *SSU911* gene (*SSU911/SSU301*) expressed 15-fold more mRNA than the intact *SSU911* gene. They also showed that this effect was at the level of transcription, not mRNA stability. Whether this effect is a function of a positive influence of the 3' region of *SSU301* or a negative effect of the 3' region of *SSU911* is not known. The 5' region of *SSU301* fused to the 3' region of *SSU911* expresses at a level about 25-fold higher than the intact *SSU911* gene. Adding the *SSU301* 5' sequences found between −285 and −204 to *SSU911* showed the same increase in transcription. This is the region that is shared between *SSU301* and the other highly expressed petunia *rbcS* gene, *SSU611*. A *rbcS* gene from *Nicotiana plumbaginifolia* (Poulsen *et al.*, 1986) and two of the five tomato *rbcS* genes (Sugita *et al.*, 1987) have sequences with homology to the *SSU301* −285 to −204 region. Finally, the region between −1 and −178 can be exchanged between *SSU301* and *SSU911* with no effect. This region is highly conserved among all petunia *rbcS* genes and contains homology to box II, a sequence conserved in the *rbcS* genes from all dicotyledonous plants so far examined (Kuhlemeier *et al.*, 1987a).

3.4. Structure of the *rbcS* Promoter—Summary

The *rbcS* promoter contains a number of regulatory elements that respond to light- and developmentally mediated signals. A general description of this promoter can be made from the studies discussed above. The region immediately upstream from the transcriptional start site is capable of low-level, light-specific promoter function. Further upstream is a region which contains positive and negative regulatory elements and which can provide full light-regulated expression of the *rbcs* gene in mature leaves. Finally, another element located further upstream is needed for full developmental regulation of the *rbcS* gene. The redundancy of elements found in the *rbcS* promoter allows for the fine tuning of gene expression to both environmental and developmental signals.

4. LIGHT-DEPENDENT TRANSLATIONAL CONTROL OF GENE EXPRESSION

Light-dependent control of translation of mRNA into protein is another means of light regulation of gene expression in plants. Translational control of gene expression has been implicated in a number of different plant systems, including *Euglena* (Miller *et al.*, 1983), amaranth (Berry *et al.*, 1985, 1986), *Spirodela* (Fromm *et al.*, 1985), pea (Inamine *et al.*, 1985), *Volvox* (Kirk and Kirk, 1985), barley (Klein and Mullet, 1986), and spinach (Deng and Gruissem, 1987).

The accumulation of the polypeptides for the LSU and SSU of Rubisco are regulated in part by control of translation. Berry *et al.* (1985, 1986) examined this regulation in amaranth and found that in seedlings grown in the dark for 8 days, the amount of mRNA for the LSU and SSU of Rubisco is low, but detectable, and accumulates dramatically 24 hr after illumination. Protein synthesis of the LSU and SSU polypeptides, as measured by incorporation of [35S]methionine into protein that can be immunoprecipitated by anti-Rubisco antibodies, increases 20-fold over general protein synthesis by 4 hr postillumination. At this point there is no measurable increase in the level of mRNA for the LSU or SSU. Shifting light-grown seedlings into the dark causes a marked decrease in LSU and SSU polypeptide synthesis. Six hours following a shift to darkness, LSU and SSU protein synthesis decreased by 50-fold while total protein synthesis had decreased only about fourfold. Six hours after shifting to the dark there was no measurable change in LSU or SSU mRNA levels. Berry *et al.* also demonstrated that this change in protein synthesis is not a function of a change in protein stability. This points out the complexity of the regulation of Rubisco synthesis. It is clear that this synthesis is regulated in response to light at both the transcriptional and translational level.

Fromm *et al.* (1985) showed that the 32-kDa protein product of the *psbA* gene is also light-regulated at the level of translation in the aquatic higher plant *Spirodela oligorrhiza*. Klein and Mullet (1986) demonstrated translational control of chloroplast-encoded chlorophyll-binding proteins in barley.

Although the mechanism for translational control of gene expression is not known, it clearly plays an important role in plant physiology and development. Apparently, as pointed out by Berry *et al.* (1986), for certain genes, responses to changes in the environment are mediated by translational regulation, whereas long-term changes in gene expression are mediated through changes in transcriptional activity.

5. LIGHT-DEPENDENT CHANGES IN PROTEIN STABILITY

Light can modulate gene expression by influencing the rate of protein turnover. In *Spirodel,* Mattoo *et al.* (1984) found that rates of synthesis and turnover of the 32-kDa chloroplast membrane protein are controlled by light. This protein is less stable in the light and more stable in the dark. They also showed that the degradation of the 32-kDa protein is dependent on photosynthetic electron transport. Bennett (1981) found that, in addition to the synthesis of the light-harvesting chlorophyll a/b binding protein being highly light dependent, it turns over rapidly when light-grown plants are transferred to the dark.

6. LIGHT MODULATION OF ENZYME ACTIVITY

Modulating enzyme activity provides a means for fine-tuning gene expression. As reviewed by Buchanan (1980), light can regulate the activity of a number of different enzymes by a variety of different mechanisms. Several enzymes of the reductive pentose phosphate cycle are activated by light and have little or no activity *in vivo* in the dark. These enzymes include fructose-1, 6-bisphosphatase (FBPase), sedoheptulose-1,7-bisphosphatase (SHPase), Rubisco, phosphoribulokinase (PRK), and NADP-glyceraldhyde 3-phosphate dehydrogenase (NADP-GAPD) (see Buchanan, 1980). Light regulation of three of these enzymes, Rubisco, FBPase, and PRK, will be discussed below.

The activity of Rubisco is light-regulated through the action of a competitive inhibitor, 2-carboxy-D-arabinitol-1-phosphate (CA-1-P), which accumulates in the dark (Seeman *et al.*, 1985; Gutteridge *et al.*, 1986; Berry *et al.*, 1987). This molecule has a structure very similar to that of one of the intermediates in the carboxylase reaction of Rubisco. Rubisco activity requires the addition of CO_2, in the presence of Mg^{2+}, to a lysine residue of the enzyme (Lorimer *et al.*, 1976). It is to this activated form of the enzyme that CA-1-P binds. Shifting plants into the light causes a rapid decrease in amount of CA-1-P (Seeman *et al.*, 1985). The metabolic pathways controlling the synthesis and degradation of CA-1-P, and ultimately this unique form of regulation, are not known.

FBPase (Wolosiuk and Buchanan, 1977) and PRK (Wolosiuk and Buchanan, 1978b), as well as SHPase (Breazeale *et al.*, 1978) and NADP-GAPD (Wolosiuk and Buchanan, 1978a), are regulated by the ferredoxin/thioredoxin system. Electrons are transferred to ferredoxin from chlorophyll in the light and then via ferredoxin–thioredoxin reductase to thioredoxin. Reduced thioredoxin activates these various enzymes. Recently it has been shown genetically that thioredoxin is essential for photosynthetic growth in the cyanobacteria *Anacystis nidulans* (Muller and Buchanan, 1989). PRK is regulated by the reversible thioredoxin-mediated reduction of a disulfide bond between Cys-16 and Cys-55 (Porter *et al.*, 1988) in the spinach enzyme. This sequence is conserved in all known higher plant PRKs (Milanez and Mural, 1988; Roesler and Ogren, 1988, 1990; Raines *et al.*, 1989). Cys-16 is located in a sequence with extensive homology to an ATP-binding site (Higgins *et al.*, 1986; Krieger *et al.*, 1987; Porter *et al.*, 1988). Although chloroplast FBPase has been sequenced from wheat (Raines *et al.*, 1988) and spinach (Marcus and Harrsch, 1990), the exact site of thioredoxin activation is not known. Earlier (Section 3.1) the light-dependent regulation of FBPase and PRK transcription was discussed. Regulation of these essential enzymes by the ferredoxin/thioredoxin system provides a further level of diurnal fine tuning for plant metabolism. A further illustration of the complexity of this regulatory network is the fact that the ferredoxin gene is itself phytochrome- (and therefore light-) regulated (Kaufman *et al.*, 1986).

7. LIGHT AND DEVELOPMENT

The recent isolation of a mutant of *A. thaliana* that develops as a light-grown plant, even when grown in the dark, promises to lead to new insights into the role of light in plant development (Chory *et al.*, 1989). Plants that are homozygous for this allele (*det1*) display many light-dependent characteristics of wild-type plants when grown in the dark, including the accumulation of mRNAs for several light-regulated nuclear genes. The recessive nature of the *det1* mutation suggests that in the absence of light, there is negative control on leaf development. Chory *et al.* (1989) suggest a model where the primary role of light on developmental gene expression is mediated by the initiation of leaf development.

8. CONCLUSIONS

Life on earth is dependent on the interaction of light and green plants. In fact, one can view the biosphere as a by-product of photosynthesis. Since it plays such a major role in the life of plants, it is not surprising that they have evolved a number of ways of sensing and responding to light. Light not only provides the energetic requirements of the plant, through photosynthesis, but also provides important signals to the plant's physiology and development. Enzymes involved in photosynthesis and carbon fixation are regulated by light at every possible level, and their activity is fine-tuned to meet the metabolic needs of the plant. Using the tools currently available and building on our present knowledge, we can anticipate that our understanding of the molecular events involved in the response of plants to light will increase as will our ability to put this knowledge to use.

ACKNOWLEDGMENTS. This work was supported by the Office of Health and Environmental Research, U.S. Department of Energy, under Contract DE-AC05-840R21400 with the Martin Marietta Energy Systems, Inc.

9. REFERENCES

Abe, H., Yamamoto, K. T., Nagatani, A., and Furuya, M., 1985, Characterization of green tissue-specific phytochrome isolated immunochemically from pea seedlings, *Plant Cell Physiol.* **26:**1387–1399.

Aoyagi, K., Kuhlemeier, C., and Chua, N-H., 1988, The pea *rbcS-3A* enhancer-like element directs cell-specific expression in transgenic tobacco, *Mol. Gen. Genet.* **213:**179–185.

Bennett, J., 1981, Biosynthesis of the light-harvesting chlorophyll *a/b* protein, *Eur. J. Biochem.* **118:**61–70.

Berry, J. O., Nikolau, B. J., Carr, J. P., and Klessig, D. F., 1985, Transcriptional and post-transcriptional regulation of ribulose 1,5 bisphosphate carboxylase gene expression in light- and dark-grown amaranth cotyledons, *Mol. Cell. Biol.* **5:**2238–2246.

Berry, J. O., Nikolau, B. J., Carr, J. P., and Klessig, D. F., 1986, Translational regulation of light-induced ribulose 1,5 bisphosphate carboxylase gene expression in amaranth, *Mol. Cell. Biol.* **6:**2347–2353.

Berry, J. A., Lorimer, G. H., Pierce, J., Seeman, J. R., and Meek, J., 1987, Isolation, identification, and synthesis of 2-carboxyarabinitol 1-phosphate, a diurnal regulator of ribulose-bisphosphate carboxylase activity, *Proc. Natl. Acad. Sci. USA* **84:**734–738.

Berry-Lowe, S. L., McKnight, T. D., Shah, D. M., and Meagher, R. B., 1982, The nucleotide sequence, expression, and evolution of one member of a multigene family encoding the small subunit of ribulose-1,5-bisphosphate carboxylase in soybean, *J. Mol. Appl. Genet.* **1:**483–498.

Boffey, S. A., Ellis, J. R., Sellden, G., and Leech, R. M., 1979, Chloroplast division and DNA synthesis in light-grown wheat leaves, *Plant Physiol.* **64:**502–505.

Breazeale, V. D., Buchanan, B. B., and Wolosiuk, R. A., 1978, Chloroplast sedoheptulose 1,

7-bisphosphatase: Evidence for regulation by the ferredoxin/thioredoxin system, *Z. Naturforsch., Teil C.* **33**:521–528.

Broglie, R., Coruzzi, G., Lamppa, G., Keith, B., and Chu, N-H., 1983, Structural analysis of nuclear genes coding for the precursor to the small subunit of wheat ribulose-1,5-bisphosphate carboxylase, *Biotechnology* **1**:55–61.

Broglie, R., Coruzzi, G., Fraley, R. T., Rogers, S. G., Horsch, R. B., Niedermeyer, J. G., Fink, C. L., Flick, J. S., and Chua, N-H., 1984, Light-regulated expression of a pea ribulose-1, 5-bisphosphate carboxylase small subunit gene in transformed plant cells, *Science* **224**:838–843.

Buchanan, B. B., 1980, Role of light in the regulation of chloroplast enzymes, *Annu. Rev. Plant Physiol.* **31**:341–374.

Chory, J., Peto, C., Feinbaum, R., Pratt, L., and Ausubel, F., 1989, Arabidopsis thaliana mutant that develops as a light-grown plant in the absence of light, *Cell* **58**:991–999.

Christensen, A. H., and Quail, P. H., 1989, Structure and expression of a maize phytochrome-encoding gene, *Gene* **85**:381–390.

Colbert, J. T., Hershey, H. P., and Quail, P. H., 1985, Phytochrome regulation of phytochrome mRNA abundance, *Plant Mol. Biol.* **5**:91–101.

Coruzzi, G., Broglie, R., Edwards, C., and Chua, N-H., 1984, Tissue-specific and light-regulated expression of a pea nuclear gene encoding the small subunit of ribulose-1,5-bisphosphate carboxylase, *EMBO J.* **3**:1671–1679.

Cretin, C., Luchetta, P., Joly, C., Miginiac-Maslow, M., Decottignies, P., Jacquot, J-P., Vidal, J., and Gadal, P., 1988, Identification of a cDNA clone for sorghum leaf malate dehydrogenase (NADP), *Eur. J. Biochem.* **174**:497–501.

Cuozzo, M., Kay, S. A., and Chua, N-H., 1988, Regulatory circuits of light-responsive genes, in *Temporal and Spatial Regulation of Plant Genes* (D. P. S. Verma and R. B. Goldberg, eds.), pp. 131–146, Springer-Verlag, Vienna.

Dean, C., and Leech, R. M., 1982, Genome expression during normal leaf development, *Plant Physiol.* **69**:904–910.

Dean, C., van den Elzen, P., Tamaki, S., Dunsmuir, P., and Bedbrook, J., 1985, Differential expression of the eight genes for the petunia ribulose bisphosphate carboxylase small subunit multi-gene family, *EMBO J.* **4**:3055–3061.

Dean, C., van den Elzen, P., Tamaki, S., Black, M., Dunsmuir, P., and Bedbrook, J., 1987a, Molecular characterization of the *rbcS* multi-gene family of *Petunia* (Mitchell), *Mol. Gen. Genet.* **206**:465–474.

Dean, C., Favreau, M., Dunsmuir, P., and Bedbrook, J., 1987b, Confirmation of the relative expression levels of the *Petunia* (Mitchell) *rbcS* genes, *Nucleic Acids Res.* **15**:4655–4668.

Dean, C., Favreau, M., Bond-Nutter, D., Bedbrook, J., and Dunsmuir, P., 1989a, Sequences downstream of translation start regulate quantitative expression of two petunia *rbcS* genes, *Plant Cell* **1**:201–208.

Dean, C., Favreau, M., Bedbrook, J., and Dunsmuir, P., 1989b, Sequences 5' to translation start regulate expression of petunia *rbcS* genes, *Plant Cell* **1**:209–215.

Deng, X-W., and Gruissem, W., 1987, Control of plastid gene expression during development: The limited role of transcriptional regulation, *Cell* **49**:379–387.

Dynan, W. S., and Tjian, R., 1985, Control of eukaryotic messenger RNA synthesis by sequence-specific DNA-binding proteins, *Nature* **316**:774–778.

Edwards, J. W., and Coruzzi, G. M., 1989, Photorespiration and light act in concert to regulate the expression of the nuclear gene for chloroplast glutamine synthetase, *Plant Cell* **1**:241–248.

Ellis, R. J., 1981, Chloroplast proteins: Synthesis, transport, and assembly, *Annu. Rev. Plant Physiol.* **32**:111–137.

Fluhr, R., and Chua, N-H., 1986, Developmental regulation of two genes encoding ribulose-bisphosphate carboxylase small subunit in pea and transgenic petunia plants: Phytochrome response and blue-light induction, *Proc. Natl. Acad. Sci. USA* **83:**2358–2362.

Fluhr, R., Kuhlemeier, C., Nagy, F., and Chua, N-H., 1986, Organ-specific and light-induced expression of plant genes, *Science* **232:**1106–1112.

Fraley, R. T., Rogers, S. G., Horsch, R. B., Sanders, P. R., Flick, J. S., Adams, S. P., Bittner, M. L., Brand, L. A., Fink, C. L., Fry, J. S., Galluppi, G. R., Goldberg, S. B., Hoffmann, N. L., and Woo, S. C., 1983, Expression of bacterial genes in plant cells, *Proc. Natl. Acad. Sci. USA* **80:**4803–4807.

Fromm, H., Devic, M., Fluhr, R., and Edelman, M., 1985, Control of *psbA* gene expression: In mature *Spirodela* chloroplasts light regulation of 32-kd protein synthesis is independent of transcription level, *EMBO J.* **4:**291–295.

Furuya, M., ed., 1987, *Phytochrome and Photoregulation in Plants*, Proceedings of the Yamada Conference XVI, Academic Press, Tokyo.

Gallagher, T. F., and Ellis, R. J., 1982, Light-stimulated transcription of genes for two chloroplast polypeptides in isolated pea leaf nuclei, *EMBO J.* **1:**1493–1498.

Gamble, P. E., Sexton, T. B., and Mullet, J. E., 1988, Light-dependent changes in *psbD* and *psbC* transcripts of barley chloroplasts: Accumulation of two transcripts maintain *psbD* and *psbC* translation capability in mature chloroplasts, *EMBO J.* **7:**1289–1297.

Giuliano, G., Pichersky, E., Malik, V. S., Timko, M. P., Scolnik, P. A., and Cashmore, A. R., 1988, An evolutionarily conserved protein binding sequence upstream of a plant light-regulated gene, *Proc. Natl. Acad. Sci. USA* **85:**7089–7093.

Green, P. J., Kay, S. A., and Chua, N-H., 1987, Sequence-specific interactions of a pea nuclear factor with light-responsive elements upstream of the *rbcS-3A* gene, *EMBO J.* **6:**2543–2549.

Green, P. J., Yong, M-H., Cuozzo, M., Kano-Murakami, Y., Silverstein, P., and Chua, N-H., 1988, Binding site requirements for pea nuclear protein factor GT-1 correlate with sequences required for light-dependent transcriptional activation of the *rbcS-3A* gene, *EMBO J.* **7:**4035–4044.

Gutteridge, S., Parry, M. A. J., Burton, S., Keys, A. J., Mudd, A., Feeney, J., Servaites, J. C., and Pierce, J., 1986, A nocturnal inhibitor of carboxylation in leaves, *Nature* **324:**274–276.

Hershey, H. P., Barker, R. F., Idler, K. B., Lissemore, J. L., and Quail, P. H., 1985, Analysis of cloned cDNA and genomic sequences for phytochrome: Complete amino acid sequences for two gene products expressed in etiolated *Avena*, *Nucleic Acids Res.* **13:**8543–8559.

Hershey, H. P., Barker, R. F., Idler, K. B., Murray, M. G., and Quail, P. H., 1987, Nucleotide sequence and characterization of a gene encoding the phytochrome polypeptide from *Avena*, *Gene* **61:**339–348.

Higgins, C. F., Hiles, I. D., Salmond, G. P. C., Gill, D. R., Downie, J. A., Evans, I. J., Holland, I. B., Gray, L., Buckel, S. D., Bell, A. W., and Hermodson, M. A., 1986, A family of related ATP-binding subunits coupled to many distinct biological processes in bacteria, *Nature* **323:**448–450.

Hoober, J. K., 1987, The molecular basis of chloroplast development, in *The Biochemistry of Plants*, Vol. 10 (M. D. Hatch and N. K. Boardman, eds.), pp. 1–74, Academic Press, New York.

Inamine, G., Nash, B., Weissbach, H., and Brot, N., 1985, Light regulation of the synthesis of the large subunit of ribulose-1,5-bisphosphate carboxylase in peas: Evidence for translational control, *Proc. Natl. Acad. Sci. USA* **82:**5690–5694.

Kaufman, L. S., Roberts, L. L., Briggs, W. R., and Thompson, W. F., 1986, Phytochrome control of specific mRNA levels in developing pea buds, *Plant Physiol.* **81:**1033–1038.

Kay, S. A., Keith, B., Shinozaki, K., Chye, M-L., and Chua, N-H., 1989, The rice phytochrome gene: Structure, autoregulated expression, and binding of GT-1 to a conserved site in the 5' upstream region, *Plant Cell* **1:**351–360.

Kendrick, R. E., and Kronenberg, G. H. M., eds., 1986, *Photomorphogenesis in Plants*, Martinus Nijhoff, Dordrecht.

Kirk, M., and Kirk, D. L., 1985, Translational regulation of protein synthesis, in response to light, at a critical stage of *Volvox* development, *Cell* **41**:419–428.

Klein, R. R., and Mullet, J. E., 1986, Regulation of chloroplast-encoded chlorophyll-binding protein translation during higher plant chloroplast biogenesis, *J. Biol. Chem.* **261**:11138–11145.

Klein, R. R., and Mullet, J. E., 1987, Control of gene expression during higher plant chloroplast biogenesis, *J. Biol. Chem.* **262**:4341–4348.

Klein, R. R., and Mullet, J. E., 1990, Light-induced transcription of chloroplast genes, *J. Biol. Chem.* **265**:1895–1902.

Krieger, T. J., Mende-Mueller, L., and Miziorko, H. M., 1987, Phosphoribulokinase: isolation and sequence determination of the cysteine-containing active-site peptide modified by 5′-*p*-fluoro-sulphoyl-benzoyladenosine, *Biochim. Biophys. Acta* **915**:112–119.

Kuhlemeier, C., Fluhr, R., Green, P. J., and Chua, N-H., 1987a, Sequences in the pea *rbcS-3A* gene have homology to constitutive mammalian enchancers but function as negative regulatory elements, *Genes Dev.* **1**:247–255.

Kuhlemeier, C., Green, P. J., and Chua, N-H., 1987b, Regulation of gene expression in higher plants, *Annu. Rev. Plant Physiol.* **38**:221–257.

Kuhlemeier, C., Cuozzo, M., Green, P. J., Goyvaerts, E., Ward, K., and Chua, N-H., 1988, Localization and conditional redundancy of regulatory elements in *rbcS-3A*, a pea gene encoding the small subunit of ribulose-bisphosphate carboxylase, *Proc. Natl. Acad. Sci. USA* **85**:4662–4666.

Kuhlemeier, C., Strittmatter, G., Ward, K., and Chua, N-H., 1989, The pea *rbcs-3A* promoter mediates light responsiveness but not organ specificity, *Plant Cell* **1**:471–478.

Lamppa, G. K., Morelli, G., and Chua, N-H., 1985, Structure and developmental regulation of a wheat gene encoding the major chlorophyl a/b binding protein, *Mol. Cell. Biol.* **5**:1370–1378.

Lissemore, J. L., and Quail, P. H., 1988, Rapid transcriptional regulation by phytochrome of the genes for phytochrome and chlorophyll a/b binding protein in *Avena sativa*, *Mol. Cell. Biol.* **8**:4840–4850.

Lissemore, J. L., Colbert, J. T., and Quail, P. H., 1987, Cloning of cDNA for phytochrome from etiolated *Cucurbita* and coordinate photoregulation of the abundance of two distinct phytochrome transcripts, *Plant Mol. Biol.* **8**:485–496.

Lorimer, G. H., Badger, M. R., and Andrews, T. J., 1976, The activation of ribulose-1,5-bisphosphate carboxylase by carbon dioxide and magnesium ions. Equilibria, kinetics, a suggested mechanism, and physiological implications, *Biochemistry* **15**:529–536.

Maniatis, T., Goodbourn, S., and Fisher, J., 1987, Regulation of inducible and tissue-specific gene expression, *Science* **236**:1237–1245.

Marcus, F., and Harrsch, P. B., 1990, Amino acid sequence of spinach chloroplast fructose-1,6-bisphosphatase, *Arch. Biochem. Biophys.* **279**:151–157.

Mattoo, A. K., Hoffman-Falk, H., Marder, J. B., and Edelman, M., 1984, Regulation of protein metabolism: Coupling of photosynthetic electron transport to *in vivo* degradation of the rapidly metabolized 32-kilodalton protein of the chloroplast membranes, *Proc. Natl. Acad. Sci. USA* **81**:1380–1384.

Mazur, B. J., and Chui, C-H., 1985, Sequence of a genomic DNA clone for the small subunit of ribulose bis-phosphate carboxylase-oxygenase from tobacco, *Nucleic Acids Res.* **13**:2373–2386.

Milanez, S., and Mural, R. J., 1988, Cloning and sequencing of cDNA encoding the mature form of phosphoribulokinase from spinach, *Gene* **66**:55–63.

Miller, M. E., Jurgenson, J. E., Reardon, E. M., and Price, C. A., 1983, Plastid translation *in organello* and *in vitro* during light-induced development in *Euglena, J. Biol. Chem.* **258:**14478–14484.

Morelli, G., Nagy, F., Fraley, R. T., Rogers, S. G., and Chu, N-H., 1985, A short conserved sequence is involved in the light-inducibility of a gene encoding the ribulose 1,5-bisphosphate carboxylase small subunit of pea, *Nature* **315:**200–204.

Muller, E. G. D., and Buchanan, B. B., 1989, Thioredoxin is essential for photosynthetic growth, *J. Biol. Chem.* **264:**4008–4014.

Mullet, J. E., and Klein, R. R., 1987, Transcription and RNA stability are important determinants of higher plant chloroplast RNA levels, *EMBO J.* **6:**1571–1579.

Nagy, F., Kay, S. A., and Chua, N-H., 1988, Gene regulation by phytochrome, *Trends Genet.* **4:**37–42.

Nelson, T., Harpster, M. H., Mayfield, S. P., and Taylor, W. C., 1984, Light-regulated gene expression during maize leaf development, *J. Cell Biol.* **98:**558–564.

Odell, J. T., Nagy, F., and Chua, N-H., 1985, Identification of DNA sequences required for activity of a plant promoter: the CaMV promoter, *Nature* **313:**810–812.

Oelmuller, R., and Mohr, H., 1985, Mode of coaction between blue/UV light and light absorbed by phytochrome in light-mediated anthocyanin formation in the milo (*Sorghum vulgare* Pers.) seedlings, *Proc. Natl. Acad. Sci. USA* **82:**6124–6128.

Otto, V., Schafer, E., Nagatani, A., Yamamoto, K. T., and Furuya, M., 1984, Phytochrome control of its own synthesis in *Pisum sativum, Plant Cell Physiol.* **25:**1579–1584.

Porter, M. A., Stringer, C. D., and Hartman, F. C., 1988, Characterization of the regulatory thioredoxin site of phophoribulokinase, *J. Biol. Chem.* **263:**123–129.

Poulsen, C., and Chua, N-H., 1988, Dissection of 5′ upstream sequences for selective expression of *Nicotiana plumbaginifolia rbcS-8B* gene, *Mol. Gen. Genet.* **214:**16–23.

Poulsen, C., Fluhr, R., Kauffman, J. M., Boutry, M., and Chua, N-H., 1986, Characterization of an *rbcS* gene from *Nicotiana plumbaginifolia* and expression of an *rbcS*-CAT chimeric gene in homologous and heterologous nuclear background, *Mol. Gen. Genet.* **205:**193–200.

Pratt, L. H., 1986, Localization within the plant, in *Photomorphogenesis in Plants* (R. E. Kendrick and G. H. M. Kronenberg, eds.), pp. 61–82, Martinus Nijhoff, Dordrecht.

Quail, P. H., Schaefer, E., and Marme, D., 1973, Turnover of phytochrome in pumpkin cotyledons, *Plant Physiol.* **52:**124–127.

Quail, P. H., Gatz, C., Hershey, H., Jones, A., Lissemore, J. L., Parks, B. M., Sharrock, R. A., Barker, R. F., Idler, K., Murray, M. G., Koorneef, M., and Kendrick, R. E., 1987, Molecular biology of phytochrome, in *Phytochrome and Photoregulation in Plants* (M. Furuya, ed.), pp. 23–37, Academic Press, Tokyo.

Raines, C. A., Llyod, J. C., Longstaff, M., Bradley, D., and Dyer, T., 1988, Chloroplast fructose-1,6-bisphosphatase: The product of a mosaic gene, *Nucleic Acids Res.* **16:**7931–7942.

Raines, C. A., Longstaff, M., Llyod, J. C., and Dyer, T. A., 1989, Complete coding sequence of wheat phosphoribulokinase: Developmental and light-dependent expression of the mRNA, *Mol. Gen. Genet.* **220:**43–48.

Roesler, K. R., and Ogren, W. L., 1988, Nucleotide sequence of spinach cDNA encoding phosphoribulokinase, *Nucleic Acids Res.* **16:**7192.

Roesler, K. R., and Ogren, W. L., 1990, *Chlamydomonas reinhardtii* phospribulokinase, *Plant Physiol.* **93:**188–193.

Sato, N., 1988, Nucleotide sequence and expression of the phytochrome gene in *Pisum sativum:* Differential regulation by light of multiple transcripts, *Plant Mol. Biol.* **11:**697–710.

Seeman, J. R., Berry, J. A., Freas, S. M., and Krump, M. A., 1985, Regulation of ribulose bisphosphate carboxylase activity *in vivo* by a light-modulated inhibitor of catalysis, *Proc. Natl. Acad. Sci. USA* **82:**8024–8028.

Sharrock, R. A., and Quail, P. H., 1989, Novel phytochrome sequences in *Arabidopsis thaliana:*

Structure, evolution, and differential expression of a plant regulatory photoreceptor family, *Genes Dev.* **3**:1745–1757.

Sharrock, R. A., Lissmore, J. L., and Quail, P. H., 1986, Nucleotide and amino acid sequence of a *Cucurbita* phytochrome cDNA clone: Identification of conserved features by comparison with *Avena* phytochrome, *Gene* **47**:287–295.

Shimazaki, Y., and Pratt, L. H., 1985, Immunochemical detection with rabbit polyclonal and mouse monoclonal antibodies of different pools of phytochrome from etiolated and green *Avena* shoots, *Planta (Berl.)* **164**:333–344.

Shropshire, W., and Mohr, H., eds., 1983, Photomorphogensis. *Encyclopedia of Plant Physiology*, New Series, Vol. 16A and 16B, Springer-Verlag, Berlin.

Silverthorne, J., and Tobin, E. M., 1984, Demonstration of transcriptional regulation of specific genes by phytochrome action, *Proc. Natl. Acad. Sci. USA* **81**:1112–1116.

Silverthorne, J., and Tobin, E. M., 1987, Phytochrome regulation of nuclear gene expression, *BioEssays* **7**:18–23.

Simpson, J., Schell, J., Van Montagu, M., and Herrera-Estrella, L., 1986, Light-inducible and tissue-specific pea *lhcp* gene expression involves an upstream element combing enhancer- and silencer-like properties, *Nature* **323**:551–554.

Sugita, M., and Gruissem, W., 1987, Developmental, organ-specific, and light-dependent expression of the tomato ribulose-1,5-bisphosphate carboxylase small subunit gene family, *Proc. Natl. Acad. Sci. USA* **84**:7104–7108.

Sugita, M., Manzara, T., Pichersky, E., Cashmore, A., and Gruissem, W., 1987, Genomic organization, sequence analysis and expression of all five genes encoding the small subunit of ribulose-1,5-bisphosphate carboxylase/oxygenase from tomato, *Mol. Gen. Genet.* **209**:247–256.

Timko, M. P., Kausch, A. P, Castresana, C., Fassler, J., Herrera-Estrella, L., Van den Broeck, G., Van Montagu, M., Schell, J., and Cashmore, A. R., 1985, Light regulation of plant gene expression by an upstream enhancer-like element, *Nature* **318**:579–582.

Tobin, E. M., and Silverthorne, J., 1985, Light regulation of gene expression in higher plants, *Annu. Rev. Plant Physiol.* **36**:569–593.

Tokuhisa, J. G., Daniels, S. M., and Quail, P. H., 1985, Phytochrome in green tissue: Spectral and immunochemical evidence for distinct molecular species of phytochrome in light grown *Avena sativa* L, *Planta (Berl.)* **164**:321–332.

Ueda, T., Pichersky, E., Malik, V. S., and Cashmore, A. R., 1989, Level of expression of the tomato *rbcS-3A* gene is modulated by a far upstream promoter element in a developmentally regulated manner, *Plant Cell* **1**:217–227.

Wimpee, C. F., Stiekma, W. J., and Tobin, E. M., 1983, Sequence heterogeneity in the RuBP carboxylase small subunit gene family of *Lemna gibba*, in *Plant Molecular Biology*, UCLA Symposium on Molecular and Cellular Biology, New Series, vol. 12 (R. B. Goldberg, ed.), pp. 391–401, Alan R. Liss, New York.

Wolosiuk, R. A., and Buchanan, B. B., 1977, Thioredoxin and glutathione regulate photosynthesis in chloroplasts, *Nature* **266**:565–567.

Wolosiuk, R. A., and Buchanan, B. B., 1978a, Activation of chloroplast NADP linked glyceraldehyde-3-phosphate dehydrogenase by the ferredoxin/thioredoxin system, *Plant Physiol.* **61**:669–671.

Wolosiuk, R. A., and Buchanan, B. B., 1978b, Regulation of chloroplast phosphoribulokinase by the ferredoxin/thioredoxin system, *Arch. Biochem. Biophys.* **189**:97–101.

Chapter 10

Genetic Manipulation of Photosynthetic Processes in Plants

Tristan A. Dyer

1. INTRODUCTION

As photosynthesis is the primary path by which high-energy organic molecules enter the biosphere, it would seem to be self-evident that we should try to optimize this process in our efforts to fulfill our requirements of such substances. However, this may not necessarily just involve an effort to increase the maximum rate at which this process occurs in ideal conditions. Also, it may be related to improving photosynthetic efficiency in suboptimal conditions of light, temperature, or water supply so that crop plants with particularly favorable characteristics can be grown in hotter or colder, wetter or drier conditions, or at different latitudes. Perhaps it may also be desirable to extend the period during which a photosynthetic organ remains viable or, alternatively, remobilizes its components quickly so that, once it is past its peak efficiency, its reservoir of valuable nitrogenous compounds is reused to greater advantage elsewhere. These issues could be particularly relevant in the situation in which we now find ourselves, in which there could be rapid climatic change (White, 1990).

Plants grow in a wide range of environments and therefore it is not surpris-

Tristan A. Dyer   Molecular Genetics Department, Cambridge Laboratory, John Innes Centre for Plant Science Research, Norwich, NR4 7UJ, United Kingdom.

Subcellular Biochemistry, Volume 17: Plant Genetic Engineering, edited by B. B. Biswas and J. R. Harris. Plenum Press, New York, 1991.

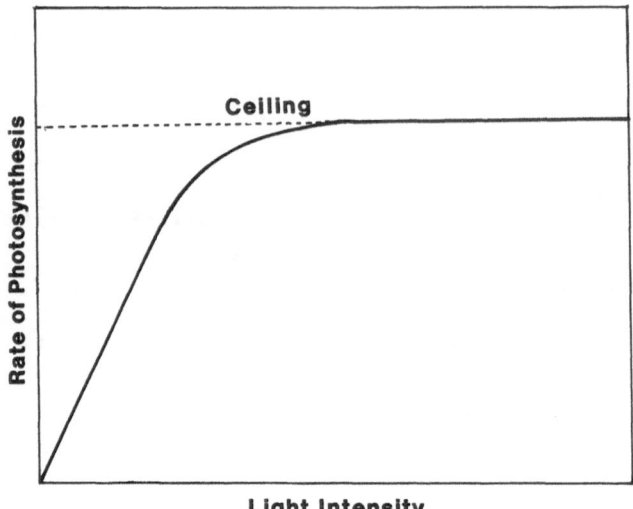

FIGURE 1. Response of leaf photosynthesis to changes in light intensity.

ing to find that there is much genetic variability in their photosynthetic properties (Austin, 1989). Although the basic biochemistry of this process is essentially the same in all plants, they have evolved a range of strategies that make them suitable for particularly ecological niches. In order to explore these differences, it is necessary to have methods of measuring and comparing their respective photosynthetic efficiencies. This is achieved by seeing how much of the energy of incident light is converted into chemical energy. Photosynthetic CO_2 fixation or O_2 evolution is measured at a range of light intensities. Essentially what is found is that at low light intensities, leaves may be very efficient, but that as the intensity of light is increased, the efficiency of photosynthesis gradually becomes less until a point is reached when negligible further increase in photosynthesis occurs with further increase in amount of light (Figure 1).

If the reasons for this decrease in efficiency were simple, there would be a reasonable chance of our being able to alter them by genetic manipulation. Presumably what one would wish to achieve is plants that would remain as efficient as possible for as long as possible as light intensity is increased.

2. LIMITING FACTORS

A concept central to the study of photosynthetic efficiency is that of limiting factors. At low light intensity, for example, the main limiting factor is apparently the rate of absorbtion of photons. In this phase, the rate of CO_2 fixation is proportional to light intensity and the slope of the line relating the two is a measure of quantum efficiency. Photosynthetic rate increases linearly until other factors become progressively limiting and eventually determine the maximum rate at which this process can take place. The underlying limitation determining this "ceiling" is thought to be the catalytic capacity of the reactions of the photosynthetic carbon reduction cycle. When this limitation occurs, excess light energy that is absorbed must be dissipated, otherwise damage to the light absorption apparatus would occur.

One way of increasing photosynthetic efficiency therefore might be to increase the quantum efficiency with which plants absorb light and a second might be to try to raise the ceiling by making the carbon reduction process more efficient. A third strategy might be to increase the sharpness of the transition from the limitation imposed by light absorption efficiency to that imposed by rate of CO_2 fixation, that is, to increase the "light utilization capacity" of plants.

Two principal methods come to mind of ways to determine which specific chloroplast constituents may be important in limiting the efficiency of photosynthesis. One may either compare the components of closely related plants that differ in their photosynthetic efficiency or make specific changes in these components by genetic manipulation. The two methods are complementary in that the first approach may indicate what constituent may be worth manipulating and the second may provide confirmatory evidence that the differences observed between naturally occurring genotypes may indeed contribute to differences in photosynthetic efficiency. What one is, in fact, searching for are tight correlations between changes in specific components (either quantitative or qualitative) and corresponding changes in photosynthetic performance.

Initially, therefore, it would seem that one of the more useful ways in which transgenic plants can be utilized is by altering the amounts of specific components of their photosynthetic apparatus. By then studying the resulting changes in the flow of metabolites, it should be possible to get an estimate of the contribution of particular catalytic components to overall control. Furthermore, this type of analysis can be carried out in a range of environmental conditions so that a comprehensive profile can be built up of the role of each component in the whole process. Central to this approach is the isolation and characterization of the coding sequences for components of the photosynthetic apparatus and subsequently the construction of transgenic plants with either increased or decreased amounts of specific components.

2.1. Selecting Photosynthetic Components for Manipulation

A series of criteria have been proposed for identifying components that control regulatory reactions in a metabolic pathway such as photosynthesis (Newsholme and Start, 1973; Rolleston, 1972). They are:

1. The enzyme should possess the appropriate regulatory properties, such as allosteric regulation or substrate cooperativity.
2. The enzyme should be present in amounts so that its activity is not greatly in excess of the required flux through the pathway and may in fact constitute a bottleneck.
3. The enzyme should catalyze a nonequilibrium or irreversible reaction.
4. There should be characteristic changes of the substrate concentration that are the reciprocal to the changes of flux.

When these criteria are applied to photosynthesis, a series of problems emerge (Stitt, 1989). First, the photosynthetic carbon reduction (PCR) pathway, for instance, contains several components that meet these criteria, such as Rubisco, chloroplastic fructose-1,6-bisphosphatase, sedoheptulose-1,7-bisphosphatase, and phosphoribulokinase. Second, the distinction between equilibrium and nonequilibrium reactions is somewhat arbitrary because it is not obvious when a reaction becomes irreversible. Also marked changes in individual reactants are generated in equilibrium reactions involving pairs of substrates or products. For example, during photosynthetic phosphoglycerate reduction, a decrease in the supply of ATP and/or NADPH will lead to an increase in the phosphoglycerate/triose phosphate ratio. This will, in turn, influence other reactions because high levels of phosphoglycerate inhibit Rubisco and phosphoribulokinase while low triose phosphate will restrict the activity of other enzymes involved in the regeneration of ribulose bisphosphate. Thus control actually involves the interplay between several components.

2.2. Kinetic Models of Photosynthesis

Considerable effort has been expended in developing models relating the activities of various components in the regulation of photosynthesis. Such models are of help in trying to establish priorities when trying to decide which components to concentrate on first. Obviously, an exhaustive analysis would require that all the components be studied, and it is to be hoped that eventually their contributions will all be evaluated. However, a large number of these are involved. In the PCR cycle alone, there are 13 enzymes acting on 16 metabolites in an intricate network of reactions. Furthermore, the dynamic and regulatory properties of such a complex system cannot be simply established by intuitive reasoning but require detailed analysis by mathematical modeling backed up by reliable experimentation.

In one of the first models (Farquhar et al., 1980), an attempt was made to integrate knowledge concerning the functioning of the biochemical components of photosynthetic carbon assimilation in C_3 plants. The conclusion was that there are two key parameters in trying to estimate the rate of CO_2 assimilation by leaves: the carboxylation capacity of the leaf and its electron transport capacity.

This model has received much attention and has been updated several times, for example by Gutschick (1984), Kirschbaum and Farquhar (1984), and Harley *et al.* (1985). However, this type of model has recently been challenged by Farazdaghi and Edwards (1988a,b), who suggested an alternative mechanistic model of photosynthesis based on the kinetic characteristics of Rubisco, which are now known in some detail. Light and CO_2 are considered the substrates for a hypothetical "photosynthetic enzyme." As a consequence, both substrates are depicted as colimiting the rate under most relevant physiological conditions. In contrast, the Farquhar *et al.* (1980) model assumes that coupling between the sequential reaction of photosynthesis is such that colimitation does not occur or is minimized. These differences in the models have important consequences for relating the capacity of individual steps (such as the activity of Rubisco detected in a leaf) to the observed rate of photosynthesis. Collatz *et al.* (1990) recently defended the Farquhar *et al.* (1980) model, claiming that the newer model is misleading and that it is often necessary to sacrifice "correctness" for "usefulness" in model building.

Other models specifically for the PCR cycle have been described (e.g., Giersch *et al.,* 1990). The analysis of Laisk *et al.* (1989) was concerned primarily with the question of whether oscillations that have been observed in photosynthesis can be related to phosphate cycling between cytosol and chloroplast stroma. However, the model is not limited to the analysis of oscillating behavior but is a mathematical description of the kinetics of the PCR cycle together with sucrose and starch synthesis. Therefore, it can be used for analysis of very different types of experimental data.

Woodrow (1986) also used a model of the cycle as a means of quantifying the degree to which the elements of the photosynthetic system contribute to flux control in order to obtain a quantitative hierarchy of importance of the constituent enzymes (see Section 3). His conclusion was that Rubisco may be a major controller of CO_2 fixation but that certain other enzymes may also play an important role. In particular, the light-activated enzymes of the stroma, such as fructose-1,6-bisphosphatase, sedoheptulose-1,7-bisphosphatase, and phosporibulokinase, which constitute a link between the electron-transport system and the PCR cycle, may be important in this context. Similar theoretical analyses have been made (Pettersson and Ryde-Pettersson, 1988, 1989). These authors have highlighted the fact that maximum PCR-cycle activity is controlled mainly by the catalytic capacities of ATP synthase and sedoheptulose-1,7-bisphosphase. Depending on the supply of phosphate external to the chloroplast, PCR cycle activity may be controlled by the phosphate translocator. Surprisingly, they conclude that Rubisco has no significant regulatory influence on the cycle under the conditions of light and CO_2 saturation and that, other than its substrate, it does not control the concentration of any cycle intermediate either.

A mathematical model of electron transport has also been formulated (Laisk and Walker, 1989). It reveals an absolute necessity for a control mechanism that dissipates the surplus light energy reaching photochemically active photosystem II (PS II) centers. At the same time, the excitation states of PS II and photosystem I (PS I) must be quantitatively balanced in such a way that overreduction or overenergization of the electron transport chain is avoided.

From this comparison of models it seems that there are a number of differences between them in what they predict for important controlling factors in photosynthesis. It is to be hoped that by flux control analysis, particularly with the aid of transgenic plants, these differences will be resolved and that a truly predictive model for photosynthesis will result that will be both "useful" and mechanistically "correct."

Models are also being formulated in which photosynthesis is considered in a wider context. For example, optimal leaf photosynthetic capacity was considered in terms of the utilization of the natural light environment. One model predicts the advantage of low photosynthetic capacity under light-limited conditions in habitats where leaf cost is high. Evans (1987) has, in fact, made a study of what happens in leaves of plants grown under different irradiances. He found that photosynthetic rate was proportional to both cytochrome f content and ATP synthase activity and that the ratio of these to one another stayed constant across all the growth irradiance treatments. The content of PS II reaction centers varied only to a small extent while the content of PS I reaction centers was unaltered. Adaptation to low irradiance was associated with a reduction in electron transport components and an increase in light-harvesting chlorophyll a/b protein such that the amount of chlorophyll per unit of thylakoid protein nitrogen increased. The converse was true with high irradiance. These results are, in general, consistent with the findings of many others (see Evans, 1987) and emphasize the importance of factors such as nitrogen cost rather than just relative photosynthetic rates (Takenaka, 1989).

3. CONTROL THEORY IN RELATION TO PHOTOSYNTHESIS

Carbon assimilation needs all the enzymes and electron-transport carriers engaged in this process. There are no secondary or alternative loops that may bypass blockages, if they occur. Therefore, every component in the system must be equally important. However, when considering the matter of limiting factors, one must ask whether some components of the system may have excess activity compared to others and are therefore less likely to exert control of the overall process. Alternatively, perhaps, the system cannot afford the luxury of overcapacity because, for example, of the usual shortage in amounts of nitrogenous compounds in plants.

The extent to which any one component may contribute to the overall process may be revealed if the activity of the component is changed. To do this, it may be reduced in amount or inactivated with an inhibitor or, alternatively, increased in amount or its activity improved. The consequences of any such change are then established by measuring the change in flux of metabolites or electrons in the system.

A formula relating these changes to the control exerted by the component on the overall process was devised by Kacser and Burns (1973) and further described by Kacser and Porteous (1987). It is as follows:

$$\frac{\delta Jm/Jm}{\delta E/E} = C^{Jm}{}_{E}$$

where C is the sensitivity coefficient or "control" coefficient of the component, Jm is the flux through the process, and E is the activity of a specified component. In a simple system, C may change from zero (when it exerts no control) to one (when it exerts full control). Usually in a multicomponent pathway, control is exerted by several components but to differing extents. Also, depending on their characteristics, different components may exert control at different stages when environmental conditions vary (Woodrow et al., 1990). As discussed above, this is likely to happen in photosynthesis.

One of the most difficult things to do in making these measurements is to vary the concentration or activity of a component while keeping all the others constant. Several methods have been used to achieve this. One is to use a mutant in which a particular activity is absent (Kruckeberg et al., 1989; Neuhaus et al., 1989). A second is to use selective inhibitors (Heber et al., 1988), and a third is to use transgenic plants in which the amount of a single component is changed (M. Stitt, unpublished results).

The main problem with mutants is that they usually result in the complete loss of a component. If the component is a crucial element of a pathway, then the mutant may be lethal. Therefore, one cannot expect to find mutants of many steps in photosynthesis, and those that do occur have pleiotrophic effects and prevent autotrophic growth.

Many difficulties are involved in their use of inhibitors. There is primarily the problem of specificity—few inhibitors are completely selective. Then there is the difficulty of application. Most inhibitors can be used only in cell-free systems because it is difficult to deliver them satisfactorily to their targets (see Heber et al., 1988).

The use of transgenic plants potentially offers the particular advantage of being able to increase and decrease the quantity óf a particular component. An increase is achieved by inserting more copies of its coding sequence into plants under the control of a powerful promoter. Decreases are achieved using constructs that produce antisense RNA (Rodermel et al., 1988), ribozyme (Cotten, 1990), or the like, which result in reduced amounts of the component being synthesized.

One problem with this approach is that alteration of one component could affect many others, especially if they are physically associated in any way. Certainly this is true for the core components of the thylakoid complexes, where loss of one component may result in abolition of the entire complex (Erickson *et al.*, 1986). Also, it takes a long time to isolate a particular coding sequence, characterize it, and then reinsert it into a plant in an appropriate construction. Such plants then have to be exhaustively characterized biochemically to ensure that the construct is being expressed and is having the desired effect and that the test plants are genetically uniform. Despite these difficulties, the transgenic plant approach appears to offer appreciable advantages in the measurement of control coefficients.

4.   ISOLATION AND MANIPULATION OF CODING SEQUENCES

A prerequisite to the manipulation of any protein component of the photosynthetic machinery is that one should have isolated and characterized its coding sequence. In most instances, this is most easily achieved if the protein itself has been isolated and antibodies raised to it. These can then be used to screen expression libraries or to see whether the products of an *in vitro* transcription–translation system contains the antigens sought. The whole process is, of course, complicated by the fact that some of the proteins are coded for by the nuclear genome and others by the chloroplast genome and therefore have to be treated quite differently.

4.1.   Location and Alteration of Genes for Chloroplast Proteins

Current knowledge of the complete sequences of several chloroplast genomes and partial sequence data for many more makes it relatively easy now to isolate specific coding sequences from any higher-plant chloroplast DNA. Most of the core components of the thylakoid, ATP synthase, and cytochrome b_6/f complexes are chloroplast encoded (see Table I) and have been studied in detail. However, there are still a number of unidentified coding sequences in chloroplast DNA that may also code for photosynthetic components but are as yet unrecognized (Ohyama *et al.*, 1988).

The literature relating to each gene referred to in Table I is cited by Hallick (1989).

Until fairly recently, it was far easier to isolate the coding sequences for chloroplast-encoded rather than nuclear-encoded chloroplast proteins. However, this is no longer the case. Using cDNA expression libraries, the coding sequence for most known nuclear-encoded thylakoid proteins involved in photosynthesis have now been isolated, and the number of similar sequences isolated for stromal enzymes involved in this process is also rapidly increasing (Raines *et al.*, 1990).

As a direct consequence of these studies, we now know far more than we

Table I
Designation and Location of Genes for Thylakoid and Polypeptides Involved in Photosynthesis

Gene	Location	Gene product
Photosystem I (PS I)		
psa	ctDNA	P700 apoprotein, subunit 1a
psaB	ctDNA	P700 apoprotein, subunit 1b
psaC	ctDNA	9-kDa Fe-S polypeptide
psaD	nucDNA	Ferredoxin-binding, subunit III
psaE	nucDNA	18–20 kDa subunit IV
psaF	nucDNA	Plastocyanin-binding subunit III
psaG	nucDNA	P35
psaH	nucDNA	10–12 kDa subunit VI
psaI	ctDNA	PSI-I polypeptide
psaJ	ctDNA	PSI-J polypeptide
psaK	nucDNA	PSI-K polypeptide ("P37")
Photosystem II (PS II)		
psbA	ctDNA	D_1-reaction center polypeptide
psbB	ctDNA	"CP47" chlorophyll apoprotein
psbC	ctDNA	"CP43" chlorophyll apoprotein
psbD	ctDNA	D_2-reaction center polypeptide
psbE	ctDNA	Cytochrome b_{559} α-subunit
psbF	ctDNA	Cytochrome b_{559} β-subunit
psbH	ctDNA	10-kDa phosphoprotein
psbI	ctDNA	4.8-kDa I-polypeptide
psbJ	ctDNA	(Reserved)
psbK	ctDNA	3.9-kDa K-polypeptide
psbL	ctDNA	PSII-L polypeptide
psbM	ctDNA	PSII-M polypeptide
psbN	ctDNA	PSII-N polypeptide
psbO	ctDNA	33-kDa polypeptide of O.E.C.[a]
psbP	ctDNA	23-kDa polypeptide of O.E.C.
psbQ	ctDNA	16-kDa polypeptide of O.E.C.
psbR	ctDNA	10-kDa polypeptide
Photosynthetic electron transport		
petA	ctDNA	Cytochrome f apoprotein
petB	ctDNA	Cytochrome b_6 apoprotein
petC	nucDNA	Rieske Fe-S polypeptide, subunit III
petD	ctDNA	Subunit IV
petE	nucDNA	Plastocyanin
petF	nucDNA	Ferredoxin
petG	ctDNA	"*orf*37" gene product, subunit V
petH	nucDNA	Ferredoxin-NADPH reductase(FNR)
petI	nucDNA	Flavodoxin
Genes for thylakoid ATP synthase components		
atpA	ctDNA	α-subunit F_1
atpB	ctDNA	β-subunit F_1

(continued)

Table I (*Continued*)

Gene	Location	Gene product
atpC	nucDNA	γ-subunit F_1
atpD	nucDNA	δ-subunit F_1
atpE	ctDNA	ϵ-subunit F_1
atpF	ctDNA	Subunit I F_0
atpG	nucDNA	Subunit II F_0
atpH	ctDNA	Subunit III F_0
atpI	ctDNA	Subunit IV F_0

[a]O.E.C. = oxygen-evolving complex

did before about the primary structure of these proteins, and from a mechanistic point of view, photosynthetic studies have benefitted enormously as a result. What we are now faced with is trying to use these sequences to make interesting transformants.

The actual technology of doing so is at least straightforward as far as certain higher plants are concerned and for nuclear genes. However, the range of plants that can be transformed easily is still relatively restricted, so one does not have a free choice as to which one can be manipulated. The constraints apply particularly to the cereals, in which there has been only limited success (Potrykus, 1990).

4.2. Plastid Transformation

As far as the genes encoded in the chloroplast DNA are concerned, these have been much more difficult to alter by transformation. The first report of chloroplast transformation was the result of cocultivation of *Nicotiana tobacum* and *Agrobacterium* (De Block *et al.*, 1985). Some of the transformants obtained showed maternal inheritance of the introduced chloramphenicol acetyltransferase activity. Although this activity was localized in the chloroplast, analysis of the integrated DNA indicated that it was not stably maintained in the chloroplast genome.

Because the instability of the introduced DNA was thought to be one of the major problems in getting chloroplast transformation, more sophisticated vectors were devised. Chloroplast promoters were fused to selection markers to ensure a high level of expression (van Grinsven *et al.*, 1986). A further step was to include in the vectors plastid DNA sequences that would allow homologous recombination without disturbing functional regions. However, such constructs when introduced by *Agrobacterium* cocultivation only become integrated into the nuclear DNA (Cornelissen *et al.*, 1987).

Hope has recently been revived in this area by reports of chloroplast genome transformation in the alga *Chlamydomonas rheinhardtii*. Transforming DNA was introduced into these organelles using microprojectiles that were fired at high velocity into the cells (Boynton *et al.*, 1988; Blowers *et al.*, 1989). Similar transformation experiments have been carried out by J-D. Rochaix and his colleagues (Department of Molecular Biology, University of Geneva, Switzerland). However, the chloroplast of *Chlamydomonas* has some unique properties. This organism has a single, large chloroplast and there are several well-characterized mutants of the chloroplast DNA. Higher plants have many much smaller chloroplasts, their chloroplasts each have many copies of their genome, and there are only a few known mutants in their organelle DNA that could be rescued. Furthermore, there is a lack of chloroplast-specific selection markers. These must be dominant and confer resistance to herbicides or antibiotics that inhibit chloroplast processes. This would make it possible to select for the few organelles that would be transformed in a large untransformed population.

Other methods of DNA delivery to organelles have been tried as well. These include microinjection of the DNA into protoplasts and then punching holes through the chloroplast membranes with a laser. Weber *et al.* (1989) were able to recover putative herbicide-resistant calli in this way. However, during regeneration the cells lost their resistant phenotype. Attempts are also being made to transfer intact chloroplasts by microinjection (see Haring and De Block, 1990). If this proves successful, then it may be possible to infiltrate the transforming DNA into the chloroplasts *en route*. A further possibility might be to chemically link a peptide sequence to one end of the DNA, which would target it into chloroplasts. Such a procedure has been used successfully to get DNA into mitochondria (Vestweber and Schatz, 1989). So far this way of getting DNA organelles has only been demonstrated to work for small, 24-bp DNA oligonucleotides.

In the case of mitochondria, it is possible to introduce genes from this organelle into the nucleus. Provided that the gene transferred in this way is spliced to a sequence that would code for a signal peptide which would target its product to the organelle, it can continue to function effectively in its new location (Nagley and Devenish, 1989). A similar approach has been attempted using the coding sequence for the large subunit of Rubisco, but so far without success. However, it has been shown by Gatenby *et al.* (1988) that this protein, synthesized *in vitro* with a transit peptide, is taken up and processed to its right size when taken up by chloroplasts. It would appear therefore that, although we are not yet sure of why certain genes are in the chloroplast genome and others are in the nucleus, we can be optimistic that it will be possible to change in this way the location of some genes so that they can be manipulated more easily.

5. PROGRESS IN THE MANIPULATION OF SPECIFIC GENES

5.1. Rubisco

Of all the proteins that are components of the photosynthetic apparatus, Rubisco has received by far the most attention with a view to changing its properties in order to increase its efficiency (see Andrews and Lorimer, 1987; Pierce, 1988; Gutteridge, 1990). The main hope is to find ways of improving the ratio between its carboxylation and oxygenation activities so that less of the photosynthate is lost by way of photorespiration in C_3 plants. However, it would also be advantageous if the enzyme could be made kinetically more efficient because at present it may have to make up more than 50% of the total soluble protein of leaves so that there is enough of it.

Unfortunately, several problems are involved in the manipulation of this enzyme. One is that its large subunit carries the catalytic site and this subunit is chloroplast encoded. Therefore, it would be very difficult to alter it *in vivo* even if we knew what changes to make. Another difficulty is that it is as yet not possible for the higher plant enzyme to be assembled properly *in vitro* or in a bacterial expression system. This makes it impossible to change its sequence by *in vitro* mutagenesis and study the consequences of making such a change on its properties. A third difficulty is that the enzyme from higher plants is the most efficient that has been detected so far with respect to the relative rate at which it carries out the carboxylation and oxygenation reactions (Parry *et al.*, 1987). Therefore, there is little indication as yet as to how the efficiency of the enzyme could be improved in this respect in plants of greatest economic interest. Also, it has not yet become clear from crystallographic studies what structural features of Rubisco contribute to determining the specificity of the reactions it catalyzes. The catalytic site itself appear to be highly conserved in structure, especially with regard to the amino acids that bind the substrate (Gutteridge, 1990), yet Rubiscos from various sources may differ appreciably in their kinetic properties. One of the most promising approaches to the study of the regulation of its activities results from work with *Chlamydomonas* mutants (Chen and Spreitzer, 1989). Alteration of particular amino acids not involved in substrate binding has been shown to be important in determining activity.

We have recently been attempting to increase the amount of enzyme present by placing a coding sequence for the small subunit under the control of a strong promoter (C. A. Raines, M. Longstaff, S. Payne, and T. A. Dyer, unpublished work). This was done in the hope that if the amount of this subunit was increased, the overall amount of the enzyme would increase also. We have some evidence to suggest this may be possible. Further evidence to suggest that the total amount of this enzyme can be manipulated via the small subunit comes from the work of Rodermel *et al.* (1988). By placing an antisense construction of this in a plant by transformation, it was possible to substantially reduce the amount of Rubisco present. This had a pronounced phenotypic effect.

5.2. Other PCR-Cycle Components

Recently, the coding sequences for several of the enzymes that catalyze the reactions of the PCR cycle have been isolated and characterized. These include the sequences of phosphoglycerate kinase (Longstaff *et al.*, 1989), glyceraldehyde phosphate dehydrogenase (Shih *et al.*, 1986; Martin and Cerff, 1986; Brinkmann *et al.*, 1989), fructose-1,6-bisphosphatase (Raines *et al.*, 1988), and phosphoribulokinase/ribose-5-phosphate kinase (Raines *et al.*, 1989; Milanez and Mural, 1988; Roesler and Ogren, 1988, 1990). We also have a putative coding sequence for sedoheptulose-1,7-bisphosphatase (C. A. Raines, J. C. Lloyd, and T. A. Dyer, unpublished work). In addition, the coding sequence of the phosphate translocator has been determined (Flügge *et al.*, 1989). Although this protein is not directly involved in the PCR itself, its activity directly affects its operation. The coding sequences for all these enzymes are therefore now available for manipulation in transgenic plants and aspects of this work are now in progress (J. Knight, A. Loins, J. C. Gray, and T. A. Dyer, unpublished work). The results of studies such as this will be of particular interest in respect to the fructose-1,6-bisphosphatase, sedoheptulose-1,7-bisphosphatase, and phosphoribulokinase, as all these enzymes are highly regulated and catalyze reactions that are essentially irreversible. The phosphate translocator may also have an important regulatory role, as discussed above (Section 2.2).

5.3. Thylakoid Proteins

Although most of the coding sequences for known thylakoid proteins involved in photosynthesis have now been isolated, relatively few of them have been manipulated in any way. This is, in part, because so many of them are chloroplast encoded (see Table I) and also because they are very highly conserved in sequence and therefore there is little variation in their structure that is interesting from a functional point of view.

The protein that has received most attention with a view to altering its properties is the PS II core protein, D_1. This is because a single amino acid change in it sequence renders a plant resistant to atrazine-type herbicides. This is apparently because of decreased biding of these herbicides, which usually displace plastoquinone from the site in which it receives electrons in the PS II complex (Michel *et al.*, 1986). However, the resulting plants are less viable than the wild types in most instances (Radosevich *et al.*, 1982). Other introduced mutations of D_1 protein and its pair, the D_2 protein, have led to significant advances in the basic understanding of photosynthetic processes (Debus *et al.*,

1988; Vermaas *et al.*, 1988). These mutations were made in the proteins of a cyanobacterium rather than of a higher plant because of the relative ease with which such alterations can be made in these organisms.

The main protein complexes of thylakoids that change in amount in plants grown in different conditions are ATP synthase, the cytochrome b_6/f complex, and the light-harvesting chlorophyll a/b proteins of the photosystems (see Evans, 1987; De la Torre and Burkey, 1990). These are therefore obvious targets for manipulation. The main problem in achieving this with the light-harvesting proteins is that in most plants they are coded for by multigene families and this might complicate attempts to alter them. The ATP synthase and cytochrome complex have both chloroplast and nuclear-encoded components. It might be quite easy to reduce the total amount of each of these complexes by introducing an antisense construct to one of the nuclear-encoded components, as was done for Rubisco. A further possibility is to put particular nuclear-coding sequences under the control of different promoters so that their expression is changed, thus preventing the usual type of acclimation from occurring when growth conditions are changed (Prioul and Reyes, 1987; De la Torre and Burkey, 1990; Sage *et al.*, 1989; Yelle *et al.*, 1989).

6. RELATIONSHIP BETWEEN PHOTOSYNTHESIS AND YIELD

It appears that in most cases there is no association, and in some cases, even a negative association, between rate of photosynthesis per unit area of individual leaves and the yield of field and foliage crops (Evans, 1975; Charles-Edwards, 1978; Ozbun, 1978). From this it has been concluded that yield is determined largely by genetic variation in partitioning of photosynthate (Gifford and Evans, 1981). Other studies suggest that yields are limited by other factors, such as leaf area index, leaf orientation, and leaf aging, which have a major influence on radiation harvesting. These factors may mask potential effects on yield due to variation in photosynthesis per leaf area (Nelson, 1988).

According to Nasyrov (1978), it should be possible genetically to increase leave photosynthesis and this would increase canopy photosynthesis. If sinks are carbohydrate limited and if the partitioning coefficients of the various sinks remain constant, then yield of both economic and noneconomic parts should increase proportionally. The concept has not been tested yet because of the lack of comparable plants that differ only in the photosynthetic capacity. However, it has been tested indirectly by manipulating environmental conditions during seed fill. Altering radiation (Satterlee and Koller, 1984) or CO_2 concentration (Hardman and Brun, 1971; Kramer, 1981; Kimball, 1983) changes leaf photosynthesis and influences economic yields.

When photosynthesis is increased by environmental manipulation, more reproductive sinks may be added to the plants or the number of grains per ear or kernel weight may be increased. With root crops such as sugar beet, late-season CO_2 enrichment increases both root size and sugar yield. The question remains, however, whether the number and activity of sinks is limiting or whether additional photosynthate stimulates the activity of sinks (Nelson, 1988).

6.1. Extrapolating Single Leaf to Canopy Photosynthesis

From what is known it seems that little genetic variation exists for quantum yield of photosynthesis, but stress conditions such as those induced by drought or low temperature can reduce quantum efficiency. The main photosynthetic processes that show genetic variation are apparently associated with the reactions of the carbon reduction cycle. If this is so, then genotype differences in photosynthesis may not be evident unless radiation exceeds some threshold value (Fig. 1). As a consequence, most crop plants situated in a canopy would be radiation limited until midmorning and then again in midafternoon. This would allow only a few hours for using genetic differences in photosynthetic rate. Furthermore, in dicots at least, the majority of the radiation is absorbed in the upper portion of the canopy (Loomis *et al.*, 1971), so only a few leaves in the upper part of the canopy would express the genetic differences in photosynthesis.

A model has been proposed to relate leaf photosynthesis and growth rate in a sward of C_3 grasses (Charles-Edwards, 1978). According to Charles-Edwards' calculations, in which several assumptions had to be made, increasing the leaf photosynthetic rate by 50% would increase the dry matter production rate by only 13% in a British summer with high radiation and by only 3% during the winter when radiation is low. Other researchers have concluded that about 40–50% of the genetic differences in leaf photosynthesis are realized in canopy photosynthesis even in high-light environments (Sinclair and de Wit, 1976).

A substantial increase in leaf photosynthesis would almost certainly require a much larger investment by the plant in proteins involved in this process. The possibility of our being able to improve the efficiency of the carbon reduction cycle by improving the kinetic properties of the individual enzymes involved seems at present to be rather remote, and therefore the cycle could only be speeded up by increasing the total amount of enzymes. We already know enough about the enzymes to suspect that even for one with a relatively large control coefficiency such as Rubisco, massive extra amounts of this protein would need to be synthesized in order for it to have any significant effect on carbon uptake. Selecting for higher Rubisco levels in breeding programs has been suggested as a breeding objective, but so far the results of such attempts have been disappointing (Secor *et al.*, 1982; Hobbs and Mahon, 1985). However, using laboratory-grown transgenic plants, we have attempted to achieve the same objective with a more encouraging outcome (T. A. Dyer, C. A. Raines, and M. Longstaff, unpublished work). Nevertheless it seems doubtful whether such an increase in a field-grown plant would be of great benefit. As nitrogen seems to be one of the main limiting constituents in the environment, investment of a large proportion of it in Rubisco might be offset by reduced amounts of other enzymes.

In view of the fact that increased efficiency of photosynthesis at low light intensities is likely to be of greater significance than those at higher intensities, probably we should not give up trying to improve the quantum efficiency of plants despite the formidable difficulties involved. These include the fact that plants may be working near the theoretical maximum anyway, leaving little scope for improvement. However, some reports suggest that the cytochrome b_6/f complex and ATP synthase may be limiting (Heber *et al.*, 1988), so boosted levels of these might be beneficial. Furthermore, the balance in amount between light harvesting and core complex seems to be much affected by conditions in which plants are grown (Evans, 1987), so it may be possible to manipulate this ratio to good effect. Another avenue that appears to be worth exploring is to minimize the adverse effects of stress conditions that lead to reduced quantum efficiency. Certainly there must be vast genetic variation controlling such a parameter that could be exploited. However, much molecular biology remains to be done in order to identify the components involved so that their coding sequences can be isolated and manipulated.

6.2. Duration of Leaf Photosynthesis

One of the most promising ways of trying to increase biomass production is to try to delay the senescence of leaves. The feasibility of doing this has already been demonstrated by Frey and his colleagues. They were able to transfer a "stay-green" character from *Avena sterilis* into a breeding line of oats (*Avena sativa*). This resulted in only a marginal increase in maximum photosynthetic rate (Brinkman and Frey, 1978), but the progeny of the cross had a higher growth rate (Takeda and Frey, 1976) and were higher yielding by 15–20%, largely owing to extended leaf duration.

Fortunately, it seems that it may be relatively easy to influence the longevity of leaves by manipulation of cytokinins in leaves. This hormone appears to be the major senescence-retarding constituent in plants and its role in leaves is particularly important. A wide variety of studies (for example, Lamattina *et al.*, 1987) have shown that leaf senescence is usually correlated with a decrease in cytokinin levels, and indeed it is possible to delay leaf senescence by the external application of this hormone.

We have recently shown (C. M. Smart, S. Schofield, M. W. Bevan, and T. A. Dyer, unpublished work) that it is possible to delay leaf senescence in transgenic plants that have inserted into them under the control of a heat shock promoter the *ipt* gene of *Agrobacterium tumefaciens*. The product of this gene catalyzes the transfer of an isopentenyl group to adenine in the formation of one of the intermediates in the synthesis of a range of cytokinins (Morris, 1986). If the expression of this gene can be appropriately regulated with a senescence-specific promoter so that it affects only leaf senescence and does not upset normal development, then we stand a good chance of being able to regulate the decline of photosynthesis in leaves. Nelson (1988) cites delayed senescence of oats as one of the most promising methods of increasing yield, especially if it can be achieved in the grain filling period.

6.3. Photorespiration

Another way of trying to increase the photosynthesis of crops with C_3-type metabolism would be to reduce or eliminate photorespiration. This is a process by which a large proportion of the photosynthate may be lost. It stems from the fact that Rubisco catalyzes oxygenation as well as carboxylation of its substrate ribulose 1,5-bisphosphate. The ratio of these reactions, known as the specificity factor, does differ in the enzymes from different sources but is most favorable in the enzyme of higher plants (Parry *et al.*, 1987) so there is little indication yet of how it could be improved. C_4-type plants are able to overcome this deficiency in their Rubisco by increasing the CO_2 concentration in the cells in which the enzyme is active using a CO_2 pump. Possibly this mechanism could be introduced into some C_3 plants too, although it may be necessary to change the expression of a whole battery of genes in order to do so (Moore, 1982). Alternatively, there may be ways of diverting the oxygenated derivatives of ribulose-1,5-bisphosphate back into normal metabolism before the rerelease of CO_2 takes place, but ways of doing this have yet to be identified.

One benefit from the increase in level of CO_2 that is currently taking place (White, 1990) is that it will decrease photorespiration (Keys, 1986) with an increase in yield as a consequence (Kimball, 1983). However, plants acclimate when grown in high CO_2 for long periods by having a reduction in Rubisco levels, thus offsetting the benefits of reduced photorespiration (Sage *et al.*, 1989). Certainly it would be interesting to study the performance of plants in which this acclimation was prevented perhaps by inserting genes into them to determine the amount of Rubisco present that was not down-regulated in high CO_2.

7. CONCLUSION

The overall impression after consideration of the available information is that photosynthesis and its control is an immensely complicated process. Therefore, attempts to find easy explanations of how it contributes to plant productivity are likely to be unsuccessful. Because of this, there seems to be ample justification for simply changing, both qualitatively and quantitatively, various components of the photosynthetic apparatus and then evaluating the effect of this. This will at least provide additional parameters by which models can be evaluated to see how faithfully they can predict what in fact happens when such components are altered. Furthermore, they may result in plants with desirable properties without necessarily providing a rigorous explanation of how a particular effect is achieved. However, biochemists who are attracted by the prospect of making transgenic plants with altered properties should be under no illusions as to how much hard work is required to make the appropriate plants and to then determine biochemically, physiologically, and agronomically the significance of the changes made.

8. REFERENCES

Andrews, T. J., and Lorimer, G. H., 1987, Rubisco: Structure, mechanisms and prospects for improvement, in *The Biochemistry of Plants*, Vol. 10 (M. D. Hatch, ed.), pp. 131–218, Academic Press, Orlando, FL.

Austin, R. B., 1989, Genetic variation in photosynthesis, *J. Agr. Sci. Camb.* **112**:287–294.

Blowers, A. D., Bogorad, L., Shark, K. B., and Sanford, J. C., 1989, Studies on *Chlamydomonas* chloroplast transformation: foreign DNA can be stably maintained in the chromosome, *Plant Cell* **1**:123–132.

Boynton, J. E., Gillham, N. W., Harris, E. H., Hosler, J. P., Johnson, A. M., Jones, A. R., Randolf-Anderson, B. L., Robertson, D., Klein, T. M., Shark, K. B., and Sanford, J. C., 1988, Chloroplast transformation in *Chlamydomonas* with high velocity microprojectiles, *Science* **240**:1534–1538.

Brinkman, M. A., and Frey, K. J., 1978, Flag leaf physiological analysis of oat isolines that differ in grain yield from their recurrent parents, *Crop Sci.* **18**:69–73.

Brinkmann, H., Cerff, R., Salomon, M., and Soll, J., 1989, Cloning and sequence analysis of cDNAs encoding the cytosolic precursors of subunits GapA and GapB of chloroplast glyceraldehyde-3-phosphate dehydrogenase from pea and spinach, *Plant Mol. Biol.* **13**:81–94.

Charles-Edwards, D. A., 1978, An analysis of the photosynthesis and productivity of vegetable crops in the United Kingdom, *Ann. Bot.* **42**:717–731.

Chen, Z., and Spreitzer, R. J., 1989, Chloroplast intragenic suppression enhances the low CO_2/O_2 specificity of mutant ribulose bisphosphate carboxylase/oxygenase, *J. Biol. Chem.* **264**:3051–3053.

Collatz, G. J., Berry, J. A., Farquhar, G. D., and Pierce, J., 1990, The relationship between the Rubisco reaction mechanism and models of photosynthesis, *Plant Cell Environ.* **13**:219–225.

Cornelissen, M. J., De Block, M., Van Montagu, M., Leemans, J., Schreier, P. H., and Schell, J., 1987, Plastid transformation. A progress report, in *Plant DNA Infectious Agents* (J. Hohn and J. Schell, eds.), pp. 311–321, Springer-Verlag, Vienna.

Cotten, M., 1990, The *in vivo* application of ribozymes, *TIBTECH* **8**:174–178.

De Block, M., Schell, J., and Van Montagu, M., 1985, Chloroplast transformation by *Agrobacterium tumefaciens*, *EMBO J.* **6**:1367–1372.

De la Torre, W. R., and Burkey, K. O., 1990, Acclimation of barley to changes in light intensity: Photosynthetic electron transport activity and components, *Photosyn. Res.* **24**:127–136.

Debus, R. J., Barry, B. A., Sithole, I., Babcock, G. T., and McIntosh, L., 1988, Directed mutagenesis indicates that the donor to P^+_{680} in photosystem II is tyrosine-161 of the D1 polypeptide, *Biochemistry* **27**:9071–9074.

Erickson, J. M., Rahire, M., and Malnöe, P., Girard-Bascou, J., Pierre, Y., Bennoun, P., and Rochaix, J-D., 1986, Lack of the D2 protein in a *Chlamydomonas* reinhardtii *psb*D mutant affects photosystem II stability and D1 expression, *EMBO J.* **5**:1745–1754.

Evans, J. R., 1987, The relationship between electron transport components and photosynthetic capacity in pea leaves grown at different irradiances, *Aust. J. Plant Physiol.* **14**:157–170.

Evans, L. T., 1975, *Crop Physiology*, Cambridge University Press, New York.

Farazdaghi, H., and Edwards, G. E., 1988a, A mechanistic model for photosynthesis based on the multisubstrate ordered reaction of ribulose-1,5-bisphosphate carboxylase, *Plant Cell Environ.* **11**:789–798.

Farazdaghi, H., and Edwards, G. E., 1988b, A model for photosynthesis and photorespiration in C_3 plants based on the biochemistry and stoichiometry of the pathways, *Plant Cell Environ.* **11**:799–809.

Farquhar, G. D., von Caemmerer, S., and Berry, J. A., 1980, A biochemical model of photosynthetic CO_2 assimilation in leaves of C_3 species, *Planta* **149**:78–90.

Flügge, U. I., Fischer, K., Gross, A., Sebald, N., Lottspeich, F., and Eckerskorn, C., 1989, The triose phosphate-3-phosphoglycerate-phosphate translocator from spinach chloroplasts: nucleotide sequence of a full-length cDNA clone and import of the *in vitro* synthesized precursor protein into chloroplasts, *EMBO J.* **8**:39–46.

Gatenby, A. A., Lubben, T. H., Ahlquist, P., and Keegstra, K., 1988, Imported large subunits of ribulose bisphosphate carboxylase/oxygenase but not imported β-ATP synthase subunits are assembled into holoenzyme in isolated chloroplasts, *EMBO J.* **7**:1307–1314.

Giersch, G., Lämmel, D., and Farquhar, G., 1990, Control analysis of photosynthetic CO_2 fixation, *Photosyn. Res.* **24**:151–165.

Gifford, R. M., and Evans, L. T., 1981, Photosynthesis, carbon partitioning and yield, *Annu. Rev. Plant Physiol.* **32**:485–509.

Gutschick, V. P., 1984, Photosynthetic model of C_3 leaves incorporating CO_2 transport, propagation of radiation and biochemistry. I Kinetics and their parameterization, *Photosynthetica* **18**:549–568.

Gutteridge, S., 1990, Limitations of the primary events of CO_2 fixation in photosynthetic organisms: the structure and mechanism of Rubisco, *Biochim. Biophys. Acta* **1015**:1–14.

Hallick, R. B., 1989, Proposals for the naming of chloroplast genes. II. Update of the nomenclature of genes for thylakoid membrane polypeptides, *Plant Mol. Biol. Rep.* **7**:266–275.

Hardman, L. L., and Brun, W. A., 1971, Effect of atmospheric carbon dioxide enrichment at different developmental stages on growth and yield components of soybeans, *Crop Sci.* **11**:886–888.

Haring, M. A., and De Block, M., 1990, New roads towards chloroplast transformation in higher plants, *Physiol. Plant* **79**:218–220.

Harley, D. C., Weber, J. A., and Gates, D. M., 1985, Interactive effects of light, leaf temperature, CO_2 and O_2 on photosynthesis in soy bean, *Planta* **165**:249–263.

Heber, U., Neimanis, S., and Dietz, K-J., 1988, Fractional control of photosynthesis by the Q_B protein, the cytochrome f/b_6 complex and other components of the photosynthetic apparatus, *Planta* **173**:267–274.

Hobbs, S. L. A., and Mahon, J. D., 1985, Inheritance of chlorophyll content, ribulose-1,5-bisphosphate carboxylase activity, and stomatal resistance in peas, *Crop Sci.* **25**:1031–1034.

Kacser, H., and Burns, J. A., 1973, The control of flux, *Sym. Soc. Exp. Biol.* **27**:67–104.

Kacser, H., and Porteous, J. W., 1987, Control of metabolism: What do we have to measure? *TIBS* **12**:5–14.

Keys, A. J., 1986, Rubisco: Its role in photorespiration, in *Proceedings of a Conference on Ribulose Bisphosphate Carboxylase-Oxygenase* (R. J. Ellis and J. C. Gray, eds.), pp. 325–336, *Philos. Trans. R. Soc. Lond. B, London*.

Kimball, B. A., 1983, Carbon dioxide and agricultural yield: an assemblage and analysis of 430 prior observations, *Agron. J.* **75**:779–788.

Kirschbaum, M. U. F., and Farquhar, G. D., 1984, Temperture dependence of whole-leaf photosynthesis in *Eucalyptus paciflora* Sieb ex Spreng, *Aust. J. Plant Physiol.* **11**:519–538.

Kramer, P. J., 1981, Carbon dioxide concentration, photosynthesis, and dry matter production, *BioScience* **31**:29–33.

Kruckeberg, A. L., Neuhaus, H. E., Feil, R., Gottlieb, L. D., and Stitt, M., 1989, Decreased-activity mutants of phosphoglucoisomerase in the cytosol and chloroplast of *Clarkia xantiana*, *Biochem. J.* **261**:457–467.

Laisk, A., and Walker, D. A., 1989, A mathematical model of electron transport. Thermodynamic necessity for photosystem II regulation: "Light stomata," *Proc. R. Soc. Lond. B* **237**:417–444.

Laisk, A., Eichelmann, H., Oja, V., Eatherall, A., and Walker, D. A., 1989, A mathematical model of the carbon metabolism in photosynthesis. Difficulties in explaining oscillations by fructose 2,6-bisphosphate regulation, *Proc. Roy. Soc. Lond. B* **237**:389–415.

Lamattina, L., Anchoverri, V., Conde, R. D., and Lezica, R. P., 1987, Quantification of the kinetic effects on protein synthesis and degradation in senescing wheat leaves, *Plant Physiol.* **83**:497–499.

Longstaff, M., Raines, C. A., McMorrow, E. M., Bradbeer, J. W., and Dyer, T. A., 1989, Wheat phosphoglycerate kinase: Evidence for recombination between the genes for the chloroplastic and cytosolic enzymes, *Nucleic Acids Res.* **17**:6569–6580.

Loomis, R. S., Williams, W. A., and Hall, A. E., 1971, Agricultural productivity, *Annu. Rev. Plant Physiol.* **22**:431–468.

Martin, W., and Cerff, R., 1986, Prokaryotic features of a nuclear-encoded enzyme, cDNA sequences for chloroplast and cytosolic glyceraldehyde-3-phosphate dehydrogenases from mustard (*Sinapsis alba*), *Eur. J. Biochem.* **159**:323–331.

Michel, H., Epp, O., and Deisenhofer, J., 1986, Pigment-protein interactions in the photosynthetic reaction centre from *Rhodopseudomonas viridis*, *EMBO J.* **5**:2445–2451.

Milanez, S., and Mural, R. J., 1988, Cloning and sequencing of cDNA encoding the mature form of phosphoribulokinase from spinach, *Gene* **66**:55–63.

Moore, P. D., 1982, Evolution of photosynthetic pathways in flowering plants, *Nature* **295**:647–648.

Morris, R. O., 1986, Genes specifying auxin and cytokinin biosynthesis in phytopathogens, *Annu. Rev. Plant Physiol.* **37**:509–538.

Nagley, P., and Devenish, R. J., 1989, Leading organellar proteins along new pathways: The relocation of mitochondrial and chloroplast genes to the nucleus, *TIBS* **14**:31–35.

Nasyrov, Y. S., 1978, Genetic control of photosynthesis and improving crop productivity, *Plant Physiol.* **29**:215–237.

Nelson, C. J., 1988, Genetic associations between photosynthetic characteristics and yield: Review of the evidence, *Plant Physiol. Biochem.* **26**:543–554.

Neuhaus, H. E., Kruckeberg, A. L., Feil, R., and Stitt, M., 1989, Reduced-activity mutants of phosphoglucose isomerase in the cytosol and chloroplast of *Clarkia xantiana*. II. Study of the mechanisms which regulate photosynthate partitioning, *Planta* **178**:110–122.

Newsholme, E. A., and Start, C., 1973, *Regulation in Metabolism*, Wiley, New York.

Ohyama, K., Kochi, T., Sano, T., and Yamada, Y., 1988, Newly identified groups of genes in chloroplasts, *TIBS* **13**:19–22.

Ozbun, J. L., 1978, Photosynthetic efficiency and crop production, *Hort. Sci.* **13**:678–679.

Parry, M. A. J., Schmidt, C. N. G., Cornelius, M. J., Millard, B. N., Burton, S., Gutteridge, S.,

Dyer, T. A., and Keys, A. J., 1987, Variations in properties of ribulose-1,5-bisphosphate carboxylase from various species related to differences in amino acid sequences, *J. Exp. Bot.* **38:**1260–1271.

Pettersson, G., and Ryde-Pettersson, U., 1988, A mathematical model of the Calvin photosynthesis cycle, *Eur. J. Biochem.* **175:**661–672.

Pettersson, G., and Ryde-Pettersson, U., 1989, Dependence of the Calvin cycle on kinetic parameters for the interaction of non-equilibrium cycle enzymes with their substrates, *Eur. J. Biochem.* **186:**683–687.

Pierce, J., 1988, Prospects for manipulating the substrate specificity of ribulose bisphosphate carboxylase/oxygenase, *Physiol. Plant.* **72:**690–698.

Potrykus, I., 1990, Gene transfer to cereals: An assessment, *BioTechnology* **8:**535–542.

Prioul, J. L., and Reyss, A., 1987, Acclimation of ribulose bisphosphate carboxylase and mRNAs to changing irradiance in adult tobacco leaves. Differential expression in LSU and SSU mRNA, *Plant Physiol.* **84:**1238–1243.

Radosevich, S. R., Sims, J., and Holt, J. S., 1982, Physiological responses and fitness of susceptible and resistant weed biotypes to triazine herbicides, in *Herbicide Resistance in Plants* (H. M. le Baron and J. Gressel, eds.), pp. 163–183, Wiley, New York.

Raines, C. A., Lloyd, J. C., Longstaff, M., Bradley, D., and Dyer, T. A., 1988, Chloroplast fructose-1,6-bisphosphatase: The product of a mozaic gene, *Nucleic Acids Res.* **16:**7931–7942.

Raines, C. A., Longstaff, M., Lloyd, J. C., and Dyer, T. A., 1989, Complete coding sequence of wheat phosphoribulokinase: Developmental and light-dependent expression of the mRNA, *Mol. Gen. Gent.* **220:**43–48.

Raines, C. A., Lloyd, J. C., and Dyer, T. A., 1990, Molecular biology of the C_3 photosynthetic carbon reduction cycle, *Photosyn. Res.* **27:**1–14.

Rodermel, S. R., Abbott, M. S., and Bogorad, L., 1988, Nuclear–organelle interactions: Nuclear antisense gene inhibits ribulose bisphosphate carboxylase enzyme levels in transformed tobacco plants, *Cell* **55:**673–681.

Roesler, K. R., and Ogren, W. L., 1988, Nucleotide sequence of spinach cDNA encoding phosphoribulokinase, *Nucleic Acids Res.* **16:**7192.

Roesler, K. R., and Ogren, W. L., 1990, *Chlamydomonas reinhardtii* phosphoribulokinase, *Plant Physiol.* **93:**188–193.

Rolleston, F. S., 1972, A theoretical background to the use of measured intermediates in the study of the control of intermediary metabolism, *Curr. Top. Cell Reg.* **5:**47–75.

Sage, R. F., Sharkey, T. D., and Seemann, J. R., 1989, Acclimation of photosynthesis to elevated CO_2 in five C_3 species, *Plant Physiol.* **89:**590–596.

Satterlee, L. D., and Koller, H. R., 1984, Response of soybean fruit respiration to changes in whole plant light and CO_2 environment, *Crop Sci.* **24:**1007–1010.

Secor, J., McCarty, D. R., Shibles, R., and Green, D. E., 1982, Variability and selection for leaf photosynthesis in advanced generations of soybeans, *Crop Sci.* **22:**255–259.

Shih, M-C., Lazar, G., and Goodman, H. M., 1986, Evidence in favour of the symbiotic origin of chloroplasts: Primary structure and evolution of tobacco glyceraldehyde-3-phosphate dehydrogenases, *Cell* **47:**73–80.

Sinclair, T. R., and de Wit, C. T., 1976, Analysis of the carbon and nitrogen limitations of soybean yield, *Agron. J.* **68:**319–324.

Stitt, M., 1989, Control of sucrose synthesis: estimation of free energy changes, investigation of the contribution of equilibrium and non-equilibrium reactions and estimation of elasticities and flux control coefficients, in *Techniques and New Developments in Photosynthesis Research* (J. Barber and R. Malkin, eds.), pp. 365–391, Plenum Press, New York.

Takeda, K., and Frey, K. J., 1976, Contributions of vegetative growth rate and harvest index to grain yield of progenies from *Avena sativa* × *A. sterilis* crosses, *Crop Sci.* **16:**817–821.

Takenaka, A., 1989, Optimal leaf photosynthetic capacity in terms of utilising a natural light environ-
    ment, *J. Theor. Biol.* **139:**517–529.
Van Grinsven, M. Q. J. M., Haring, M. A., De Haas, J. M., Nijkamp, H. J. J., and Kool, A. J.,
    1986, Identification, cloning and transfer of chloroplast genes of *Petunia hybrida,* in *Genetic
    Manipulation in Plant Breeding* (W. Horn, C. J. Jensen, W. Odenbach, and W. Schieder, eds.),
    pp. 859–866, Eucarpia, de Gruyter, Berlin.
Vermaas, W. F. J., Rutherford, A. W., and Hansson, Ö, 1988, Site-directed mutagenesis in pho-
    tosystem II of the cyanobacterium *Synechocystis* sp. PCC 6803: Donor D is a tyrosine residue in
    the D2 protein, *Proc. Natl. Acad. Sci. USA* **85:**8477–8481.
Vestweber, D., and Schatz, G., 1989, DNA-protein conjugates can enter mitochondria via the protein
    import pathway, *Nature* **338:**170–172.
Weber, G., Monajembashi, S., Greulich, K-O., and Wolfrum, J., 1989, Uptake of DNA in chlo-
    roplasts of *Brassica napus L* facilitated by UV-laser micro beam, *Eur. J. Cell Biol.* **49:**73–79.
White, R. M., 1990, The great climate debate, *Sci. Am.* **263:**18–25.
Woodrow, I. E., 1986, Control of the rate of photosynthetic carbon dioxide fixation, *Biochim.
    Biophys. Acta* **851:**181–192.
Woodrow, I. E., Ball, J. T., and Berry, J. A., 1990, Control of photosynthetic carbon dioxide fixation
    by the boundary layer, stomata and ribulose-1,5-bisphosphate carboxylase/oxygenase, *Plant
    Cell Environ.* **13:**339–347.
Yelle, S., Beeson, R. C., Trudel, M. J., and Gosselin, A., 1989, Acclimation of two tomato species
    to high atmosphere CO_2. II. Ribulose-1,5-bisphosphate carboxylase/oxygenase and phos-
    phoenolypyruvate carboxylase, *Plant Physiol.* **90:**1473–1477.

Chapter 11

Transfer RNA Involvement in Chlorophyll Biosynthesis

Gary P. O'Neill, Dieter Jahn, and Dieter Söll

1. INTRODUCTION

Porphyrin-containing molecules, such as hemes and chlorophylls, are key components of respiratory and photosynthetic metabolism. The tetrapyrrole rings of these molecules are formed via a branched biosynthetic pathway from the condensation of the important five-carbon intermediate 5-aminolevulinic acid (δ-aminolevulinic acid, ALA). The central role of ALA in porphyrin biosynthesis is underscored by the observation that it is a major point for regulation of heme and chlorophyll synthesis in prokaryotes, yeast, algae, plants, and in avian and mammalian cells (Beale and Weinstein, 1990; Jordan, 1990; Labbe-Bois and Labbe, 1990). The biosynthesis of ALA has recently attracted considerable research interest because of the discovery, initially made in plants, that one pathway for its biosynthesis requires an unusual involvement of a transfer RNA (tRNA). In this review we shall focus on various aspects of the required tRNA

Abbreviations used: ALA, δ-aminolevulinic acid; Glu, glutamate; GluRS, glutamyl-tRNA synthetase; GluTR, glutamyl-tRNA reductase; GSA, glutamate 1-semialdehyde; KAP, 7-keto-8-aminopelargonic acid; ORF, open reading frame; PAP, pyridoxamine 5'-phosphate; PLP, pyridoxal 5'-phosphate.

Gary P. O'Neill, Dieter Jahn, and Dieter Söll   Department of Molecular Biophysics and Biochemistry, Yale University, New Haven, Connecticut 06511.

Subcellular Biochemistry, Volume 17: Plant Genetic Engineering, edited by B. B. Biswas and J. R. Harris. Plenum Press, New York, 1991.

(e.g., its characterization, role in the pathway, regulation, and involvement in other metabolic processes), and the enzymes involved in this tRNA-dependent formation of ALA. Most of the progress in this area has involved plant experimental systems; however, it has become increasingly apparent in the last year that the tRNA-dependent formation of ALA is also widely distributed among the eubacterial and archaebacterial kingdoms. Since the enzymes and the mechanisms involved in the tRNA-dependent formation of ALA appear to be remarkably similar in plants and bacteria, we shall also discuss the recent rapid progress made in several bacterial systems. Other reviews have dealt with the development of this research area from a historical perspective and in relation to the overall process of porphyrin biosynthesis (Beale, 1990; Beale and Weinstein, 1990; Castelfranco and Beale, 1983; Jordan, 1990; Kannangara, 1991; O'Neill and Söll, 1990a; von Wettstein, 1991).

2. PATHWAYS OF δ-AMINOLEVULINIC ACID BIOSYNTHESIS

2.1. The "Shemin" Pathway (ALA from Succinyl-CoA and Glycine)

The first pathway of ALA formation, described in the 1950s, involves the condensation of succinyl-coenzyme A and glycine catalyzed by the enzyme ALA synthase and is often referred to as the Shemin pathway (Gibson *et al.*, 1958; Kikuchi *et al.*, 1958; Shemin and Russell, 1953). This pathway has been shown to function in a limited number of eubacteria, yeast, and in avian and mammalian cells (Beale and Weinstein, 1990; Jordan, 1990; Labbe-Bois and Labbe, 1990). Except for one report of ALA synthase activity in the mitochondria of *Euglena gracilis* (Beale *et al.*, 1981; Weinstein and Beale, 1983), all attempts to detect this activity in plants and other photosynthetic eukaryotes have failed. The inability to detect ALA synthase activity in plants prompted the search for alternate routes of ALA formation.

2.2. The tRNA-Dependent C$_5$-Pathway (ALA from Glutamate)

The first clue that an alternate route of ALA biosynthesis existed in plant chloroplasts was deduced from experiments in which the fate of radioactive carbon atoms derived from exogenous compounds such as glycine, succinate, α-ketoglutarate, and glutamate was examined *in vivo* (Beale and Castelfranco, 1973, 1974; Meller *et al.*, 1975). These experiments demonstrated that the five-carbon backbones of glutamate and α-ketoglutarate, and not the carbon skeletons of glycine or succinate, were transformed intact into ALA. This transformation, often designated the C$_5$-pathway, was originally described in greening tissues from several plant species (Beale and Castelfranco, 1973, 1974; Beale *et al.*,

1975; Meller *et al.*, 1975) and has been extended by *in vivo* radiotracer studies to many photosynthetic and nonphotosynthetic organisms, including a number of higher plant species, red and green algae, and cyanobacteria (Beale and Weinstein, 1990). The physiological relevance of this pathway was shown by radiotracer experiments and other techniques, such as ^{13}C-NMR analysis, which have demonstrated that all the carbon and nitrogen atoms in plant tetrapyrroles are derived from ALA formed via the C_5-pathway (Beale and Weinstein, 1990; Castelfranco and Beale, 1983; Oh-Hama *et al.*, 1982; Porra *et al.*, 1983; Schneegurt and Beale, 1986).

Direct biochemical analysis of the *in vitro* conversion of glutamate into ALA proved difficult to study as the ALA-forming activity in plant or algal chloroplast extracts was extremely labile (Bruyant and Kannangara, 1987; Kannangara *et al.*, 1988). Several major advances in the biochemical elucidation of the C_5-pathway were made by G. Kannangara, S. Gough, and their co-workers at the Carlsberg Research Laboratory during the last decade. By using a combination of physicochemical and affinity chromatographic techniques, they were able to isolate three soluble activities that together catalyzed the *in vitro* conversion of glutamate into ALA (Wang *et al.*, 1981). The three soluble activities were originally proposed to be glutamate-1-phosphate kinase, glutamate-1-phosphate dehydrogenase, and glutamate-1-semialdehyde aminotransferase (Wang *et al.*, 1981). The participation of the putative aminotransferase was supported by its partial purification from barley chloroplasts and the demonstration that this enzyme fraction efficiently converted synthetic glutamate-1-semialdehyde (GSA) into authentic ALA (Houen *et al.*, 1983; Kannangara and Gough, 1978; Wang *et al.*, 1981).

A surprising observation was made by the Carlsberg research group in 1984 when one of the three soluble components required for *in vitro* ALA formation was identified as RNA in extracts from chloroplasts of barley and *Chlamydomonas reinhardtii* (Huang *et al.*, 1984; Kannangara *et al.*, 1984). This conclusion was based on the ability of RNase A and snake venom phosphodiesterase to greatly reduce *in vitro* ALA synthesis and the partial purification by high-performance liquid chromatography (HPLC) of an RNA fraction required for ALA synthesis (δ-ALA RNA). It was also observed that glutamate could be covalently bound to the δ-ALA RNA, which led to the hypothesis that "the activation of glutamate is accomplished by a tRNA-like molecule" that "is similar to tRNA^{Glu} in that it can accept glutamate" (Huang *et al.*, 1984; Kannangara *et al.*, 1984). A dehydrogenase activity was proposed that would reduce the activated glutamate to yield GSA (Kannangara *et al.*, 1984).

The transfer RNA nature of the δ-ALA RNA was established (Schön *et al.*, 1986) after purification of the active RNA species from the δ-ALA RNA fraction previously prepared from barley chloroplasts (Kannangara *et al.*, 1984). Sequence determination showed a primary sequence that could be folded into the

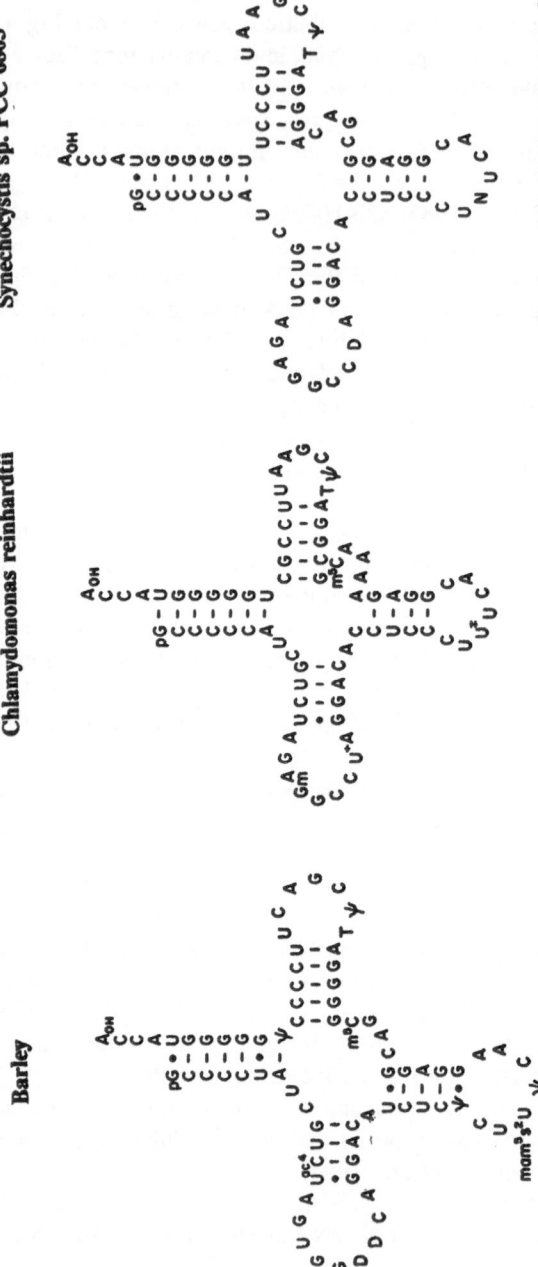

FIGURE 1. Nucleotide sequences of tRNA[Glu] species active in the C5-pathway of ALA formation from the chloroplasts of barley (Schön *et al.*, 1986) and *C. reinhardtii* (O'Neill *et al.*, 1990) and the cyanobacterium *Synechocystis* 6803 (O'Neill *et al.*, 1988).

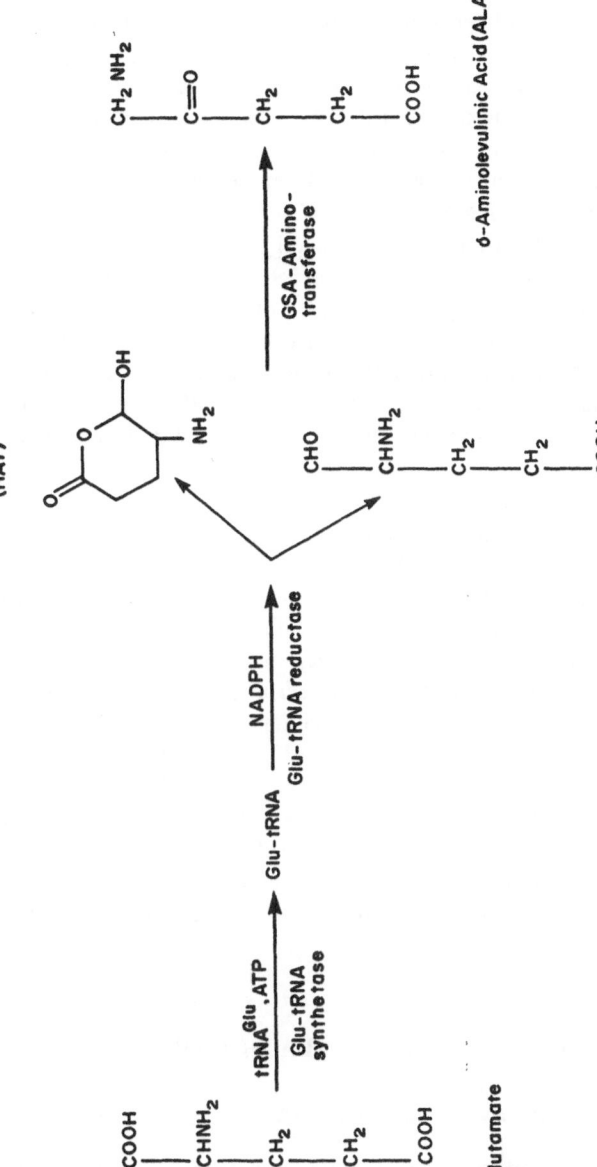

FIGURE 2. C_5-pathway of ALA formation (Kannangara *et al.*, 1988; Jordan, 1990; Schön *et al.*, 1986).

typical cloverleaf structure of tRNAs (Figure 1). The anticodon sequence UUC' identified the tRNA as a $tRNA^{Glu}$, which was in agreement with the fact that the pure δ-ALA RNA could be charged with glutamate *in vitro* (Schön *et al.*, 1986).

Based on the accumulated information, a reaction mechanism for the C_5-pathway of ALA formation was proposed (Kannangara *et al.*, 1988; Schön *et al.*, 1986) (Figure 2). In the first step of glutamyl-tRNA synthetase (GluRS) attaches glutamate to tRNA and generates $Glu-tRNA^{Glu}$ via a classical ATP- and Mg^{2+}-dependent aminoacylation reaction. The next step involves the reduction of the "activated" carboxyl group at the C_1 position of glutamate by an NADPH-dependent reductase, Glu-tRNA reductase (GluTR), to yield GSA. In the final step, ALA is formed by transamination catalyzed by GSA-aminotransferase. Although this model remains essentially correct, several aspects of the C_5-pathway remain incompletely characterized. For example, only recently has the purification of the fourth solubile component, the GluTR, been achieved (Chen *et al.*, 1990a). In addition, the exact chemical nature of the GluTR product and the reaction mechanisms of both the GluTR and the GSA aminotransferase are controversial (Beale and Weinstein, 1990; Jordan, 1990; Kannangara, 1991; O'Neill and Söll, 1990a).

2.3. Phylogenetic Distribution of the C_5-Pathway

The tRNA-dependent C_5-pathway of ALA synthesis exists in the chloroplasts of all algal and plant species tested thus far, and recent studies have demonstrated its presence in a wide variety of photosynthetic and nonphotosynthetic eubacteria, such as *Escherichia coli* and *Bacillus subtilis*, and archaebacteria (Avissar and Beale, 1989a; Avissar *et al.*, 1989; Friedmann and Thauer, 1986; Li *et al.*, 1989a; Oh-Hama *et al.*, 1986a, 1986b, 1988; O'Neill *et al.*, 1988, 1989; Rieble and Beale, 1988; Rieble *et al.*, 1989). These reports have been summarized recently in an excellent review (Beale and Weinstein, 1990). It is clear that the C_5-pathway occurs in many different prokaryotic groups with widely disparate metabolic properties. For example the C_5-pathway occurs in both photosynthetic and nonphotosynthetic bacteria, aerobes, facultative anaerobes, and strict anaerobes.

3. RNA INVOLVEMENT IN THE C_5-PATHWAY

3.1. Evidence for the Involvement of RNA

Several experimental approaches have been taken to demonstrate the requirement for $tRNA^{Glu}$ and the participation of $Glu-tRNA^{Glu}$ in the C_5-pathway of ALA formation. RNA involvement is readily demonstrated by the sensitivity of *in vitro* ALA-forming activity in crude cell extracts to RNase A or phosphodiesterase treatment (Huang *et al.*, 1984; Kannangara *et al.*, 1984). Conversely, a concentration-dependent stimulation of *in vitro* ALA synthesis has also been demonstrated by addition to cell extracts of a crude RNA fraction prepared by phenol extraction. The ALA-supporting capacity of such an RNA fraction is completely resistant to treatment by both proteases and deoxyribonucleases (Kannangara *et al.*, 1984). The tRNA nature of δ-ALA RNA is also supported by experiments in which heterologous, but well-characterized, tRNAs are used to support *in vitro* ALA formation. For example, it was shown that *E. coli* tRNA$_2^{Glu}$ can complement *Chlamydomonas* enzyme fractions in *in vitro* ALA formation (Huang and Wang, 1986).

3.2. Evidence for the Involvement of tRNA^{Glu}

The most direct approach to characterize the δ-ALA RNA is by its purification and sequence analysis (O'Neill *et al.*, 1988, 1990; Schön *et al.*, 1986). Because the δ-ALA RNA is initially obtained as a small fraction of a complex mixture of many other tRNA and small-molecular-weight RNA molecules, a combination of high-resolution purification techniques is required to obtain homogeneous δ-ALA RNA. Purification of the required RNA species is followed by assaying for its ability to support *in vitro* ALA formation and also by glutamate acceptor activity (see below). The purification techniques have included chlorophyllin-(or heme) Sepharose affinity chromatography, tRNA^{Phe} anticodon affinity chromatography, ion-exchange chromatography, HPLC separations on reversed-phase columns, and electrophoretic separations through denaturing polyacrylamide gels (Chen *et al.*, 1990b; Kannangara *et al.*, 1984; O'Neill *et al.*, 1988; Schneegurt and Beale, 1988; Schön *et al.*, 1986; Wang *et al.*, 1981). Whatever purification approach is taken, it is critical that the isolated RNA species be carefully analyzed for contamination by other glutamate-accepting RNAs and their ability to support ALA formation *in vitro*. For example, in barley different tRNA^{Glu} species exist in the cytoplasm, mitochondria, and chloroplasts (Peterson *et al.*, 1988; Schön *et al.*, 1986, 1988). The number of glutamate-accepting RNA species is further complicated by the "mischarging" of chloroplastic tRNA^{Gln} with glutamate by the chloroplastic glutamyl-tRNA synthetase (see below and Schön *et al.*, 1988; Schön and Söll, 1988b). Thus, when barley chloroplast tRNA is fractionated by HPLC, three glutamate-accepting RNA species are resolved into one tRNA^{Glu} species capable of supporting ALA formation and two tRNA^{Gln} species inactive in ALA formation (Schön *et al.*, 1986, 1988).

3.3. Structure of the tRNA^{Glu}

Sequence analysis of purified δ-ALA RNAs from barley (Schön *et al.*, 1986), *C. reinhardtii* (O'Neill *et al.*, 1990), and the cyanobacterium *Synechocystis* 6803 (O'Neill *et al.*, 1988) has shown that all three δ-ALA RNAs are *bone fide* tRNA^{Glu} species. Apart from the barley and *Chlamydomonas* tRNA species, no other chloroplast tRNA^{Glu} has been sequenced at the RNA level. Sequence comparison of barley, *C. reinhardtii*, and *Synechocystis* 6803 tRNA^{Glu} species has revealed several notable features (Figure 1). All three tRNAs possess a primary sequence of 76 nucleotides with the UUC anticodon expected of a tRNA^{Glu} species. Their sequences are highly homologous to each other and to the known chloroplast tRNA^{Glu} genes of tobacco, pea, broad bean, liverwort, spinach, wheat, and *Euglena gracilis,* and also to *E. coli* and the cyanelle of *Cyanophora paradoxa* (Berry-Lowe, 1987; Evrard *et al.,* 1988; Hollingsworth and Hallick, 1982; Holschuh *et al.,* 1984; Komine *et al.,* 1990; Kuntz *et al.,* 1984; Ohme *et al.,* 1985; Ohyama *et al.,* 1986; Quigley and Weil, 1985; Rasmussen *et al.,* 1984; Shinozaki *et al.,* 1986). All of these tRNAs, except *E. coli* tRNA^{Glu}, possess an A:U base pair in position 53:61 (numbering according to Sprinzl *et al.,* 1989), which is a highly conserved G:C base pair in all other elongator tRNAs (Schön and Söll, 1988b; Sprinzl *et al.,* 1989). This unusual $A_{53}:U_{61}$ base pair may be important in tRNA recognition by barley Glu-tRNA reductase (Peterson *et al.,* 1988; Schön and Söll, 1988a,b). A $G_{53}:C_{61}$ base pair is formed in the barley chloroplast tRNA^{Gln} and barley cytoplasmic tRNA^{Glu}, two glutamate-accepting tRNAs that are incapable of supporting ALA formation *in vitro* (Peterson *et al.,* 1988; Schön *et al.,* 1988). The modified nucleotide 5-methylaminomethyl-2-thiouridine (mam^{5}s^{2}U) is found in the first position of the anticodon of tRNA^{Glu} species; this is the same nucleotide found in eubacterial tRNAs (Sprinzl *et al.,* 1989). The cytoplasmic tRNA^{Glu} species from *Saccharomyces cerevisiae* and *Schizosaccharomyces pombe* carry a differently modified uridine, 5-methoxycarbonylmethyl-2-thiouridine, in this position (Sprinzl *et al.,* 1989). The occurrence of mam^{5}s^{2}U in chloroplast tRNAs supports the endosymbiontic theory of the prokaryotic origin of chloroplasts (Whatley and Whatley, 1981; Woese, 1987).

3.4.   Glu-tRNA^{Glu} as an Intermediate

The typical tRNA^{Glu} structure of the δ-ALA RNA and its capacity to be charged with glutamate by glutamyl-tRNA synthetase strongly support the idea that in analogy with other aminoacyl-tRNAs, the covalent linkage between the tRNA and glutamate is through an aminoacyl bond between the terminal adenosine of tRNA^{Glu} and the C_1-carboxyl (α-carboxyl) group of glutamate. Enzymatic removal of the 3'-terminal CCA of the δ-ALA RNA abrogates its capacities to be charged with glutamate and to support ALA formation *in vitro*. Both these capabilities can be restored to the tRNA by enzymatic readdition of the 3'-CCA end (Schön *et al.,* 1986). The intermediacy of Glu-tRNA^{Glu} has been demonstrated by showing that Glu-tRNA^{Glu} prepared in a separate reaction using purified components (i.e., *E. coli* GluRS and *E. coli* tRNA^{Glu}) can be used as a substrate for *in vitro* ALA formation (Avissar and Beale, 1988; Chen *et al.*, 1990a; Huang and Wang, 1986).

3.5. Genes Coding for the tRNA^{Glu}

Genes coding for tRNA^{Glu} species known to be active in ALA formation have been cloned from the chloroplast genomes of barley and *C. reinhardtii* and from *Synechocystis* 6803 (Berry-Lowe, 1987; O'Neill *et al.*, 1990; O'Neill and Söll, 1990b). Hybridization of δ-ALA RNA fragments to nuclear and chloroplast DNA from barley showed that the RNA was encoded in the chloroplast genome. This was confirmed by sequence analysis of a chloroplast DNA fragment encoding this gene (Berry-Lowe, 1987). This work also showed that there is only one gene for tRNA^{Glu} in the barley chloroplast genome (Berry-Lowe, 1987). The *C. reinhardtii* chloroplast genome is unusual in that it contains two copies of the tRNA^{Glu} gene (*trnE1* and *trnE2*), one of which (*trnE2*) is adjacent to the gene encoding the elongation factor Tu (Baldauf and Palmer, 1990; O'Neill *et al.*, 1990). Both copies of the *Chlamydomonas* tRNA^{Glu} genes are bounded by inverted repeats and insertion sequence elements, suggesting that one of the tRNA^{Glu} genes arose through gene duplication.

Many other *trnE* genes that encode tRNA^{Glu} species presumably active in ALA formation have been isolated from the chloroplasts of *Euglena gracilis*, spinach, broad bean, tobacco, wheat, pea, and liverwort, and from the cyanelle of *Cyanophora paradoxa* (Berry-Lowe, 1987; Evrard *et al.*, 1988; Hollingsworth and Hallick, 1982; Holschuh *et al.*, 1984; Komine *et al.*, 1990; Kuntz *et al.*, 1984; Ohme *et al.*, 1985; Ohyama *et al.*, 1986; Quigley and Weil, 1985; Rasmussen *et al.*, 1984; Shinozaki *et al.*, 1986). The 3'-CCA end of the tRNA^{Glu} is not encoded by the *Synechocystis* 6803 or any of the chloroplast tRNA^{Glu} genes. Our current knowledge shows that all chloroplast genomes, with the exception of *Chlamydomonas* (O'Neill *et al.*, 1990), contain only one tRNA^{Glu} gene (Berry-Lowe, 1987; Ohyama *et al.*, 1986; Shinozaki *et al.*, 1986). Therefore, tRNA^{Glu} is a dual-function molecule that is used in both protein and chlorophyll synthesis.

In most chloroplast genomes the tRNA^{Glu} gene is associated and probably cotranscribed (see below) with other tRNA genes, the most frequent arrangement being *trnE-trnY-trnD* (Figure 3). Inspection of the gene structure of the various cloned t-DNA sequences suggests that there does not appear to be any arrangement of the chloroplast tRNA^{Glu} genes conserved through evolution (Figure 3).

3.6. Transcription of tRNA^{Glu} Chloroplast Genes

In vivo analysis of the transcription of chloroplast tRNA^{Glu} genes from barley, spinach, and tobacco has revealed that these genes are cotranscribed with

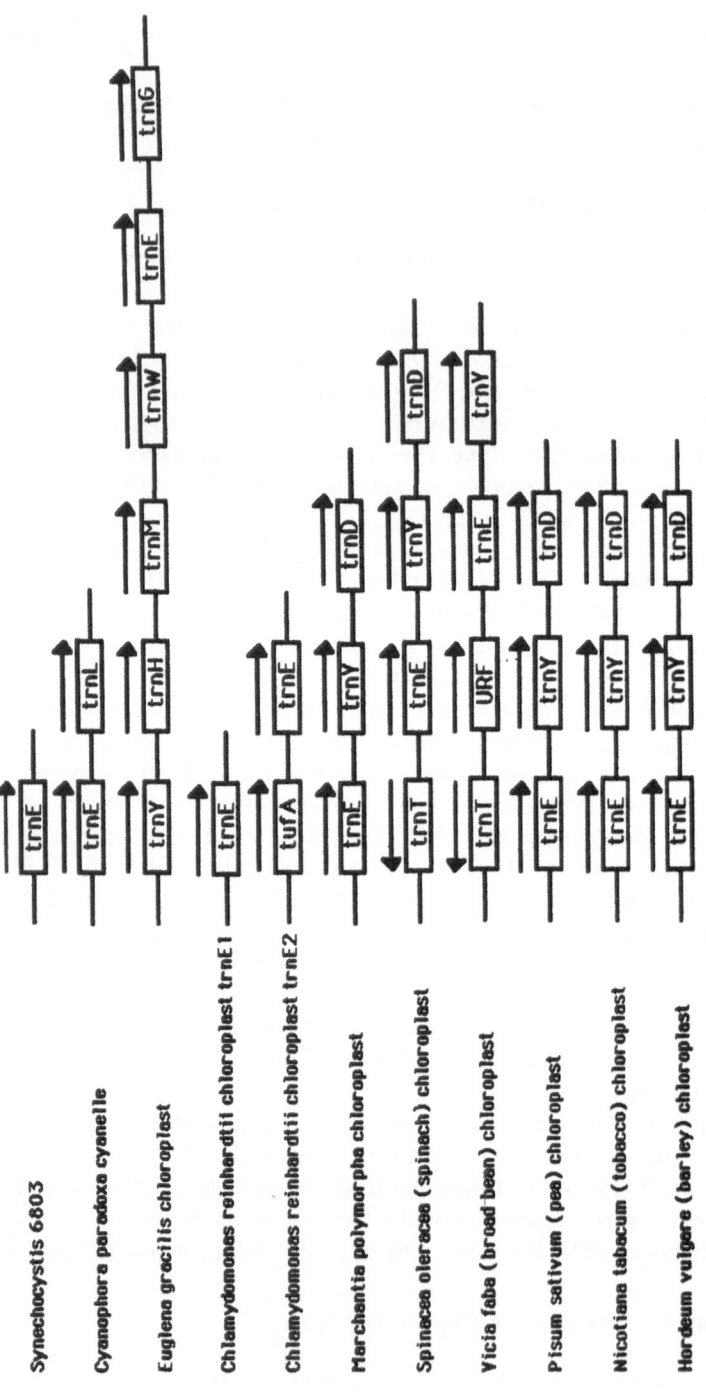

FIGURE 3. Structures of the sequenced *trnE* gene regions in oxygenic photosynthetic organisms. The organizations for 11 tDNA segments for tRNA^Glu from the cyanobacteria *Synechocystis* 6803, the cyanelle of *Cyanophora paradoxa*, and from the chloroplasts of the phytoflagellate *Euglena gracillis*, the green algae *Chlamydomonas reinhardtii*, the liverwort *Marchantia polymorpha*, and various higher plants are shown. Arrows indicate the direction of transcription. *tufA*, elongation factor Tu. The sources of the sequences are noted in the text.

other tRNA genes (Berry-Lowe, 1987; Holschuh *et al.*, 1984; Ohme *et al.*, 1985), while inspection of the gene structure for chloroplast *trnE* in pea, broad bean, *Marchantia polymorpha*, and *Euglena gracilis* strongly suggests that *trnE* is cotranscribed with other tRNA genes (Figure 3). Thus, in barley *trnE* is transcribed with two other tRNA genes into a precursor RNA (tRNA^{Glu}-tRNA^{Tyr}-tRNA^{Asp}), which is similar to the ones found in tobacco, spinach, and pea (Berry-Lowe, 1987; Holschuh *et al.*, 1984; Ohme *et al.*, 1985; Rasmussen *et al.*, 1984). In addition, the barley study showed that transcription of this tRNA gene cluster was not light-inducible (Berry-Lowe, 1987).

As mentioned above, the *Chlamydomonas* chloroplast genome contains two identical copies of tRNA^{Glu} (Baldauf and Palmer, 1990; O'Neill *et al.*, 1990). We have established a *Chlamydomonas*-based *in vitro* transcription system, based on earlier work (Gruissem *et al.*, 1983), to investigate the expression of the tRNA^{Glu} genes. Analysis of the *C. reinhardtii trnE1* gene sequence indicated that the cloned gene possessed its own transcriptional promotor and terminator elements. In the *Chlamydomonas in vitro* transcription system, the gene was highly transcribed by chloroplast RNA polymerase from its own promotor. Thus, the *C. reinhardtii* transcription system can be used to establish whether the two *trnE* genes differ in the nature of their transcription.

3.7. Possible Role of tRNA as Cofactor in ALA Formation

In the porphyrin biosynthetic pathway the biosynthetic steps after ALA formation are energetically favorable, requiring decarboxylations, oxidations, and aromatic ring formation (Jordan, 1990). Therefore, the source of energy to drive their biosynthesis is derived from the linkage energy stored in ALA. When ALA is formed via the Shemin pathway, the succinyl-CoA bond provides the bond energy. In analogy, the activated aminoacyl bond of Glu-tRNA^{Glu} also supplies the required energy for the synthesis of hemes and chlorophylls. Of course, this use of a high-energy aminoacyl bond to drive a biosynthetic process is one of the most often used in biology, i.e., protein biosynthesis.

3.8. Dual Role for tRNA^{Glu} in Protein Biosynthesis and ALA Formation

In the chloroplast genomes of tobacco (Shinozaki *et al.*, 1986), barley (Berry-Lowe, 1987), and *M. polymorpha* (Ohyama *et al.*, 1986), and also in *Synechocystis* 6803 (O'Neill and Söll, 1990b) only one *trnE* gene is found. Thus, in these organisms the product of this gene must provide tRNA^{Glu} that functions in both protein and ALA biosynthesis. However, in *Chlamyodomonas* two tRNA^{Glu} genes are present in the chloroplast genome, raising the possibility that different genes may encode tRNA destined for either protein or ALA biosynthesis. It is also possible that posttranscriptional nucleotide modifications of the tRNA also give rise to different tRNA^{Glu} species, which may be preferentially used in either protein or ALA formation. The question of a dual function for tRNA^{Glu} has been addressed in studies in the cyanobacterial species *Synechocystis* 6803 and the green algae *C. reinhardtii* (O'Neill and Söll, 1990b; Schneegurt *et al.*, 1988; G. P. O'Neill and D. Söll, unpublished observations). In these studies purified and precharged [^{14}C]Glu-tRNA species that had been shown to be active in ALA formation were tested for their ability to support *in vitro* protein synthesis. For *in vitro* protein synthesis both homologous *Synechocystis* 6803 cell extracts and heterologous *E. coli* cell extracts have been employed. The two studies using *Synechocystis* tRNAs concluded that both isoaccepting *Synechocystis* tRNA^{Glu} species supported protein and ALA synthesis in both homologous and heterologous protein synthesizing systems (O'Neill and Söll, 1990b; Schneegurt *et al.*, 1988). The two *C. reinhardtii* chloroplast tRNA^{Glu} species, which have identical primary sequences (O'Neill *et al.*, 1990), also have equivalent capacities to participate in an *E. coli*–derived *in vitro* protein synthesis system (G. P. O'Neill, unpublished data). These experiments support the dual role of the ALA RNA in protein and ALA formation.

4. ENZYMOLOGY

4.1. Enzyme Assays

Enzymatic assays for the complete Glu → ALA conversion or individual assays for GluRS, GluTR, and GSA aminotransferase are available. The complete conversion of glutamate into ALA is conveniently measured in crude cell extracts or in a reconstituted system using either glutamate or [^{14}C]glutamate as substrate (Mauzerall and Granick, 1956; Wang *et al.*, 1981; Weinstein and Beale, 1985). The choice of buffer used for preparation of the cell extract and for assaying the complete C_5 enzyme assay is critical. N-Tris[hydroxymethyl] methylglycine (commonly known as tricine) has been used almost exclusively; tris[hydroxymethyl]aminomethane (commonly known as trizma) buffers completely inhibit ALA formation via the C_5-pathway. Additions required for the complete Glu → ALA conversion include ATP, Mg^{2+}, NADPH, and levulinic acid. ALA formed in the cell extract accumulates due to the inhibition by levulinic acid of ALA dehydratase, the next enzyme in the porphyrin biosynthetic pathway (Jordan, 1990). The reaction products Glu, GSA, and ALA can be separated either by reversed-phase HPLC or by chromatography over Dowex A50WX8 (Kannangara and Gough, 1978; Mau *et al.*, 1987; Wang *et al.*, 1981). If [^{14}C]Glu is the substrate and reversed-phase separation of the reaction products is employed, the products are often sufficiently pure at this stage to be quantitated by liquid scintillation counting. The quantitation of nonradioactive ALA requires forming ALA-pyrrole by condensing ALA with ethylacetoacetate, reaction of ALA-pyrrole with Ehrlich's reagent (p-aminobenzaldehyde), and spectrophotometric determination of the colored product (Mauzerall and Granick, 1956; Urata and Granick, 1963). Thin-layer and paper chromatographic analytical techniques are also often used for definitive identification of ALA-pyrrole (Gough and Kannangara, 1976; Kannangara and Gough, 1977; Wang et al., 1981).

The individual assays for each of the C_5 enzymes require considerably more effort in the preparation of the substrates, which are not commercially available. GluRS is assayed in the routinely used aminoacylation assay (Lapointe et al., 1985). In the presence of ATP and Mg^{2+}, GluRS will esterify radioactive glutamate onto tRNA to yield Glu-tRNA, which can be quantitated as acid-insoluble material.

The direct assay for GluTR measures the conversion of Glu-tRNA into GSA and requires the preparation of Glu-tRNA. In the presence of a source of reducing power such as NADPH or NADH, the GluTR catalyzes the conversion of Glu-tRNA into free GSA and discharged tRNA (Chen et al., 1990a; Huang and Wang, 1986; Jahn et al., 1991c). The GSA is purified by reversed-phase HPLC or ion-exchange chromatography on Dowex 1 or Dowex A50WX8 and then quantitated directly (if [^{14}C]Glu-tRNA was used) (Chen et al., 1990a; Jahn et al., 1990c). Alternatively, the GSA can be converted into ALA by adding GSA aminotransferase and then quantitated using the colorimetric assay for ALA-pyrrole described above (Chen et al., 1990a). The use of a highly purified GluRS was found to be essential for the preparation of [^{14}C]Glu-tRNA^{Glu} in order to obtain low backgrounds in the GluTR reaction (Jahn et al., 1991c).

The assay for GSA aminotransferase requires preparation of the substrate GSA. Of the two reported procedures, synthesis via the ozonolysis of 4-vinyl-4-aminobutyric acid is the most rapid and gives high yields (Gough et al., 1989; Kannangara and Schouboe, 1985). Synthetic GSA is stable in its hydrated form at pH values less than 3 but is notoriously unstable under physiological pH conditions, with a half-life of 3–4 min at pH 8.0 (Hoober et al., 1988). At pH 6.5–7.0, GSA is sufficiently stable to permit the assay of GSA aminotransferase, which remains active within this pH range (Hoober et al., 1988). Depending on the source of the enzyme, either pyridoxal 5'-phosphate or pyridoxamine 5'-phosphate has been reported to act as cofactor. In certain systems, a cofactor requirement cannot be shown or a complete dependence for the required cofactor has been demonstrated (Avissar and Beale, 1989b; Grimm et al., 1989; Hoober et al., 1988; Houghton et al., 1989; Jahn et al., 1991a).

4.2. Purification of the Enzymes of the C_5-Pathway

Much effort has been directed to the purification of the C_5 enzymes, mainly from barley chloroplasts and C. reinhardtii (Bruyant and Kannangara, 1987; Chang et al., 1990; Chen et al., 1990a,b; Grimm et al., 1989; Wang et al., 1981,

1984; Weinstein *et al.*, 1987). The difficulties in the purification of these enzymes have mostly been concerned with the lability of the GluTR, the apparent low concentration of GluTR in cell extracts, and the difficulty in separating active GluTR from GluRS (Chen *et al.*, 1990a; Kannangara *et al.*, 1988; Wang *et al.*, 1984). The affinity and ion-exchange chromatographic resins that have proven particularly successful have exploited the capacities of the GluRS and the GluTR to bind either ATP, NADPH, or tRNA. These resins include nucleotide-analog affinity matrices, phosphocellulose, and various anion and cation ion-exchange resins. A difficult obstacle in purification of the C_5 enzymes is separation of enzymatically active GluTR from GluRS, which has recently been achieved by gradient elution of a diethylaminoethyl-cellulose chromatographic column (Chen *et al.*, 1990a,b), serial affinity chromatography on 2',5'-ADP agarose and Reactive Blue 2-Sepharose (Weinstein *et al.*, 1987), and immunoaffinity chromatography utilizing an anti-GluRS antibody (Bruyant and Kannangara, 1987). Storage and stabilization of the enzyme fractions are important; the enzyme activities are stable at $-80°C$ but are rapidly inactivated at 2–4°C (Weinstein *et al.*, 1987). Glycerol at a concentration of 1 M or higher stabilizes the enzyme activities (Weinstein *et al.*, 1987).

4.3. Structural Features and Catalytic Properties of the Enzymes

4.3.1. Glutamyl-tRNA Synthetase

The GluRS has been purified and characterized from the chloroplasts of wheat (Ratinaud *et al.*, 1983), barley (Bruyant and Kannangara, 1987), and *C. reinhardtii* (Chang *et al.*, 1990; Chen *et al.*, 1990b). The apparent native molecular weights of the enzymes and their quaternary structures are unusually diverse. For *C. reinhardtii* GluRS two different experimental analyses have been obtained. One study reports a monomeric structure of 62,000 Da (Chen *et al.*, 1990b), in contrast to a second investigation which observed a homodimeric structure with subunits of 32,500 Da (Chang *et al.*, 1990). Further experiments are needed to resolve these differences. The monomeric *C. reinhardtii* GluRS structure is similar to the homologous enzymes from *E. coli* and *B. subtilis*, which are active as monomers of 53,810 and 65,000 Da (Breton *et al.*, 1986; Lapointe *et al.*, 1985; Lapointe and Söll, 1972; Proulx *et al.*, 1983; Proulx and Lapointe, 1985). The barley and wheat GluRS enzymes are homodimers with subunits of 54,000 and 56,000, respectively (Bruyant and Kannangara, 1987; Ratinaud *et al.*, 1983). Since only one chloroplastic GluRS was detected in barley, wheat, and *C. reinhardtii*, this enzyme must provide charged Glu-tRNA for both protein synthesis and ALA formation. In addition, the chloroplastic GluRS has another metabolic role in that it is required for the production of Gln-tRNA[Gln] for protein biosynthesis by a natural misacylating mechanism (Chen *et al.*, 1990b; Jahn *et al.*, 1990b; Schön *et al.*, 1988). This mechanism involves misacylation of tRNA^{Gln} with glutamate by GluRS and subsequent transamidation of Glu-tRNA^{Gln} to the correctly acylated Gln-tRNA^{Gln} (Jahn *et al.*, 1990b; Wilcox and Nirenberg, 1968). Like other aminoacyl-tRNA synthetases, GluRS has exquisite capabilities to recognize and discriminate between different tRNAs; e.g., the barley chloroplast GluRS can charge chloroplast tRNA^{Glu} but not the homologous cytoplasmic tRNA^{Glu} species (Peterson *et al.*, 1988; Schön and Söll, 1988a,b).

There is evidence that the GluRS enzymes from *Chlamydomonas* and *Scenedesmus obliquus* are negatively regulated by heme and protochlorophyllide *in vitro* (Chang *et al.*, 1990; Dörnemann *et al.*, 1989). These findings suggest that GluRS is the first enzyme of the C_5-pathway, although its product, Glu-tRNA^{Glu}, is a dual-purpose molecule functioning in both protein and ALA formation.

4.3.2. Glutamyl-tRNA Reductase

The enzymic conversion by GluTR of Glu-tRNA^{Glu} to GSA and discharged tRNA is the least understood, but the most unique and unusual, step in the C_5-pathway. The enzyme catalyzes a novel reduction reaction by an unknown reaction mechanism that requires a specific charged tRNA and NADPH. This step of the C_5-pathway provides a major regulatory point for the C_5-pathway and the biosynthesis of heme and chlorophyll (Beale, 1990; Beale and Weinstein, 1990). In the literature GluTR is frequently referred to as a dehydrogenase. We prefer the term "reductase," as the product, GSA, is the consequence of a reduction of glutamate (it could be termed NADPH:Glu-tRNA oxidoreductase). GluTR was first purified from *C. reinhardtii* chloroplasts (Chen *et al.*, 1990a). The monomeric enzyme has an apparent molecular mass of 130 kDa. Only Glu-tRNA^{Glu} as substrate and NADPH as cofactor are required for GluTR to form free GSA. Thus, GluTR alone is sufficient for conversion of Glu-tRNA^{Glu} to GSA. While the major product of the reduction reaction using pure GluTR was identified as GSA by its comigration with synthetic GSA on reversed-phase HPLC and by its ability to act as substrate for the GSA aminotransferase, its chemical structure was not determined (Chen *et al.*, 1990a). As discussed below, the structure of GSA is somewhat controversial (Beale and Weinstein, 1990; Jordan, 1990). In addition to GSA, at least two other products of the GluTR reaction have been observed by HPLC and thin-layer chromatographic analysis in *Chlamydomonas* (Chen *et al.*, 1990a), *Cyanidium caldarium* (Houghton *et al.*, 1989), and *E. coli* (G. O'Neill, unpublished observations). These products cannot be converted to ALA by the aminotransferase and may represent intermediates in the GluTR step (Houghton *et al.*, 1989; Jahn *et al.*, 1991c). Two different mechanisms for the Glu-tRNA reductase-catalyzed two-electron reduction of the α-carboxyl group of glutamate have been proposed (Chen *et al.*, 1990a; Kannangara, 1991). One mechanism (Chen *et al.*, 1990a) is formally analogous to the back reaction of glyceraldehydephosphate dehydrogenase where the activated (by phosphorylation) carboxyl group of 3-phosphoglycerate is reduced to 3-phosphoglyceraldehyde in the presence of NADH (see, e.g., Stryer, 1988). This reaction proceeds through an acyl enzyme intermediate. The similarity with Glu-tRNA reductase is obvious; the enzyme reduces a carboxyl group of glutamate (activated by aminoacyl-tRNA formation) to GSA in the presence of NADPH. A significantly different mechanism was proposed (Kannangara, 1991) which requires several steps that are analogous to reactions in protein biosynthesis. In the first step a complex is formed between Glu-tRNA^{Glu}, GTP, Glu-tRNA synthetase, and Glu-tRNA reductase. This complex is then a substrate for an additional enzyme, similar to peptidyl transferase. The reaction mechanism would involve a nucleophilic attack by the α-amino group of glutamate on the esterified carboxyl carbon of the Glu-tRNA^{Glu}, yielding the intermediate azeridinone propionate. In the final step, the NADPH-dependent GluTR would reduce the azeridinone propionate to GSA. These reaction mechanisms and intermediates proposed (Chen *et al.*, 1990a; Kannangara, 1991) are so different that it should not be long before the correct mechanism is established. It should be emphasized that although additional accessory proteins may enhance the overall conversion of Glu-tRNA^{Glu} to GSA, the purified *C. reinhardtii* Glu-tRNA reductase catalyzes this reaction requiring only the substrate Glu-tRNA and NADPH (Chen *et al.*, 1990a).

GluTR has the ability to discriminate between different charged tRNAs, and often the tRNA specificities of GluTR and the homologous GluRS are markedly different. Thus, the barley chloroplast GluTR can recognize and use as substrate charged barley chloroplast tRNA^{Glu}, but not *E. coli* Glu-tRNA^{Glu}, mischarged barley chloroplastic Glu-tRNA^{Gln}, or barley cytoplasmic Glu-tRNA^{Glu} (Kannangara *et al.*, 1984; Peterson *et al.*, 1988). In contrast, barley chloroplastic GluRS accepts all these tRNA substrates except barley cytoplasmic tRNA^{Glu}. On the other hand, the *Chlamydomonas* GluTR can recognize *E. coli* tRNA^{Glu} in addition to its homologous tRNA (Huang and Wang, 1986; Chen *et al.*, 1990b). As discussed elsewhere (Schön and Söll, 1988a,b), the $A_{53}:U_{61}$ base pair of tRNA^{Glu} may be important in tRNA discrimination displayed by the Glu-tRNA reductase.

Currently genes encoding the plant chloroplastic GluTR's have not been isolated. Our only information on genes for this interesting enzyme comes from results in the C_5-pathway-containing eubacteria *E. coli*, *Salmonella typhimurium*, and *B. subtilis* (Avissar and Beale, 1989a; Li *et al.*, 1989a; O'Neill *et al.*, 1989). In these organisms, isolation of genes by complementation cloning of ALA-requiring mutants (termed *hemA* mutants) has proven highly successful (Avissar and Beale, 1990; Drolet *et al.*, 1989; Elliott, 1989; Li *et al.*, 1989b;

Petricek *et al.*, 1990; Verkamp and Chelm, 1989). It was proposed (Avissar and Beale, 1989a) that the *hemA* gene in *E. coli* encodes the GluTR or a component of it. This conclusion was based on the ability of partially purified *Chlorella* reductase to restore ALA biosynthetic activity to extracts prepared from the ALA-requiring *hemA* mutant strain of *E. coli*. The best evidence that the *hemA* gene is the reductase is the recent demonstration (Avissar and Beale, 1990) that a cloned *Chlorobium vibrioforme* DNA fragment complemented the *E. coli hemA* mutation in *trans* and conferred a different tRNA specificity to the *in vitro* GluTR activity of the transformed strains. Unfortunately, there are no data yet on the enzymatic activity of the purified natural or recombinant forms of the *hemA* gene product.

The deduced amino acid sequences of the putative bacterial GluTR genes show a high degree of identity (Drolet *et al.*, 1989; Elliott, 1989; Li *et al.*, 1989b; Petricek *et al.*, 1990; Verkamp and Chelm, 1989). The *hemA* gene product has a molecular mass of 46 kDa in *E. coli* and *S. typhimurium* and 51 kDa in *B. subtilis*. Amino acid sequence comparisons of the three bacterial sequences reveals a highly conserved 31-amino-acid sequence, which is predicted to be a β-α-β nucleotide binding fold, which may be involved in the binding of NADPH (Petricek *et al.*, 1990). The bacterial glutamyl-tRNA reductases are less than half the size of the purified *C. reinhardtii* GluTR (Chen *et al.*, 1990a). But when GluTR activity was purified from aerobically grown *E. coli* (Jahn *et al.*, 1991c), two proteins were found that convert *in vitro E. coli* Glu-tRNA^{Glu} to GSA. One protein is a monomer of 85 kDa molecular mass, while the other one separates well from the first activity and is a monomeric 45-kDa protein. Interestingly, both activities were found to copurify over several chromatographic steps with *E. coli* glutamyl-tRNA synthetase. It may be a coincidence that the combined molecular mass of the two GluTR activities in *E. coli* equals that of the single enzyme (130 kDa) of *Chlamydomonas* (Chen *et al.*, 1990a).

It has been reasoned, based on the known chloroplast genome sequences, that the plant GluTR must be nuclear-encoded and imported into the chloroplast (Kannangara *et al.*, 1988). Comparison of the DNA and deduced amino acid sequences of the GluTR-containing *hemA* operon of *B. subtilis* to the chloroplast genomes of *Marchantia polymorpha* and tobacco indicates that in addition to the δ-ALA tRNA^{Glu}, at least one protein component of the plant C$_5$-pathway is chloroplast encoded (G. O'Neill, unpublished results). In *B. subtilis* the *hemA*-encoded GluTR precedes an open reading frame (ORF), called ORF2, in an operon containing two other genes in porphyrin synthesis (Petricek *et al.*, 1990). Mutations in either the *B. subtilis hemA* or ORF2, a gene coding for a 30-kDa peptide, lead to ALA auxotrophy (Petricek *et al.*, 1990). The *B. subtilis* ORF2 peptide is 35–46% identical to hypothetical proteins deduced from the chloroplast genomes of tobacco [termed hypothetical 35.5-kDa protein (ORF 313)] and of *Marchantia polymorpha* [termed hypothetical 35-kDa protein (ORF 320)]

(G. O'Neill, unpublished results). The two chloroplast ORFs and the *B. subtilis* ORF2 also display striking homology at the amino acid level to NADH-ubiquinone oxidoreductases, cytochrome C oxidase, and cytochrome B (G. O'Neill, unpublished results).

4.3.3. GSA Aminotransferase

The final step of the C_5-pathway is an isomerization of GSA catalyzed by the GSA aminotransferase (Kannangara and Gough, 1978; Wang *et al.*, 1981), which has been purified from barley and *C. reinhardtii* chloroplasts and the cyanobacterium *Synechococcus* (Grimm *et al.*, 1989; Jahn *et al.*, 1991). The barley enzyme was found to be a dimer of two identical 46,000-Da subunits, while the aminotransferases from *Synechococcus* and *C. reinhardtii* were active as monomers of 46,000 Da and 43,000 Da, respectively (Grimm *et al.*, 1989; Jahn *et al.*, 1991).

As mentioned above, the chemical nature of the substrate for the aminotransferase is uncertain. Based on *in vitro* and *in vivo* evidence, it was established as GSA (Gough *et al.*, 1989; Grimm *et al.*, 1989; Hoober *et al.*, 1988; Houen *et al.*, 1983; Kannangara and Gough, 1978; Kannangara and Schouboe, 1985; Wang *et al.*, 1981). The first line of evidence that GSA is an intermediate involved its synthesis and the demonstration that synthetic GSA could be enzymatically converted to ALA (Wang *et al.*, 1981). Two protocols for synthesizing GSA have been reported (Gough *et al.*, 1989; Houen *et al.*, 1983), and in both cases the synthetic GSA was a suitable substrate for the GSA-aminotransferase. A second line of evidence establishing GSA as an intermediate was the isolation of natural GSA from gabaculin-treated barley leaf tissue (Kannangara and Schouboe, 1985). Gabaculin (3-amino 2,3-dihydrobenzoic acid) is a potent inhibitor of certain ω-aminotransferases; gabaculin treatment apparently leads to the accumulation of GSA, the substrate for the GSA-aminotransferase (Kannangara and Schouboe, 1985; Hoober *et al.*, 1988; Jahn *et al.*, 1991a). The structures of both the synthetic and the natural GSA were confirmed by ^{1}H-NMR and ^{13}C-NMR spectroscopy, infrared, and mass spectrometric analysis (Gough *et al.*, 1989; Hoober *et al.*, 1988; Houen *et al.*, 1983; Kannangara and Schouboe, 1985). However, the role of GSA as an intermediate in the C_5-pathway and the proposed structure has been questioned by several groups (Beale and Weinstein, 1990; Jordan, 1990). The basis for the controversy is the extreme instability of an α-aminoaldehyde such as GSA, which should rapidly polymerize. Another possible product of the GluTR, suggested by Jordan (Jordan, 1990), is 2-hydroxy,3-aminotetrahydropyran-1-one (HAT; Figure 2). The possible role of HAT in the C_5-pathway is supported by the observations that synthetic HAT is more stable than GSA at physiological pH and that it can be enzymatically converted to ALA (Jordan, 1990).

Cofactor requirements for the GSA aminotransferase vary depending on the source of the enzyme. The barley enzyme does not display any cofactor requirements; the *Synechococcus-* and *Cyanidium*-derived enzymes are stimulated by pyridoxamine 5'-phosphate (PAP) while the *Chlamydomonas* enzyme activity is completely dependent on the addition of pyridoxal 5'-phosphate (PLP) to the assay (Grimm *et al.*, 1989; Hoober *et al.*, 1988; Houghton *et al.*, 1989; Jahn *et al.*, 1991). The different cofactor requirements of the GSA aminotransferases have been suggested to be due to different reaction mechanisms for these aminotransferases (Gough *et al.*, 1989; Grimm *et al.*, 1989; Hoober *et al.*, 1988; Houghton *et al.*, 1989).

Several possible mechanisms and intermediates have been postulated for the GSA aminotransferase reaction (Gough *et al.*, 1989; Grimm *et al.*, 1989; Hoober *et al.*, 1988). In the case of the cofactor-independent GSA aminotransferase from barley, a dimer of GSA is proposed as an intermediate. In this model, the enzyme catalyzes an intermolecular exchange of the C_4 amino groups between a pair of oppositely oriented GSA molecules. This model is supported by the observation that head-to-tail concentration-dependent GSA hydrate dimerization occurs (Gough *et al.*, 1989). A reaction mechanism proceeding through a GSA dimer would obviate the need for a PLP or PAP cofactor. Interestingly, the barley enzyme was inhibited by gabaculine, which inhibits transamination through its interaction with the enzyme-bound PAP (Soper and Manning, 1982). This result led to the proposal for a PAP-independent mode of gabaculin inhibition in the barley system (Gough *et al.*, 1989; Grimm *et al.*, 1989).

The cofactor-dependent monomeric GSA aminotransferases from *Chlamydomonas* and *Synechococcus* probably catalyze transaminations that are more similar to classical pyridoxal 5'-phosphate-dependent aminotransferases (Grimm *et al.*, 1989; Jahn *et al.*, 1991). In these cases the enzymatic reaction would start with the binding of the cofactor PLP to a lysine residue in the active site, followed by transfer of the C_4 amino group of GSA to PLP to yield the pyridoxamine form of the enzyme and 4,5-dioxovalerate (Breu and Dörnemann, 1988; Jahn *et al.*, 1991). This intermediate may remain bound to the enzyme or it could be replaced by another molecule of GSA or 4,5-dioxovalerate. The reaction could also proceed through another proposed intermediate, 4,5-diaminovalerate (Hoober *et al.*, 1988). Evidence has been obtained for an intermolecular transamination mechanism by following label transfer between different GSA molecules enzymatically generated from differentially labeled [13]C- and ^{15}N-glutamate molecules (Mau and Wang, 1988). The PLP-dependent reaction mechanism of the monomeric GSA aminotransferases is supported by the finding that the primary sequence of these proteins contains the highly conserved amino acid sequence LGKIIGG (Grimm, 1990; von Wettstein, 1991), which is the PLP binding site in other enzymes using this cofactor (Tanizawa *et al.*, 1989; Mukherjee and Dekker, 1990).

Genes encoding the GSA aminotransferase from barley, the cyanobacterium *Synechococcus,* and the gram-negative *E. coli* have been cloned (Grimm, 1990; von Wettstein, 1990). The barley gene was isolated using the polymerase chain reaction and oligonucleotides based on the partial amino acid sequence of the purified barley enzyme (Grimm, 1990; Grimm *et al.,* 1989). The barley precursor protein contains a 34-amino-acid long transit peptide responsible for chloroplast import. Two other genes for GSA aminotransferase from *Synechococcus* and *E. coli* have been isolated with the aid of the barley cDNA (von Wettstein, 1991). The GSA aminotransferases from barley, *Synechococcus,* and *E. coli* share 48% identity in their primary structures (von Wettstein, 1991). The homology between the barley and cyanobacterial GSA aminotransferases is 72% (von Wettstein, 1990).

4.4. Multienzyme Complex Formation

Although all three C_5-enzymes can catalyze their individual reactions independently (see above), one might expect complexes to form between the different enzymes. Complexed enzymes could directly shunt an unstable product, such as GSA, to the next enzyme. A complex of GluRS with GluTR would simplify the process of routing Glu-tRNA^{Glu} into either protein or ALA biosynthesis. One proposal (Kannangara, 1990; Kannangara *et al.,* 1988) suggests that GluRS and GluTR are complexed with Glu-tRNA and that this complex is the "substrate" for another enzyme that initiates the reduction. The only evidence to date for complex formation is in the *Chlamydomonas* system; in the absence of NADPH, stable complexes of GluTR:Glu-tRNA and GluRS:Glu-tRNA^{Glu}:GluTR can be isolated by glycerol gradient sedimentation ultracentrifugation (Chen *et al.,* 1990a; M. W. Chen and D. Jahn, unpublished observations).

5. REGULATION

The cell's demand for porphyrins varies during different metabolic states and conditions. In many organisms ALA synthesis is a key regulatory and rate-limiting step for the formation of heme and chlorophylls (Beale and Weinstein, 1990; Jordan, 1990; Labbe-Bois and Labbe, 1990; May *et al.,* 1986). Two observations regarding the regulation of the overall formation of ALA in various photosynthetic biological systems are clear: (1) upon illumination ALA synthesis is stimulated several-fold (Beale and Weinstein, 1990; Huang and Castelfranco, 1989; Kannangara and Gough, 1979; Weinstein and Beale, 1985), and (2) exogenous protoheme is a potent inhibitor of ALA synthesis in intact plastids and many of the soluble extracts tested for ALA formation (Beale, 1990; Beale and Weinstein, 1990; Castelfranco and Beale, 1983). The mode of action of both light and heme are unknown; the light-dependent stimulation was shown to be mediated through the phytochrome response (Huang *et al.*, 1989). Other intermediates in heme and chlorophyll biosynthesis, including protoporphyrin, Mg-protoporphyrin, protochlorophyllide, and chlorophyllide, at high concentration have minimal inhibitory effects on *in vitro* ALA formation and probably are not physiologically relevant regulatory effectors (Beale, 1990; Beale and Weinstein, 1990). In contrast, Mg-protoporphyrin is an effective inhibitor of ALA formation in intact plastids (Chereskin and Castelfranco, 1982). One can conceive of at least four possible targets for regulation of ALA formation in the C_5-pathway, including tRNA^{Glu}, GluRS, GluTR, and GSA-aminotransferase.

5.1. Regulation at the Level of tRNA^{Glu}

The level of tRNA^{Glu} appears not to be light-regulated, as established in studies with barley chloroplasts (Berry-Lowe, 1987) and in the cyanobacterium *Synechocystis* 6803 (O'Neill and Söll, 1990b). The study in *Synechocystis* asked the question whether there was a change in the steady-state levels of tRNA^{Glu}, Glu-tRNA^{Glu}, and GluRS under conditions where chlorophyll demand changed greatly. Under conditions where chlorophyll levels varied over a 10-fold range the level of tRNA^{Glu}, GluRS, and Glu-tRNA^{Glu} remained constant (O'Neill and Söll, 1990b). Thus, there appears to be a functional excess of charged tRNA in the cell that can accommodate the changing demands of protein and ALA biosynthesis (O'Neill and Söll, 1990b).

5.2. Regulation at the Level of GluRS

Two reports suggest that GluRS in extracts of *Chlamydomonas* (Chang *et al.*, 1990) or *Scenedesmus obliquus* (Dörnemann *et al.*, 1989) can be inhibited *in vitro* by heme or protochlorophyllide, respectively. That heme might have a regulatory effect on an aminoacyl-tRNA synthetase may be related to its well-known role in the regulation of the inhibition of protein synthesis in other eukaryotic systems (London *et al.*, 1987). As mentioned above, the level of GluRS was not affected by light in *Synechocystis* (O'Neill and Söll, 1990b).

5.3. Regulation at the Level of GluTR

Little is known about the regulation of the chloroplastic enzyme GluTR; in high concentrations heme has about a 50% inhibitory effect on the partially purified *C. reinhardtii* enzyme *in vitro* (Huang and Wang, 1986). In contrast, heme has no effect on either of the purified *E. coli* GluTR activities (Jahn *et al.*, 1991). Other studies reveal that more highly purified preparations as compared to less pure preparations require higher heme concentrations to achieve 50% inhibition, suggesting that maximal heme inhibition may be exerted through more than one component (Beale and Weinstein, 1990). Currently there is no information on transcriptional regulation of plant GluTR. Northern (RNA) blot analysis of the *E. coli* GluTR mRNA levels in cells grown aerobically and anaerobically with different reduced carbon sources did not detect any changes in the level of message; however, transcription from alternate transcriptional initiation sites was observed (Verkamp and Chelm, 1989). In analogy to ALA synthase, sophisticated regulatory mechanisms could function at that level (Labbe-Bois and Labbe, 1990; May *et al.*, 1986). The proposed regulation of the *hemA* gene product in *S. typhimurium* is intriguing. The *hemA* gene is cotranscribed with the gene (*prfA*) encoding the peptide chain release factor (RF-1) (Elliott, 1989). This protein recognizes UAG and UAA termination codons and catalyzes the release of the finished peptide chain during protein biosynthesis (Craigen *et al.*, 1985). Interestingly, the *hemA* gene ends with a UAG codon and thus requires RF-1 for correct termination. For this reason it was suggested that read-through of the *hemA* UAG codon should occur at a rate inversely proportional to the cellular concentration of RF-1, resulting in a form of autogenous regulation (Elliott, 1989).

5.4. Regulation at the Level of GSA Aminotransferase

Exposure of greening barley plastids and mature chloroplasts to light results in a several-fold increase in GSA aminotransferase activity (Kannangara and Gough, 1978). The increased aminotransferase activity is not due to increased transcription of the gene for this enzyme as its mRNA levels do not change when it is exposed to light (Grimm, 1990).

5.5. Other Questions regarding Regulation

A different consideration of how the C_5-pathway is regulated springs from the question of how tRNA[Glu] is shunted into ALA or protein biosynthesis. Clearly, the GluTR must compete with elongation factor Tu for binding of Glu-tRNA[Glu]; thus, the relative binding constants and intracellular concentrations of these proteins may be important. In this context it may be pertinent that in the *C. reinhardtii* chloroplast genome, one copy of the *trnE* gene is immediately adjacent to the gene encoding elongation factor Tu.

Still another idea for regulating the C_5-pathway was recently proposed (Kannangara, 1990). It is known that oxidation of the modified anticodon base, mam^5s^2U, leads to a tRNA which cannot be charged (Carbon *et al.*, 1965; Kannangara *et al.*, 1988). Thus, a reversible oxidation of this nucleotide may restrict the availability of active tRNA[Glu]. However, such a modification would also effect the role of tRNA[Glu] in protein biosynthesis (Agris *et al.*, 1973).

6. BIOCHEMICAL ORIGIN OF THE C₅-PATHWAY

What is the evolutionary significance of this pathway? Why is tRNA-activated Glu utilized as opposed to succinyl-CoA and glycine? It is generally believed that the C_5-pathway predates the Shemin pathway. In the early evolution of the biotic world, RNAs have been proposed to lay key structural and catalytic roles prior to the appearance of proteins. Activation of chemical groups by ligation through the acyl linkage of a bound tRNA might have provided a stable intermediate in many different biosynthetic pathways. Other examples of transformations of amino acids bound to tRNA are (1) the formation of Gln-tRNA^{Gln} by the transamidation of Glu-tRNA^{Gln} (Jahn et al., 1990b; Schön et al., 1988; Wilcox and Nirenberg, 1968), (2) the formylation of methionine upon acylation to prokaryotic initiator tRNAs (Dickerman et al., 1967; Migita and Doi, 1970), and (3) the formation of the required amino acid selenocysteine (Böck and Stadtman, 1988; Stadtman et al., 1989). This compound is formed from serine after acylation of tRNA.

The ALA synthase–catalyzed condensation of succinyl-CoA and glycine has been detected only in organisms capable of aerobic growth and possessing a complete tricarboxylic acid cycle. Many bacteria that utilize the C_5-pathway contain an incomplete tricarboxylic acid cycle owing to the absence of the 2-oxoglutarate-oxidizing enzyme (Oh-Hama et al., 1986a,b; O'Neill et al., 1989). It was suggested that the C_5-pathway was correlated with the absence of the 2-oxoglutarate-oxidizing activity since its product succinyl-CoA is the substrate for the ALA synthase. But the facultative anaerobe E. coli and the aerobe B. subtilis both contain a complete tricarboxylic acid cycle (during aerobic growth) and synthesize ALA via the C_5-pathway (Avissar and Beale, 1989a; Li et al., 1989a; O'Neill et al., 1989). Until the evolutionary development of a complete Krebs cycle to provide a ready supply of succinyl-coenzyme A, the C_5-route may have been the sole route of ALA formation.

While E. coli has not been shown to contain ALA synthase, it contains the biof gene, whose product, 7-keto-8-aminopelargonic acid (KAP) synthetase shares strong homologies (41% amino acid identity) with bacterial and vertebrate ALA synthases (Otsuka et al., 1988). The substrates and reactions catalyzed by 7-KAP synthetase and ALA synthetase are very similar. 7-KAP synthetase condenses alanine and pimelic acid-CoA through a decarboxylation of alanine to yield 7-KAP (Otsuka et al., 1988). In comparison, ALA synthase catalyzes the condensation of glycine and succinyl-CoA via a decarboxylation of glycine to yield ALA (Jordan, 1990). Both proteins are pyridoxal 5'-phosphate-dependent enzymes that catalyze the condensation of an amino acid with a CoA-activated carboxyl group of an organic acid with the release of CO_2 (Jordan, 1990; Otsuka et al., 1988). Considering the high degree of amino acid identity between these enzymes, the chemical and steric similarities of their substrates, and the reactions catalyzed, 7-KAP synthetase and ALA synthase are clearly members of the same enzyme family. The obvious question is whether *E. coli* possesses an ALA synthase that was lost during evolution or a cryptic or as yet undetected ALA synthase?

7. CONCLUSIONS

The availability of pure enzymes and the cloned genes will allow investigation of the mechanism of the biochemical transformation of glutamate into ALA. The most challenging and intriguing questions that now remain involve the mechanism of the Glu-tRNA reductase–catalyzed reduction and the regulation of ALA biosynthesis. There are two aspects to the question of regulation of ALA biosynthesis. One aspect is the regulation of the expression of the genes and their products at the transcriptional and translational levels. The second aspect of regulation concerns the mechanism that directs the Glu-tRNA^{Glu} into either protein or ALA biosynthesis. While the genes and enzymes of the C_5-pathway could conceivably be regulated by conventional effector mechanisms (i.e., transcriptional activation/repression by effector molecules, end-product allosteric feedback inhibition of the C_5-pathway enzymes by heme or chlorophyll precursors), there is as yet no known mechanism that routes or shunts a bifunctional tRNA into either protein synthesis or a nonprotein biosynthetic process.

Finally, how does the plant cell coordinate ALA synthesis and porphyrin synthesis between the heme-requiring mitochondria and the chlorophyll-requiring chloroplast?

Further genetic and biochemical studies will shed light on the role of tRNA and its role in the activation of the α-carboxyl group of glutamate prior to reduction. Because of the essential role ALA-derived porphyrins play in respiration and photosynthesis, the regulation of ALA formation and the role of tRNA in this process are at the center of the cell's energy generating and utilization systems.

ACKNOWLEDGMENT. Work in the authors' laboratory was supported by grants from the National Institutes of Health and Department of Energy.

8. REFERENCES

Agris, P., Söll, D., and Seno, T., 1973, Biological function of 2-thiouridine in *Escherichia coli* glutamic acid transfer RNA, *Biochemistry* **12**:4331–4337.
Avissar, Y. J., and Beale, S. I., 1988, Biosynthesis of tetrapyrrole pigment precursors. Formation

and utilization of glutamyl-tRNA for δ-aminolevulinic acid synthesis by isolated enzyme fractions, *Plant Physiol.* **88:**879–886.

Avissar, Y. J., and Beale, S. I., 1989a, Identification of the enzymatic basis for δ-aminolevulinic acid auxotrophy in a *hemA* mutant of *Escherichia coli, J. Bacteriol.* **171:**2919–2924.

Avissar, Y. J., and Beale, S. I., 1989b, Biosynthesis of tetrapyrrole pigment precursors, *Plant Physiol.* **89:**852–859.

Avissar, Y. J., and Beale, S. I., 1990, Cloning and expression of a structural gene from *Chlorobium vibrioforme* that complements the *hemA* mutation in *Escherichia coli, J. Bacteriol.* **172:**1656–1659.

Avissar, Y. J., Ormerod, J. G., and Beale, S. I., 1989, Distribution of δ-aminolevulinic acid biosynthetic pathways among phototrophic bacterial groups, *Arch. Microbiol.* **151:**513–519.

Baldauf, S. L., and Palmer, J. D., 1990, Evolutionary transfer of the chloroplast *tufA* gene to the nucleus, *Nature* **344:**262–265.

Beale, S. I., 1990, Biosynthesis of the tetrapyrrole pigment precursor, δ-aminolevulinic acid, from glutamate, *Plant Physiol.* **93:**1273–1279.

Beale, S. I., and Castelfranco, P. A., 1973, ^{14}C incorporation from exogenous compounds into δ-aminolevulinic acid by greening cucumber cotyledons, *Biochem. Biophys. Res. Commun.* **52:**143–149.

Beale, S. I., and Castelfranco, P. A., 1974, The biosynthesis of δ-aminolevulinic acid in higher plants. II. Formation of ^{14}C-δ-aminolevulinic acid from labeled precursors in greening plant tissues, *Plant Physiol.* **53:**297–303.

Beale, S. I., and Weinstein, J. D., 1990, Tetrapyrrole metabolism in photosynthetic organisms, in *Biosynthesis of Heme and Chlorophyll* (H. A. Dailey, ed.), pp. 287–391, McGraw-Hill, New York.

Beale, S. I., Gough, S. P., and Granick, S., 1975, Biosynthesis of δ-aminolevulinic acid from the intact carbon skeleton of glutamic acid in greening barley, *Proc. Natl. Acad. Sci. USA* **72:**2719–2723.

Beale, S. I., Foley, T., and Dzelzkalns, V., 1981, Δ-aminolevulinic acid synthase from *Euglena gracilis, Proc. Natl. Acad. Sci. USA* **78:**1666–1669.

Berry-Lowe, S., 1987, The chloroplast tRNA glutamate gene required for δ-aminolevulinate synthesis, *Carlsberg Res. Commun.* **52:**197–210.

Böck, A., and Stadtman, T. C., 1988, Selenocysteine, a highly specific component of certain enzymes is incorporated by a UGA-directed co-translational mechanism, *Biofactors* **1:**245–250.

Breton, R., Sanfacon, H., Papayannopoulos, T., Biemann, K., and Lapointe, J., 1986, Glutamyl-tRNA synthetase of *Escherichia coli*. Isolation and primary structure of the *gltX* gene and homology with other aminoacyl-tRNA synthetases, *J. Biol. Chem.* **261:**10610–10617.

Breu, V., and Dörnemann, D., 1988, Formation of 5-aminolevulinic acid via glutamate 1-semialdehyde and 4,5-dioxovalerate with participation of a RNA component in *Scenecdesmus obliquus* mutant C-2A, *Biochem. Biophys. Acta* **967:**135–140.

Bruyant, P., and Kannangara, C. G., 1987, Biosynthesis of Δ-aminolevulinate in greening barley leaves. VIII. Purification and characterization of the glutamate-tRNA ligase, *Carlsberg Res. Commun.* **52:**99–109.

Carbon, J. A., Hung, L., and Jones, D. S., 1965, A reversible oxidative inactivation of specific transfer RNA species, *Proc. Natl. Acad. Sci. USA* **53:**979–986.

Castelfranco, P. A., and Beale, S. I., 1983, Chlorophyll biosynthesis: recent advances and areas of current interest, *Annu. Rev. Plant Physiol.* **34:**241–278.

Chang, T. E., Wegmann, B., and Wang, W. Y., 1990, Purification and characterization of glutamyl-tRNA synthetase—An enzyme involved in chlorophyll biosynthesis, *Plant Physiol.* **93:**1641–1649.

Chen, M. W., Jahn, D., O'Neill, G. P., and Söll, D., 1990a, Purification of the glutamyl-tRNA reductase from *Chlamydomonas reinhardtii* involved in δ-aminolevulinic acid formation during chlorophyll biosynthesis, *J. Biol. Chem.* **265:**4058–4063.

Chen, M. W., Jahn, D., Schön, A., O'Neill, G. P., and Söll, D., 1990b, Purification and characterization of *Chlamydomonas reinhardtii* chloroplast glutamyl-tRNA synthetase, a natural misacylating enzyme, *J. Biol. Chem.* **265:**4054–4057.

Chereskin, B. A., and Castelfranco, P. A., 1982, Effects of iron and oxygen on chlorophyll biosynthesis. II. Observations on the biosynthetic pathway in isolated etiochloroplasts, *Plant Physiol.* **69:**112–116.

Craigen, W. J., Cook, R. G., Tate, W. P., and Caskey, C. T., 1985, Bacterial peptide chain release factors: Conserved primary structure and possible frameshift regulation of release factor 2, *J. Mol. Biol.* **138:**179–207.

Dickerman, H. W., Steers, E., Redfield, B. G., and Weissbach, H., 1967, Methionyl soluble ribonucleic acid transformylase. I. Purification and partial characterization, *J. Biol. Chem.* **242:**1522–1525.

Dörnemann, D., Kotzabasis, K., Richter, P., Breu, V., and Senger, H., 1989, The regulation of chlorophyll biosynthesis by the action of protochlorophyllide on glu-t-RNA-ligase, *Botan Acta* **102:**112–115.

Drolet, M., Peloquin, L., Echelard, Y., Cousineau, L., and Sasarman, A., 1989, Isolation and nucleotide sequence of the *hemA* gene of *Escherichia coli* K12, *Mol. Gen. Genet.* **216:**347–352.

Elliott, T., 1989, Cloning, genetic characterization, and nucleotide sequence of the *hemA-prfA* operon of *Salmonella typhimurium*, *J. Bacteriol.* **171:**3948–3960.

Evrard, J. L., Kuntz, M., Straus, N. A., and Weil, J. H., 1988, A class-I intron in a cyanelle tRNA gene from *Cyanophora paradoxa:* Phylogenetic relationship between cyanelles and plant chloroplasts, *Gene* **71:**115–122.

Friedmann, H. C., and Thauer, R. K., 1986, Ribonuclease-sensitive δ-aminolevulinic acid formation from glutamate in cell extracts of *Methanobacterium thermoautotrophicum*, *FEBS Lett.* **207:**84–88.

Gibson, K. D., Laver, W. G., and Neuberger, A., 1958, Initial stages in the biosynthesis of porphyrins. II. The formation of 5-ALA from glycine and succinyl CoA by particles from chicken erythrocytes, *Biochem. J.* **70:**71–76.

Gough, S. P., and Kannangara, C. G., 1976, Synthesis of δ-aminolevulinic acid by isolated plastids, *Carlsberg Res. Commun.* **41:**183–190.

Gough, S. P., Kannangara, C. G., and Block, K., 1989, A new method for the synthesis of glutamate 1-semialdehyde. Characterization of its structure in solution by NMR spectroscopy, *Carlsberg Res. Commun.* **54:**99–108.

Grimm, B., 1990, Primary structure of a key enzyme in plant tetrapyrrole synthesis: Glutamate 1-semialdehyde aminotransferase, *Proc. Natl. Acad. Sci. USA* **87:**4169–4173.

Grimm, B., Bull, A., Welinder, K. G., Gough, S. P., and Kannangara, C. G., 1989, Purification and partial amino acid sequence of the glutamate 1-semialdehyde aminotransferase of barley and *Synechococcus*, *Carlsberg Res. Commun.* **54:**67–79.

Gruissem, W., Narita, J. O., Greenberg, B. M., Prescott, D. M., and Hallick, R. B., 1983, Selective *in vitro* transcription of chloroplast genes, *J. Cell. Biochem.* **22:**31–46.

Hollingsworth, M. J., and Hallick, R. B., 1982, *Euglena gracilis* chloroplast transfer RNA transcription units. Nucleotide sequence analysis of a tRNA^Tyr-tRNA^His-tRNA^Met-tRNA^Trp-tRNA^Glu-tRNA^Gly gene cluster, *J. Biol. Chem.* **257:**12795–12799.

Holschuh, K., Bottomley, W., and Whitfield, P. R., 1984, Organization and nucleotide sequence of the genes for spinach chloroplast tRNA, *Plant Mol. Biol.* **3:**313–317.

Hoober, J. K., Kahn, A., Ash, D. E., Gough, S., and Kannangara, C. G., 1988, Biosynthesis of Δ-aminolevulinate in greening barley leaves. IX. Structure of the substrate, mode of gabaculine inhibition, and the catalytic mechanism of glutamate 1-semialdehyde aminotransferase, *Carlsberg Res. Commun.* **53**:11–25.

Houen, G., Gough, S. P., and Kannangara, C. G., 1983, Δ-aminolevulinate synthesis in greening barley V. The structure of glutamate 1-semialdehyde, *Carlsberg Res. Commun.* **48**:567–572.

Houghton, J. D., Brown, S. B., and Gough, S. P., 1989, Biosynthesis of δ-aminolevulinate in *Cyanidium caldarium:* Characterization of tRNA^Glu, ligase, dehydrogenase and glutamate 1-semialdehyde aminotransferase, *Carlsberg Res. Commun.* **54**:131–143.

Huang, L., and Castelfranco, P. A., 1989, Regulation of 5-aminolevulinic acid synthesis in developing chloroplasts. I. Effect of light/dark treatments *in vivo* and *in organello, Plant Physiol.* **90**:996–1002.

Huang, D. D., and Wang, W. Y., 1986, Chlorophyll biosynthesis in *Chlamydomonas* starts with the formation of glutamyl-tRNA, *J. Biol. Chem.* **261**:13451–13455.

Huang, D. D., Wang, W. Y., Gough, S. P., and Kannangara, C. G., 1984, Δ-Aminolevulinic acid-synthesizing enzymes need an RNA moiety for activity, *Science* **225**:1482–1484.

Huang, D. D., Bonner, B. A., and Castelfranco, P. A., 1989, Regulation of 5-aminolevulinic acid (ALA) synthesis in developing chloroplasts. II. Regulation of ALA-synthesizing capacity by phytochrome, *Plant Physiol.* **90**:1003–1008.

Jahn, D., Chen, M. W., and Söll, D., 1991a, Purification and functional characterization of glutamate-1-semialdehyde aminotransferase from *Chlamydomonas reinhardtii, J. Biol. Chem.* **266**:161–167.

Jahn, D., Kim, Y. C., Ishino, Y., Chen, M. W., and Söll, D., 1991b, Purification and functional characterization of the Glu-tRNA^Gln aminotransferase from *Chlamydomonas reinhardtii, J. Biol. Chem.* **265**:8059–8064.

Jahn, D., Michelsen, U., and Söll, D., 1991c, Two glutamyl-tRNA reductase activities in *Escherichia coli, J. Biol. Chem.* **266**:2542–2548.

Jordan, P. M., 1990, Biosynthesis of 5-aminolevulinic acid and its transformation into coproporphyrinogen in animals and bacteria, in *Biosynthesis of Heme and Chlorophylls* (H. A. Dailey, ed.), pp. 55–121, McGraw-Hill, New York.

Kannangara, C. G., 1991, Biochemistry and molecular biology of chlorophyll biosynthesis, in *Cell Culture and Somatic Cell Genetics of Plants* (L. Bogorad and I. K. Vasil, eds.), The Molecular Biology of Plastids and Mitochondria, Vol. 7B, Academic Press, New York, pp. 302–321.

Kannangara, C. G., and Gough, S. P., 1977, Synthesis of δ-aminolevulinic acid and chlorophyll by isolated plastids, *Carlsberg Res. Commun.* **42**:441–458.

Kannangara, C. G., and Gough, S. P., 1978, Biosynthesis of δ-aminolevulinate in greening barley leaves: Glutamate 1-semialdehyde aminotransferase, *Carlsberg Res. Commun.* **43**:185–194.

Kannangara, C. G., and Gough, S. P., 1979, Biosynthesis of δ-aminolevulinate in greening barley leaves. II. Induction of enzyme synthesis by light, *Carlsberg Res. Commun.* **44**:11–20.

Kannangara, C. G., and Schouboe, A., 1985, Biosynthesis of δ-aminolevulinate in greening barley leaves. VII. Glutamate 1-semialdehyde accumulation in gabaculine treated leaves, *Carlsberg Res. Commun.* **50**:179–191.

Kannangara, C. G., Gough, S. P., Oliver, R. P., and Rasmussen, S. K., 1984, Biosynthesis of δ-aminolevulinate in greening barley leaves. VI. Activation of glutamate by ligation to RNA, *Carlsberg Res. Commun.* **49**:417–437.

Kannangara, C. G., Gough, S. P., Bruyant, P., Hoober, J. K., Kahn, A., and von Wettstein, D., 1988, tRNA^Glu as a cofactor in δ-aminolevuinate biosynthesis: Steps that regulate chlorophyll synthesis, *Trends Biochem. Sci.* **13**:139–143.

Kikuchi, G., Kumar, A., Talmage, P., and Shemin, D., 1958, The enzymatic synthesis of δ-aminolevulinic acid, *J. Biol. Chem.* **233**:1214–1219.

Komine, Y., Adachi, T., Inokuchi, H., and Ozeki, H., 1990, Genomic organization and physical mapping of the transfer RNA genes in *Escherichia coli* K12, *J. Mol. Biol.* **212**:579–598.

Kuntz, M., Weil, J. H., and Steinmetz, A., 1984, Nucleotide sequence of a 2 kbp *Bam*Hl fragment of *Vicia faba* chloroplast DNA containing the genes for threonine, glutamic acid and tyrosine transfer RNAs, *Nucleic Acids Res.* **12**:5037–5047.

Labbe-Bois, R., and Labbe, P., 1990, Tetrapyrrole and heme biosynthesis in the yeast *Saccharomyces cerevisiae*, in *Biosynthesis of Heme and Chlorophyll* (H. A. Dailey, ed.), pp. 235–285, McGraw-Hill, New York.

Lapointe, J., and Söll, D., 1972, Glutamyl-tRNA synthetase of *Escherichia coli*. I. Purification and properties, *J. Biol. Chem.* **247**:4966–4974.

Lapointe, J., Levasseur, S., and Kern, D., 1985, Glutamyl-tRNA synthetase from *Escherichia coli*, *Methods Enzymol.* **113**:42–49.

Li, J. M., Brathwaite, O., Cosloy, S. D., and Russell, C. S., 1989a, 5-Aminolevulinic acid synthesis in *Escherichia coli*, *J. Bacteriol.* **171**:2547–2552.

Li, J. M., Russell, C. S., and Cosloy, S. D., 1989b, Cloning and structure of the *hemA* gene of *Escherichia coli* K12, *Gene* **82**:209–217.

London, I. M., Levin, D. H., Matts, R. L., Thomas, N. S. B., Petryshyn, R., and Chen, J. J., 1987, Regulation of protein synthesis, *Enzymes* **18**:359–380.

Mau, Y. L., and Wang, W. Y., 1988, Biosynthesis of Δ-aminolevulinic acid in *Chlamydomonas reinhardtii*, *Plant Physiol.* **86**:793–797.

Mau, Y. H., Wang, W. Y., Tamura, R. N., and Chang, T. E., 1987, Identification of an intermediate of δ-aminolevulinic acid biosynthesis in *Chlamydomonas* by HPLC, *Arch. Biochem. Biophys.* **255**:75–79.

Mauzerall, D., and Granick, S., 1956, The occurrence and determination of δ-aminolevulinic acid and porphobilinogen in urine, *J. Biol. Chem.* **219**:435–446.

May, B. K., Borthwick, I. A., Srivastava, G., Pirola, B. A., and Elliott, W. H., 1986, Control of 5-aminolevulinate synthases in animals, *Current Topics Cell. Reg.* **28**:233–262.

Meller, E., Belkin, S., and Harel, E., 1975, The biosynthesis of δ-aminolevulinic acid in greening maize leaves, *Phytochemistry* **14**:2399–2402.

Migita, L. K., and Doi, R. H., 1970, Formylation of methionyl-transfer RNA from prokaryotes and eukaryotes by *Bacillus subtilis* transformylase, *Arch. Biochem. Biophys.* **138**:457–463.

Mukherjee, J. J., and Dekker, E. E., 1990, 2-Amino-3-ketobutyrate CoA ligase of *Escherichia coli*: Stoichiometry of pyridoxal phosphate binding and location of the pyridoxal lysine peptide in the primary structure of the enzyme, *Biochem. Biophys. Acta* **1037**:24–29.

Oh-Hama, T., Seto, H., and Miyachi, S., 1986a, ^{13}C NMR evidence for bacteriochlorophyll *c* formation by the C_5 pathway in the green sulfur bacterium *Prosthecochloris*, *Eur. J. Biochem.* **159**:189–194.

Oh-Hama, T., Seto, H., and Miyachi, S., 1986b, ^{13}C-NMR evidence of bacteriochlorophyll *a* formation by the C_5 pathway in *Chromatium*, *Arch. Biochem. Biophys.* **246**:192–198.

Oh-Hama, T., Seto, H., Otake, N., and Miyachi, S., 1982, ^{13}C-NMR evidence for the pathway of chlorophyll biosynthesis in green algae, *Biochem. Biophys. Res. Commun.* **105**:647–652.

Oh-Hama, T., Stolowich, N. J., and Scott, A. J., 1988, 5-Aminolevulinic acid formation from glutamate via the C_5-pathway in *Clostridium thermoaceticum*, *FEBS Lett.* **228**:89–93.

Ohme, M., Kamogashira, T., Shinozaki, K., and Sugiura, M., 1985, Structure and cotranscription of tobacco chloroplast genes for tRNA^{Glu}(UUC), tRNA^{Tyr}(GUA), and tRNA^{Asp}(GUC), *Nucleic Acids Res.* **13**:1045–1056.

Ohyama, K., Fukuzawa, H., Kohchi, T., Shirai, H., Sano, T., Sano, S., Umesono, K., Shiki, Y., Takeuchi, M., Chang, Z., Aota, S., Inokuchi, H., and Ozeki, H., 1986, Chloroplast gene organization deduced from complete sequence of liverwort *Marchantia polymorpha* chloroplast DNA, *Nature* **322**:572–574.

O'Neill, G. P., and Söll, D., 1990a, Transfer RNA and the formation of the heme and chlorophyll precursor, 5-aminolevulinic acid, *Biofactors* **2**:227–234.

O'Neill, G. P., and Söll, D., 1990b, Expression of the *Synechocystis* sp. strain PCC 6803 tRNA^Glu gene provides tRNA for protein and chlorophyll biosynthesis, *J. Bacteriol.* **172**:6363–6371.

O'Neill, G. P., Peterson, D. M., Schön, A., Chen, M. W., and Söll, D., 1988, Formation of the chlorophyll precursor δ-aminolevulinic acid in cyanobacteria requires aminoacylation of a tRNA^Glu species, *J. Bacteriol.* **170**:3810–3816.

O'Neill, G. P., Chen, M. W., and Söll, D., 1989, Δ-Aminolevulinic acid biosynthesis in *Escherichia coli* and *Bacillus subtilis* involves formation of glutamyl-tRNA, *FEMS Microbiol. Lett* **60**:255–260.

O'Neill, G. P., Schön, A., Chow, H., Chen, M. W., Kim, Y. C., and Söll, D., 1990, Sequence of tRNA^Glu and its genes from the chloroplast genome of *Chlamydomonas reinhardtii, Nucleic Acids Res.* **18**:5893.

Otsuka, A. J., Buoncristiani, M. R., Howard, P. K., Flamm, J., Johnson, C., Yamamoto, R., Uchida, K., Cook, C., Ruppert, J., and Matsuzaki, J., 1988, The *Escherichia coli* biotin biosynthetic enzyme sequences predicted from the nucleotide sequence of the bio operon. *J. Biol. Chem.* **263**:19577–19585.

Peterson, D., Schön, A., and Söll, D., 1988, The nucleotide sequences of barley cytoplasmic transfer RNAs and structural features essential for formation of δ-aminolevulinic acid, *Plant Mol. Biol.* **11**:293–299.

Petricek, M., Rutberg, L., Schröder, I., and Hederstedt, L., 1990, Cloning and characterization of the *hemA* region of the *Bacillus subtilis* chromosome, *J. Bacteriol.* **172**:2250–2258.

Porra, R. J., Klein, O., and Wright, P. E., 1983, The proof by ^13^C-NMR spectroscopy of the predominance of the C₅ pathway over the Shemin pathway in chlorophyll biosynthesis in higher plants and the formation of the methyl ester group of chlorophyll from glycine, *Eur. J. Biochem.* **130**:509–516.

Proulx, M., and Lapointe, J., 1985, Purification of glutamyl-tRNA synthetase from *Bacillus subtilis, Methods Enzymol.* **113**:50–54.

Proulx, M., Duplain, L., Lacoste, L., Yaguchi, M., and Lapointe, J., 1983, The monomeric glutamyl-tRNA synthetase from *Bacillus subtilis* 168 and its regulatory role. Their purification, characterization, and the study of their interaction, *J. Biol. Chem.* **258**:753–759.

Quigley, F., and Weil, J. H., 1985, Organization and sequence of five tRNA genes and of an unidentified reading frame in the wheat chloroplast genome: evidence for gene rearrangements during the evolution of chloroplast genomes, *Curr. Genet.* **9**:495–503.

Rasmussen, O. F., Stummann, B. M., and Henningsen, K. W., 1984, Nucleotide sequence of a 1.1 kb fragment of the pea chloroplast genome containing three tRNA genes, one of which is located within an open reading frame of 91 codons, *Nucleic Acids Res.* **12**:9143–9153.

Ratinaud, M. H., Thomes, J. C., and Julien, R., 1983, Glutamyl-tRNA synthetases from wheat. Isolation and characterization of three dimeric enzymes, *Eur. J. Biochem.* **135**:471–477.

Rieble, S., and Beale, S. I., 1988, Transformation of glutamate to δ-aminolevulinic acid by soluble extracts of *Synechocystis* sp. PCC 6803 and other oxygenic prokaryotes, *J. Biol. Chem.* **263**:8864–8871.

Rieble, S., Ormerod, J. G., and Beale, S. I., 1989, Transformation of glutamate to 5-aminolevulinic acid by soluble extracts of *Chlorobium vibrioforme, J. Bacteriol.* **171**:3782–3787.

Schneegurt, M. A., and Beale, S. I., 1986, Biosynthesis of protoheme and heme *a* from glutamate in maize, *Plant Physiol.* **81**:965–971.

Schneegurt, M. A., and Beale, S. I., 1988, Characterization of the tRNA required for biosynthesis of 5-aminolevulinic acid from glutamate. Purification by anticodon-based affinity chromatography and determination that the UUC glutamate anticodon is a general requirement for function in ALA biosynthesis, *Plant Physiol.* **86**:497–504.

Schneegurt, M. A., Rieble, S., and Beale, S. I., 1988, The tRNA required for *in vitro* δ-aminolevulinic acid formation from glutamate in *Synechocystis* extracts, *Plant Physiol.* **88**:1358–1366.

Schön, A., and Söll, D., 1988a, Transfer RNA specificity of a mischarging aminoacyl-tRNA synthetase: Glutamyl-tRNA synthetase from barley chloroplasts, *FEBS Lett.* **228**:241–244.

Schön, A., and Söll, D., 1988b, Dual role of glutamyl-tRNA in barley chloroplasts: involvement in chlorophyll and protein biosynthesis, *Arch. Biol. Med. Ex.* **21**:467–474.

Schön, A., Krupp, G., Gough, S., Berry-Lowe, S., Kannangara, C. G., and Söll, D., 1986, The RNA required in the first step of chlorophyll biosynthesis is a chloroplast glutamate tRNA, *Nature* **322**:281–284.

Schön, A., Kannangara, C. G., Gough, S., and Söll, D., 1988, Protein biosynthesis in organelles requires misaminoacylation of tRNA, *Nature* **331**:187–190.

Shemin, D., and Russell, C. S., 1953, 5-Aminolevulinic acid, its role in the biosynthesis of porphyrins and purines, *J. Am. Chem. Soc.* **75**:4873–4875.

Shinozaki, K., Ohme, M., Tanaka, M., Wakasugi, T., Hayashida, N., Matsubayashi, T., Zaita, N., Chunwongse, J., Obokata, J., Yamaguchi-Shinozaki, K., Ohto, C., Torazawa, K., Meng, B. Y., Sugita, M., Deno, H., Kamogashira, T., Yamada, K., Kusuda, J., Takaiwa, F., Katol, A., Tohdoh, N., Shimada, H., and Sugiura, M., 1986, The complete nucleotide sequence of the tobacco chloroplast genome: its gene organization and expression, *EMBO J.* **5**:2043–2049.

Soper, T. S., and Manning, J. M., 1982, Inactivation of pyridoxal phosphate enzymes by gabaculine-correlation with enzymic exchange of beta-protons, *J. Biol. Chem.* **257**:13930–13936.

Sprinzl, M., Hartmann, T., Weber, J., Blank, J., and Zeidler, R., 1989, Compilation of tRNA sequences and sequences of tRNA genes, *Nucleic Acids Res.* **17**:1–172.

Stadtman, T. C., Davis, J. N., Zehelein, E., and Böck, A., 1989, Biochemical and genetic analysis of *Salmonella typhimurium* and *Escherichia coli* mutants defective in specific incorporation of selenium into formate dehydrogenase and tRNAs, *Biofactors* **2**:35–44.

Stryer, L., 1988, *Biochemistry*, 3rd ed., pp. 366–367, W. H. Freeman, San Francisco.

Tanizawa, K., Masu, Y., Asano, S., Tanaka, H., and Soda, K., 1989, Thermostable D-amino acid aminotransferase from a thermophilic *Bacillus* species. Purification, characterization, and active site sequence determination, *J. Biol. Chem.* **264**:2445–2449.

Urata, G., and Granick, S., 1963, Biosynthesis of α-aminoketones and the metabolism of aminoacetone, *J. Biol. Chem.* **238**:811–820.

Verkamp, E., and Chelm, B. K., 1989, Isolation, nucleotide sequence, and preliminary characterization of the *Escherichia coli* K-12 *hemA* gene, *J. Bacteriol.* **171**:4728–4735.

von Wettstein, D., 1991, Chlorophyll biosynthesis (in press).

Wang, W. Y., Gough, S. P., and Kannangara, C. G., 1981, Biosynthesis of δ-aminolevulinate in greening barley leaves. IV. Isolation of three soluble enzymes required for the conversion of glutamate to δ-aminolevulinate, *Carlsberg Res. Commun.* **46**:243–257.

Wang, W. Y., Huang, D. D., Stachon, D., Gough, S. P., and Kannangara, C. G., 1984, Purification, characterization, and fractionation of the δ-aminolevulinic acid synthesizing enzymes from light-grown *Chlamydomonas reinhardtii* cells, *Plant Physiol.* **74**:569–575.

Weinstein, J. D., and Beale, S. I., 1983, Separate physiological roles and subcellular compartments for two tetrapyrrole biosynthetic pathways in *Euglena gracilis*, *J. Biol. Chem.* **258**:6799–6807.

Weinstein, J. D., and Beale, S. I., 1985, RNA is required for enzymatic conversion of glutamate to δ-aminolevulinate by extracts of *Chlorella vulgaris*, *Arch. Biochem. Biophys.* **239**:87–93.

Weinstein, J. D., Mayer, S. M., and Beale, S. I., 1987, Formation of δ-aminolevulinic acid from glutamic acid in algal extracts. Separation into an RNA and three required enzyme components by serial affinity chromatography, *Plant Physiol.* **84**:244–250.

Whatley, J. M., and Whatley, F. R., 1981, Chloroplast evolution. *New Phytol.* **87**:233–247.

Wilcox, M., and Nirenberg, M., 1968, Transfer RNA as a cofactor coupling amino acid synthesis with that of protein, *Proc. Natl. Acad. Sci. USA* **61**:229–236.

Woese, C. R., 1987, Bacterial evolution, *Microbiol. Rev.* **51**:221–271.

Chapter 12

Biochemical and Molecular Studies on Plant Development *In Vitro*

Hans-Jörg Jacobsen

1. INTRODUCTION

The origin of plant cell and tissue culture as a technique can be traced back to the early days of plant developmental biology, when the German botanist Haberlandt in the first two decades of this century tried to understand plant development and to experimentally prove his hypothesis of plant cell totipotency (Haberlandt, 1902). His attempts to develop a methodology for obtaining controlled growth of excised plant tissues failed because of the limited knowledge of the nutritional and hormonal requirements of plant tissues and cells grown under aseptic conditions during his era, but today a vast amount of literature on *in vitro* regeneration is available. All these reports clearly show that many plant cells indeed possess the unique character of totipotency. The first scientist to grow plant tissues *in vitro* was P. R. White, who had kept tomato roots in a liquid medium since 1934 (cited in Chaleff, 1980). A great step forward in understanding plant development was made after elucidation of the role that phytohormones play in controlling tissue culture techniques. The role of auxins and cytokinins as the key

Hans-Jörg Jacobsen   Institut für Genetik, Universität Bonn, D-5300 Bonn-1, Germany.   Present address: Lehrgebiet Molekubr-genetik, FB Biologie, Universität Hannover, D-3000 Hannover, Germany.

Subcellular Biochemistry, Volume 17: Plant Genetic Engineering, edited by B. B. Biswas and J. R. Harris. Plenum Press, New York, 1991.

factors for triggering either cell proliferation (callus formation and cell suspensions) or root and shoot development was elucidated by the elegant work of Skoog and co-workers in 1950 in tobacco (cited in Skoog, 1971).

About 20–30 years ago, plant tissue and cell culture techniques were discovered and found to be promising tools for developing new strategies for crop improvement. It was expected that the techniques already available would have a great impact on the process of breeding new and better varieties in a shorter time. The validity of the new techniques was proven in subsequent years with well-known model systems such as *Nicotiana tabacum, Datura* sp., and *Daucus carota,* where plants could be regenerated almost at will from explants or even single cells and somatic hybrids could be created by protoplast fusion. But it was soon realized that not all plants are as easy to handle as these model systems with respect to their *in vitro* regenerability.

Yet, even today, the cereals and the seed legumes, to name the two most important groups of crop plants for human and animal nutrition, must be considered as being recalcitrant species with respect to *in vitro* regeneration. Thus the regeneration barrier is still the major bottleneck preventing full application of the emerged new technologies in crop improvement. Despite the numerous articles published on plant cell regeneration in the past and the vast amount of knowledge accumulated, plant tissue and cell culture is still an empirical science, since only little is known about the molecular basis of differentiation. The development of regeneration schemes is time and money consuming, and success is not at all predictable. It is my aim in this chapter to discuss some of the unsolved problems and to focus on some recent developments in plant molecular and developmental biology which are directed to furthering our understanding of the physiology and molecular biology underlying plant development *in vitro*. Special attention will be paid to recent developments in plant hormone receptor research and signal transduction systems in plants. The applied branch of plant cell and tissue culture, in particular plant biotechnology, will take advantage of a better understanding of fundamental processes in plant development. To achieve this, better cooperation between various disciplines in plant science, including plant physiology, plant molecular biology, and developmental cell biology among others, is required in the future.

2. PATHWAYS FOR PLANT REGENERATION: ORGANOGENESIS VERSUS SOMATIC EMBRYOGENESIS

Regeneration of plants from isolated cells or tissues can be achieved by subsequentially inducing organogenesis (root and shoot formation) or via the induction of somatic embryos, first discovered independently by Steward *et al.* (1958) and Reinert (1958) in carrot. This pathway can be regarded as having the best potential for simple, efficient, practical applications. Somatic embryo-genesis, the process leading from putatively single somatic cells to embryos and, subsequently, to plants, was found in many plant species to be dependent on the sole application of strong auxins at relatively high concentrations (for review, see Kysely and Jacobsen, 1990). It is interesting to note that the effect of auxin on embryo induction can be counteracted by cytokinin (Kysely and Jacobsen, 1990).

In general, somatic embryogenesis can be divided into five distinct stages (stages 0, 1, 2, globular, and heart shape), where the transition of state 0 cells to state 1 cells is dependent on the application of auxin (phase 0 or "auxin$^+$"). Embryogenic cell clusters of state 1 develop during the auxin-free phases I–III to globular and then to heart-shaped embryos (for review, see Komamine *et al.*, 1990). More simply, one may distinguish between *somatic embryo induction,* which is auxin-dependent, and *embryo development,* where auxins are inhibitory. Considerable progress has been reported with respect to processes associated with embryo development (see Ammirato, 1989, and Komamine *et al.*, 1990, for further references), but little is known about the molecular mechanisms of the induction process, which, however, in most systems such as carrot or the cereals may be difficult to access.

2.1. The Genotype Problem

Despite some recent breakthroughs in regeneration of certain important crop species via somatic embryogenesis, it should not be forgotten that by adapting and optimizing existing protocols based on trial-and-error approaches, many schemes for plant cell regeneration are restricted to a few genotypes or cultivars of a given species. Therefore, besides the culture conditions applied, the genetic background strongly influences the development of plant cells *in vitro*. It has been found by classical genetics and linkage analysis studies that the number of genes controlling regeneration apparently is rather limited (Peng and Hodges, 1989; Willman *et al.*, 1989; Nadolska-Orczyk and Malepszy, 1989; Koorneef *et al.*, 1987; Koorneef and Hanhart, 1990; Wricke, personal communication). Both nuclear and cytoplasmic genes seem to be involved in the regeneration process. Linkage analysis studies in tomato revealed the location of the two dominant genes in tomato on chromosomes 2 and 3, respectively (Koorneef, personal communication). The nature of these genes and their respective role(s) in normal plant development are completely unknown, and there is, of course, no way to call these genes "regeneration genes," because it is obvious that the actions of these genes under certain *in vitro* conditions reflect just another facet of pleiotropism. The molecular and functional characterization of these genes will give more insight with respect to the important steps for plant cell development *in vivo* and *in vitro*.

On the other hand, the effect that in the same species under certain condi-tions explants or cell lines derived from one genotype will regenerate into a plant, while those from another will not, provides the opportunity to characterize these genes at the molecular level by means of differentially screening cDNA libraries. An additional approach is provided when a genotype capable of regeneration will not regenerate under conditions where a single factor (i.e., a phytohormone) is changed. In both cases, the specific and differential behavior can be used to study, at the biochemical and molecular level, alterations being associated or correlated with the morphogenetic response.

2.2. Auxins and Somatic Embryogenesis

Auxins have been the best-studied class of plant hormones, since the discovery of the first native auxin β-indole acetic acid (IAA) by Kögl *et al.* (1932). The literature on auxin effects and their putative mode(s) of action(s) is almost uncalculable. The main effects of auxins are listed in Table I. Auxin effects can be separated into rapid (or short-term) and morphogenetic (or long-term) effects. The group of compounds exerting one or another auxin effect is rather extensive, including both natural and synthetic substances. The terminology of naming a compound an auxin depends on its auxin effects in plants or *in vitro*, rather than on structural relationships.

It is important to realize, but often neglected, that the different auxins differ considerably in their effectivity and specificity, depending on the type of auxin effect being analyzed: auxins like IAA or one of its synthetic analogs, β-naphthyl acetic acid (NAA), which both cause strong effects in cell elongation, are absolutely inactive with respect to somatic embryo induction (Kysely and Jacobsen, 1990). On the contrary, potent inducers of somatic embryogenesis such as 2,4-dichlorophenoxy acetic acid (2,4-D) or the auxin-active herbicide picloram have an effect on cell elongation only at concentrations which, in the case of picloram,

Table I
Short-Term and Long-Term Effects of Auxin

Short-term effects
  Cell elongation
  Stimulation of cytoplasma circulation
  Stimulation of cell-to-cell water transport
  Hyperpolarization of plasma membrane
  Proton secretion
Long-term effects
  Control of apical dominance
  Root induction and growth
  Flowering of long-day plants
  Callus induction
  Induction of somatic embryogenesis

are about 50 times higher than those which are optimal for IAA or NAA to stimulate cell elongation (Chang and Foy, 1983). It can therefore be concluded that plant cells should have mechanisms to distinguish between different auxins (or phytohormones in general), and it can be completely misleading to render a particular auxin inactive, if it has no effect or only weak effects in one particular assay system designed to detect rapid effects such as cell elongation. In addition, plant cells express different degrees of competence to a given hormone: While mesophyll protoplasts of pea so far cannot routinely be regenerated to plants via somatic embryogenesis, a genotype-dependent percentage of calli regenerated from protoplasts (isolated from young lateral buds of pea) can be induced to form somatic embryos and plants (Lehminger-Mertens and Jacobsen, 1989). Moreover, we found in the same study that for some genotypes, 2,4-D applied in an identical molarity as picloram appeared to be lethal, while with picloram somatic embryo formation was induced. Other genotypes reacted in the same way to either 2,4-D or picloram. In cell cultures of *Morinda* species, Zenk *et al.*(1984) found an inhibition of anthraquinone synthesis by 2,4-D and a stimulation by NAA. Spena *et al.* (1987) reported that some of the *rol* genes from the T-DNA of *Agrobacterium rhizogenes* modulate the function of cytokinins, and Shen *et al.* (1988) found a higher sensitivity of hairy roots to auxin than their normal counterparts. Both the ability to apparently distinguish between different auxins and the expression of different levels of competence in the reactions to auxin application point to a key role of signal recognition systems for auxins.

3. SIGNALS AND SIGNAL TRANSDUCTION SYSTEMS IN PLANT DEVELOPMENT

Although a bulk of literature reporting successful regeneration protocols in almost all plant genera has been accumulated in the past, a general concept for plant regeneration *in vitro* is still lacking. A number of factors influencing plant cell development in a positive or negative way in tissue culture have been reported, but in most of the cases reported, we have almost no idea whether direct or indirect, primary or secondary effects were monitored. In order to solve this particular problem, a better knowledge of specific and general signal transduction mechanisms for the various signals and triggers controlling plant development is required. In a recent elegant review, Sethi and Guha-Mukherjee (1990) discussed the available data on the role of the cell 'cycle and signal transduction pathways in *in vitro* plant cell differentiation and regeneration. The available literature indicates that in some areas there are great homologies between animal and plant systems. This is true at least for signal transduction pathways at the membrane level (Guern *et al.*, 1990), i.e., with respect to the role of some second messengers such as diacylglycerol (DG), calcium, or inositole triphosphate (IP_3). There are, however, also striking differences between the two king-doms; i.e., in plant systems cyclic nucleotides seem not to play a comparable and central role as they do in animal and fungal systems (Amrhein, 1977). On the other hand, animals do not possess light-dependent controlling mechanisms like the plants have with the phytochrome system. The plant growth regulators or phytohormones, i.e., the auxins and cytokinins, although synthesized also in animals or bacteria, exert eminent morphogenetic effects only in plants. Their relative concentrations and combinations control plant organogenesis *in vitro* and *in vivo* qualitatively (Skoog, 1971). In many species it is clear that, depending on the concentration applied, auxins induce and promote callus or root growth, whereas an excess of cytokinins, which also promote cell division or retard senescence, is able to trigger explants to develop shoots.

Plant Hormone Receptors

Signal recognition and transduction systems for plant hormones have been studied for almost 20 years (for a recent review, see Napier and Venis, 1990). As of this writing, a number of proteins have been described that fulfill most of the requirements for receptor functions (mainly on binding kinetics and specificity). With respect to auxins, binding sites with both a high affinity and great specifici-ty have been detected and characterized on membranes (mABP on plasmalemma and ER; Hertel *et al.*, 1972; Bhattacharyya and Biswas, 1978; Löbler and Klämbt, 1985a,b; Hesse *et al.*, 1989; Venis and Napier, 1990), as well as being soluble proteins in the cytoplasm and nucleus (sABP; Oostrom *et al.*, 1980; Jacobsen, 1982, 1984; Jacobsen and Hajek, 1985; Jacobsen *et al.*, 1987; Herber *et al.*, 1988; Sakai, 1985; Sakai and Hanagata, 1985; Sakai *et al.*, 1986; Paulus and Jacobsen, 1989). Reports are available of some studies in which it was demonstrated that receptor-containing cytosol fractions stimulated *in vitro* tran-scription (Sakai *et al.*, 1986; van de Linde *et al.*, 1984; Libbenga *et al.*, 1987), indicating a mechanism that might be similar to the mode of action of steroid hormones in animal and fungal systems. The stimulation of transcription was found to correlate with the concentration of auxin-binding protein, and the op-timum auxin concentration was 10^{-8} M for 2,4-D and 10^{-7} M for IAA, respec-tively (Figure 1). It was also demonstrated that auxins are bound with a high affinity in isolated soybean nuclei (Paulus and Jacobsen, 1989).

First evidence that auxin-binding proteins can induce specific transcripts *in vitro* in isolated nuclei only in the presence of auxin was presented by Kikuchi *et al.* (1989), who found an increase in the synthesis of two polypeptides after incubating isolated mung bean nuclei with auxin and sABP preparations and *in vivo* upon auxin treatment. There is, however, no evidence yet that soluble auxin receptors represent a class of transcriptional activation factors, although these factors are likely to exist in plants (Sommer *et al.*, 1990).

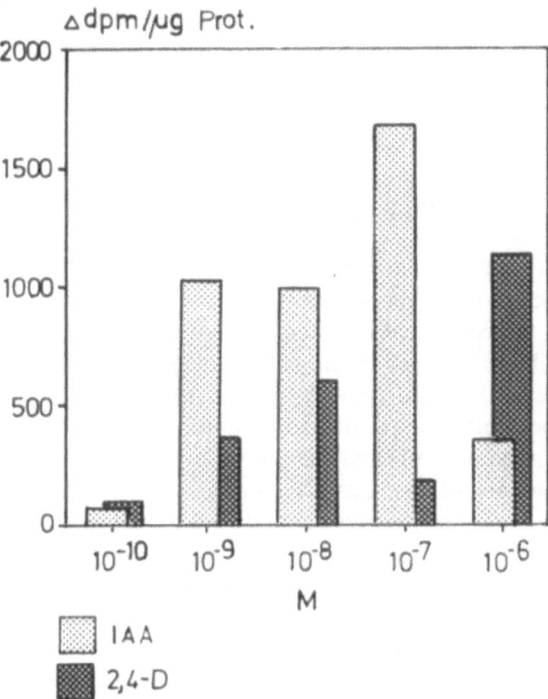

△dpm/µg Prot.

M

IAA

2,4-D

FIGURE 1. Effect of various auxin concentrations on sABP-dependent stimulation of *in vitro* transcription in isolated pea nuclei (B. Köllner, personal communication).

In soybean cell suspensions, both growth-stage-dependent modulations in the number of binding sites in cytoplasm and nucleus and shifts in their apparent binding affinities were found (Herber *et al.*, 1988; Paulus and Jacobsen, 1989). During the *lag* phase of cell growth, high-affinity binding was only detectable in the cytoplasm, while during the *log* phase it could be detected in isolated nuclei (Paulus and Jacobsen, 1989). From these data it can be concluded that in plants, obviously two different signal recognition and transduction systems for auxins are likely to exist, but it is not clear at the moment whether these pathways are linked or whether particular cells express both mechanisms at the same time. Also, it is unclear how the binding interacts with cellular and morphogenetic responses. So far, none of the hormone-binding proteins can be called a true receptor, since no plant hormone–induced response could be shown unequivocally to be dependent on hormone binding (Napier and Venis, 1990). But it is not unlikely that the membrane-located binding sites are associated indirectly with the rapid auxin effects of cell elongation, hyperpolarization of the plasmalemma, or H^+ secretion (Barbier-Brygoo *et al.*, 1989; Klämbt, 1990), while the sites found in both the cytoplasm and the nucleus may be involved in controlling long-term morphogenetic responses to auxins.

4. BIOCHEMICAL AND MOLECULAR MARKERS

4.1. Markers for Somatic Embryogenesis

The transition of cells in a tissue (or when cultured *in vitro*) from one developmental state to another or from a nonembryogenic to an embryogenic state depends on the controlled activation and repression of specific genes and, accordingly, results in changes in the pattern of gene products. These changes can be monitored and should provide some initial information about the biochemistry and molecular control behind this process.

Sung and Okimoto (1981) and Choi and Sung (1984) demonstrated specific changes in the polypeptide composition during somatic embryogenesis. In two cell lines each independently derived from two different genotypes of pea, Stirn and Jacobsen (1987) found two polypeptides that were specific for an embryo-like differentiation pattern. These proteins showed some homologies with respect to their molecular weights and their IEP to two proteins found in carrot (Sung and Okimoto, 1981). Two hybridoma clones producing embryo-specific monoclonal antibodies were isolated after a differential screening of hybridoma libraries derived from embryogenic calli and somatic embryos of pea. They recognize two polypeptides (20 and 50 kDa, respectively) specific for somatic embryos and/or embryogenic callus of pea and young globular somatic embryos of carrot (Stirn and Jacobsen, 1990). Recent results indicate that the 50-kDa protein is auxin-inducible within a short time after exposure of the tissues to auxin, and it is expressed in apical tissues, hypocotyl, and root tips of young seedlings (Stirn *et al.*, submitted). This may be of particular importance, since the application of auxin (2,4-D or picloram) was found to be the inductive event for somatic embryogenesis in pea and other legumes (Kohlenbach, 1978, 1985; Kysely and Jacobsen, 1990; Lehminger-Mertens and Jacobsen, 1989; Pickardt *et al.*, 1989; Bögre *et al.*, 1990; Jacobsen, 1991). It should be noted, however, that for none of the marker proteins reported so far could indications for a *functional* involvement in the process of somatic embryogenesis be derived from their amino acid sequences (when available) or otherwise be demonstrated. On the other hand, it can be expected that knowledge on unequivocally characterized marker proteins may be important for the early detection of morphogenetic pathways and thus speed up the process of developing regeneration protocols.

Extracellular proteins secreted to the medium have also been reported to play a key role in carrot somatic embryogenesis (de Vries *et al.*, 1988; Booij *et al.*, 1990). Particular secreted proteins have been found to interact in the restoration of somatic embryogenesis in some nonembryogenic suspension lines.

4.2. Auxin-Inducible Genes

A number of papers were published in the past decade reporting auxin-inducible genes and proteins (Alliotte *et al.*, 1989; for a review, see Key, 1989). Whether or not these genes are involved in *in vitro* differentiation processes remains unknown, since most of the data reported were obtained using intact plants or excised organs treated for some time with auxin. van der Zaal *et al.* (1987) reported auxin-inducible transcripts in tobacco cell suspensions, which accumulate prior to auxin-induced cell divisions. Genomic clones were isolated and the respective promotors analyzed in transgenic tobacco using *gus* fusions. The results show that the promotor of one of the auxin-inducible genes confers expression in meristematic tissues (Mennes *et al.*, 1990). A clone coding for a protein with a similar function in tobacco mesophyll protoplasts was also reported by Takahashi *et al.* (1989).

The auxin-inducible proteins share no obvious homology in their coding regions, but there are some interesting conserved DNA-sequence elements in the 5' regions in some of the corresponding genes, which, however, also share some degree of homology with stress-inducible genes (wounding, heat shock; An *et al.*, 1990). The proteins seem not to be species-specific, since in recent experiments using PCR we found homologous sequences to the small-auxin–up-regulated proteins (SAURs) of soybean in pea (McClure *et al.*, 1989; Mayerbacher, personal communication).

5. CONCLUSIONS

The regeneration barrier in most of the important crop plants must be overcome in order to make use of the emerging new technologies for crop improvement. The conventional trial-and-error method of developing regeneration schemes is time- and labor-consuming, so new approaches are required to further our understanding of biochemical and molecular processes of plant development *in vitro*. These new approaches include attempts leading to a better understanding of signal recognition and transduction systems in plants as well as identification and molecular characterization of genes that have been reported to be crucial in plant development. Although many factors have been found to influence plant development considerably, the major effectors for control of growth and development are the phytohormones, i.e., auxins and cytokinins. Their respective mode of action is far from clear, but there is strong evidence that receptors are the key mediators between hormone application and morphogenetic response.

6. REFERENCES

Alliotte, T., Tiré, C., Engler, G., Peleman, J., Caplan, A., van Montagu, M., and Inzé, D., 1989, An auxin-regulated gene of *Arabidopsis thaliana* encodes a DNA-binding protein, *Plant Physiol.* **89**:743–752.

Ammirato, P. V., 1989, Recent progress in somatic embryogenesis, *IAPTC Newslett.* **57**:2–16.

Amrhein, N., 1977, The current status of cyclic AMP in higher plants, *Annu. Rev. Plant Physiol.* **28**:123–132.

An, G., Costa, M. A., and Ha, S-B., 1990, Nopaline synthase promotor is wound inducible and auxin inducible, *Plant Cell,* **2**:225–233.

Barbier-Brygoo, H., Ephritikhine, G., Klämbt, D., Ghislain, M. M., and Guern, J., 1989, Functional evidence for an auxin receptor at the plasmalemma of tobacco mesophyll protoplasts, *Proc. Natl. Acad. Sci. USA* **86**:891–895.

Bhattacharyya, K., and Biswas, B. B., 1978, Membrane-bound auxin receptors from *Avena*-roots, *Ind. J. Biochem. Biophys.* **15**:445–448.

Bögre, L., Stefanov, I., Abraham, M., Somogyi, I., and Dudits, D., 1990, Differences in response to 2,4-D treatment between embryogenic and non-embryogenic lines of alfalfa, in *Progress in Plant Cellular and Molecular Biology* (H. J. H. Nijkamp, L. H. W., van der Plas, and J. van Aartrijk, eds.), pp. 427–436, Kluwer Academic Publ., Dordrecht, The Netherlands.

Booij, H., Sterk, P., Schellekens, G. A., van Kammen, A., and de Vries, S. C., 1990, Tissue and cell-specific expression of genes encoding carrot extracellular protein, in *Progress in Plant Cellular and Molecular Biology* (H. J. J. Nijkamp, L. H. W. van der Plas, and J. van Aartrijk, eds.), pp. 398–401, Kluwer Academic Publ., Dordrecht, The Netherlands.

Chaleff, R., 1980, *Somatic Cell Genetics of Higher Plants,* Cambridge University Press, Cambridge.

Chang, I. K., and Foy, C. L., 1983, Rapid growth responses of dwarf corn coleoptile sections to picloram, *Pesticide Biochem. Physiol.* **19**:203–209.

Choi, J. H., and Sung, Z. R., 1984, Two-dimensional gel analysis of carrot somatic embryogenic proteins, *Plant Mol. Biol. Rep.* **2**:19–25.

de Vries, S. C., Booij, H., Janssens, R., Vogels, R., Saris, L., LoSchiavo, F., Terzi, M., and van Kammen, A., 1988, Carrot somatic embryogenesis depends on the phytohormone-controlled presence of correctly glycosylated extracellular proteins, *Genes Dev.* **2**:462–476.

Guern, J., Ephritikhine, G., Imhoff, V., and Pradier, J. M., 1990, Signal transduction at the membrane level of plant cells, in *Progress in Plant Cellular and Molecular Biology* (H. J. J. Nijkamp, L. H. W. van der Plas, and J. van Aartrijk, eds.), pp. 466–479, Kluwer Academic Publ., Dordrecht, The Netherlands.

Haberlandt, G., 1902, Culturversuche mit isolierten Pflanzenzellen, *Sitzungsber. Akad. Wiss. Wien, Math.-Nat. Classe 111,* **1**:69–92.

Herber, B., Ulbrich, B., and Jacobsen, H-J., 1988, Modulation of soluble auxin-binding proteins in soybean cell suspensions, *Plant Cell Rep.* **7**:178–181.

Hertel, R., Thomson, K-St., and Russo, V. E. A., 1972, *In vitro* auxin binding to particulate cell fractions from maize coleoptiles, *Planta* **107**:325–340.

Hesse, T., Feldwisch, J., Balshüsemann, D., Bauw, G., Puype, M., Vanderkerckhofe, J., Löbler, M., Klämbt, D., Schell, J., and Palme, K., 1989, Molecular cloning and structural analysis of a gene from *Zea mays* (L.) coding for a putative receptor for the plant hormone auxin, *EMBO J.* **8**:2453–2461.

Jacobsen, H-J., 1982, Soluble auxin binding proteins in pea epicotyls, *Physiol. Plant* **56**:161–167.

Jacobsen, H-J., 1984, Two different soluble cytoplasmic auxin-binding sites in etiolated pea epicotyls, *Plant Cell Physiol.* **25**(6):867–873.

Jacobsen, H-J., 1991, Somatic embryogenesis in seed legumes: The possible role of soluble auxin receptors, *Isr. J. Bot.* **40**:139–143.

Jacobsen, H-J., and Hajek, K., 1985, Genotype-specific soluble auxin-binding in etiolated pea epicotyls, *Biol. Plant.* **27**(263):110–113.

Jacobsen, H-J., Hajek, K., Mayerbacher, R., and Herber, B., 1987, Soluble auxin-binding: Is there a correlation between growth-stage dependent high-affinity auxin-binding and auxin competence? in *Plant Hormone Receptors* (D. Klämbt, ed.), pp. 63–70, Springer Verlag, Berlin.

Key, J. L., 1989, Modulation of gene expression by auxin, *BioEssays* **11**(263):52–58.

Kikuchi, M., Imaseki, H., and Sakai, S., 1989, Modulation of gene expression in isolated nuclei by auxin-binding proteins, *Plant Cell Physiol.* **30**(5):765–773.

Klämbt, D., 1990, A view about the function of auxin-binding proteins at plasma membranes, *Plant Mol. Biol.* **14**:1045–1090.

Kögl, F., Haagen-Smit, A. J., and Erxleben, H., 1932, Über ein Phytohormon der Zellstreckung. Reindarstellung des Auxins aus menschlichem Harn, *Hoppe-Seylers Z. Physiol. Chem.* **228**:113–121.

Kohlenbach, H. W., 1978. Comparative somatic embryogenesis, in *Frontiers of Plant Tissue Culture 1978* (T. A. Thorpe, ed.), pp. 59–66, 4th IAPTC-Congress, Calgary.

Kohlenbach, H. W., 1985, Fundamental and applied aspects of in vitro plant regeneration by somatic embryogenesis, in In Vitro *Techniques—Propagation and Long Term Storage* (A. Schäfer-Menuhr, ed.), pp. 101–109. Martinus Nijhoff/Dr. W. Junk Publ., Dordrecht.

Komamine, A., Matsumoto, M., Tsukahara, M., Fujiwara, A., Kawahara, R., Ito, M., Smith, J., Nomura, K., and Fujimura, T., 1990, Mechansims of somatic embryogenesis in cell cultures—physiology, biochemistry and molecular biology, in *Progress in Plant Cellular and Molecular Biology* (H. J. J. Nijkamp, L. H. W. van der Plas, and J. van Aartrijk, eds.), pp. 307–313, Kluwer Academic Publ., Dordrecht, The Netherlands.

Koorneef, M., and Hanhart, C. J., 1990, The genetics of regeneration capacity in tomato, Abstr. VII IAPTC-Congress, Amsterdam, June 24–29, 1990, p. 300.

Koorneef, M., Hanhart, C. J., and Martinelli, L., 1987, A genetic analysis of cell culture traits in tomato, *Theor. Appl. Genet.* **74**:633–641.

Kysely, W., and Jacobsen, H-J., 1990, Somatic embryogenesis from pea embryos and shoot apices, *Plant Cell Tiss. Org. Cult.* **20**:7–14.

Lehminger-Mertens, R., and Jacobsen, H-J., 1989, Plant regeneration from pea protoplasts via somatic embryogenesis, *Plant Cell Rep.* **8**:379–382.

Libbenga, K. R., van Telgen, H. J., Mennes, A. M., van der Linde, P. C. G., and van der Zaal, E. J., 1987, Characterization and function analysis of a high-affinity cytoplasmic auxin-binding binding protein, in *Molecular Biology of Plant Growth Control* (J. E. Fox and M. Jacobs, eds.), pp. 229–243, Alan R. Liss, New York.

Löbler, M., and Klämbt, D., 1985a, Auxin-binding protein from coleoptile membranes of corn (*Zea mays* L.). I. Purification by immunological methods and characterization, *J. Biol. Chem.* **260**:9848–9853.

Löbler, M., and Klämbt, D., 1985b, Auxin-binding protein from coleoptile membranes of corn (*Zea mays* L.). II. Localization of a putative auxin receptor, *J. Biol. Chem.* **260**:9854–9859.

McClure, B. A., Hagen, G., Brown, C. S., Gee, M. A., and Guilfoyle, T. J., 1989, Transcription, organization and sequence of an auxin regulated gene cluster in soybean, *Plant Cell* **1**:229–239.

Mennes, A. M., Boot, C. J. M., Libbenga, K. R., van der Zaal, E. J., and Maan, A. C., 1990, IAA perception and auxin-regulated gene expression, in *Plant Growth Substances 1988* (S. B. Rood and R. P. Pharis, eds.), pp. 100–105, Springer Verlag, Berlin.

Nadolska-Orczyk, A., and Malepszy, S., 1989, *In vitro* culture of *Cucumis sativus* L., *Theor. Appl. Genet.* **78**:836–840.

Napier, R. M., and Venis, M. A., 1990, Receptors for plant growth regulators: Recent advances, *J. Plant Growth Reg.* **9**:113–126.

Oostrom, H., Kulescha, Z., van Vliet, Th. B., and Libbenga, K. R., 1980, Characterization of a cytoplasmic auxin receptor from tobacco pith callus, *Planta* **149**:44–47.

Paulus, C., and Jacobsen, H-J., 1989, Characterization of cytoplasmic auxin-binding proteins in soybean cell suspensions, DECHEMA *Biotechnol. Conf.* **3**:417–420.

Peng, J., and Hodges, T. K., 1989, Genetic analysis of plant regeneration in rice (*Oryza sativa* L.), *In Vitro Cell. Dev. Biol.* **25**(1):91–94.

Pickardt, T., Huancaruna, P., and Schieder, O., 1989, Plant regeneration via somatic embryogenesis in *Vicia narbonnensis, Protoplasma* **149**:5–10.

Reinert, J., 1958, Morphogenese und ihre Kontrolle an Gewebekulturen aus Karrotten, *Natur-wissenschaften* **45**:344–345.

Sakai, S., 1985, Auxin-binding protein in etiolated mung bean seedlings: Purification and properties of auxin-binding protein, *Plant Cell Physiol.* **26**:185–192.

Sakai, S., and Hanagata, T., 1985, Purification of an auxin-binding protein from etiolated mung bean seedlings by affinity chromatography, *Plant Cell Physiol.* **24**:685–693.

Sakai, S., Seki, J., Imaseki, H., 1986, Stimulation of RNA-synthesis in isolated nuclei by auxin-binding proteins I and II, *Plant Cell Physiol.* **27**:635–643.

Sethi, U., and Guha-Mukherjee, S., 1990, Biochemistry and molecular biology of competence of plant cell differentiation and regeneration *in vitro*—A review, *Curr. Sci.* **59**(6):308–311.

Shen, W. J., Petit, A., Guern, J., and Tempé, J., 1988, Hairy roots are more sensitive to auxin than normal roots, *Proc. Natl. Acad. Sci. USA* **85**:3417–3421.

Skoog, F., 1971, Les cultures des tissus des plantes, Colloques Internationaux CNRS, Paris, No. 193, pp. 115–135.

Sommer, H., Beltran, J-P., Huijser, P., Pape, H., Lönnig, W-E., Saedler, H., and Schwarz-Sommer, S., 1990, *Deficieus,* a homeotic gene involved in the control of flower morphogenesis in *Antirrhinum majus:* the protein shows homology to transcription factors, *EMBO J.* **9**(3):605–613.

Spena, A., Schmülling, T., Koncz, C., and Schell, J., 1987, Independent and synergistic activity of *rol* A, B and C loci in stimulating abnormal growth in plants, *EMBO J.* **6**:3891–3899.

Steward, F. C., Mapes, M. O., and Mears, K., 1958, Growth and organized development of cultured cells. II. Organization in cultures from freely suspended cells, *Am. J. Bot.* **45**:705–708.

Stirn, S., and Jacobsen, H-J., 1987, Marker proteins for embryogenic differentiation patterns in pea callus, *Plant Cell Rep.* **6**:50–54.

Stirn, S., and Jacobsen, H-J., 1990, Production and characterization of monoclonal antibodies against marker proteins for somatic embryogenesis in pea (*Pisum sativum* L.), in *Progress in Plant Cellular and Molecular Biology* (H. J. J. Nijkamp, L. H. W. van der Plas, and J. van Aartrijk, eds.), pp. 460–465, Kluwer Academic Publ., Dordrecht, The Netherlands.

Stirn, S., Altherr, S., and Jacobsen, H-J., An early marker protein for somatic embryogenesis in pea and carrot is induced by auxin (submitted).

Sung, Z. R., and Okimoto, R., 1981, Embryogenic proteins in somatic embryos of carrot, *Proc. Natl. Acad. Sci. USA* **78**:3683–3687.

Takahashi, Y., Kuroda, H., Tanaka, T., Machida, Y., Takebe, I., and Nagata, T., 1989, Isolation of an auxin-regulated gene cDNA expressed during the transition from G_0 to S-phase in tobacco mesophyll protoplasts, *Proc. Natl. Acad. Sci. USA* **86**:9279–9283.

van der Linde, P. C. G., Bouman, H., Mennes, A. M., and Libbenga, K. R., 1984, A soluble auxin-binding protein from cultured tobacco tissue stimulates RNA-synthesis *in vitro, Planta* **160**:102–106.

Plant Development *In Vitro* 277

van der Zaal, E. J., Memelink, J., Mennes, A. M., Quint, A., and Libbenga, K. R., 1987, Auxin-induced mRNA species in tobacco cell cultures, *Plant Mol. Biol.* **10**:145–157.

Venis, M. A., and Napier, R. M., 1990, Antibodies to the maize membrane auxin receptor, in *Signal Perception and Transduction in Higher Plants* (R. Ranjeva and A. M. Boudet, eds.), pp. 13–26, Springer-Verlag, Berlin.

Willman, M. R., Schroll, S. M., and Hodges, T. K., 1989, Inheritance of somatic embryogenesis and plantlet regeneration from primary (type 1) callus in maize, *In Vitro Cell. Dev. Biol.* **25**(1):95–100.

Zenk, M. H., Schulte, U., and El-Shagi, H., 1984, Regulation of anthraquinone formation by phenoxyacetic acids in *Morinda* cell cultures, *Naturwissenschaften* **71**:266.

Chapter 13

The Molecular Basis of Ethylene Biosynthesis, Mode of Action, and Effects in Higher Plants

Dominique Van Der Straeten and Marc Van Montagu

1. INTRODUCTION

Understanding the molecular basis of plant hormone biosynthesis and mode of action is one of the keys to unraveling fundamental aspects of plant development. Progress in molecular genetics has led to the recent isolation of a number of genes induced specifically by hormonal treatment. This has been reviewed for auxins (Theologis, 1986), gibberellic acid (Jacobsen and Chandler, 1987), and abscisic acid (Skriver and Mundy, 1990). Moreover, the recently developed techniques of integrated physical/RFLP maps (Hwang *et al.*, 1991) and yeast artificial chromosomes (Grill and Somerville, 1991) will lead to the isolation of

Abbreviations: ACC, 1-aminocyclopropane-1-carboxylic acid; AdoMet, *S*-adenosyl-L-methionine; AOA, aminooxyacetic acid; AVG, aminoethoxyvinylglycine; BA, benzyladenine; EFE, ethylene-forming enzyme; EMS, ethylmethanesulfonate; ER, endoplasmic reticulum; IAA, indole-3-acetic acid; Met, methionine; PLP, pyridoxal-5′-phosphate; RFLP, restriction fragment length polymorphism.

Dominique Van Der Straeten and Marc Van Montagu   Laboratorium voor Genetica, Universiteit Gent, B-9000 Gent, Belgium.

Subcellular Biochemistry, Volume 17: Plant Genetic Engineering, edited by B. B. Biswas and J. R. Harris. Plenum Press, New York, 1991.

genes encoding hormone biosynthetic enzymes, receptors, and signal transducers (Bleecker *et al.*, 1990; Hwang *et al.*, 1991; Giraudat *et al.*, 1990).

Ethylene is one of the six commonly occurring plant hormones. It is the smallest of all plant growth substances and one of the most pleiotropic of known effectors. In 1901, the Russian physiologist Neljubow discovered that this simple two-carbon atom molecule was the active compound in illuminating gas eliciting the "triple response" of etiolated pea plants. In 1934, Gane provided direct proof that a plant tissue is capable of producing ethylene. The devoted efforts of generations of plant physiologists and biochemists have led to a vast amount of information concerning the biology of this hormone, acting at concentrations between 0.1 and 1 ppm (Burg and Burg, 1967; Abeles, 1973).

Ethylene is not only known to play an important role in several stages of plant development, including germination and seedling growth, leaf and root growth, senescence, abscission, fruit ripening, flowering and sex expression, but is also generally referred to as the "stress hormone", possibly triggering the plant's defense (Abeles, 1973; Yang and Hoffman, 1984; McKeon and Yang, 1987; Moore, 1989). Ethylene production rates are influenced by itself (Riov and Yang, 1982b; Liu *et al.*, 1985b) as well as other hormones: auxins, cytokinins, and abscisic acid (Yu and Yang, 1979; Yoshii and Imaseki, 1981; McKeon *et al.*, 1982; Wright, 1980). In some cases, it might act in concert with or in opposition to another hormone, depending on the target cell type (Osborne, 1982). The first reports on the induction of genes by ethylene treatment date from the early eighties (Christoffersen and Laties, 1982; Nichols and Laties, 1984; Tucker and Laties, 1984), but the function of many of these genes is related more to the expression of phenotypes induced by ethylene than to its biosynthesis or reception. A more valuable breakthrough on the molecular aspects of ethylene has occurred within the last 2 years, with the isolation of clones encoding S-adenosyl-L-methionine synthetase (Peleman *et al.*, 1989a), 1-aminocyclopropane-1-carboxylic acid synthase (Sato and Theologis, 1989; Van Der Straeten *et al.*, 1990), a candidate ethylene-forming enzyme (Hamilton *et al.*, 1990), and a candidate ethylene receptor (Bleecker *et al.*, 1990). These and other molecular probes will facilitate our task of answering the following fundamental questions in ethylene biology: (1) Which molecular mechanisms govern ethylene formation in higher plants? This covers the study of *cis*- and *trans*-acting factors regulating temporal and spatial expression of genes encoding enzymes in ethylene biosynthesis under different conditions. (2) How can the plant perceive ethylene and transfer the signal to result in the well-characterized responses? (3) What are the molecular requirements for ethylene inducibility of a gene? In this chapter we summarize recent progress in our understanding of each of these basic questions. Since this review is devoted first to the molecular aspects of ethylene and since there exists an overwhelming amount of information concerning ethylene physiology, we have tried to limit ourselves to the key points of ethylene physiology and biochemistry relevant to the molecular data presented here.

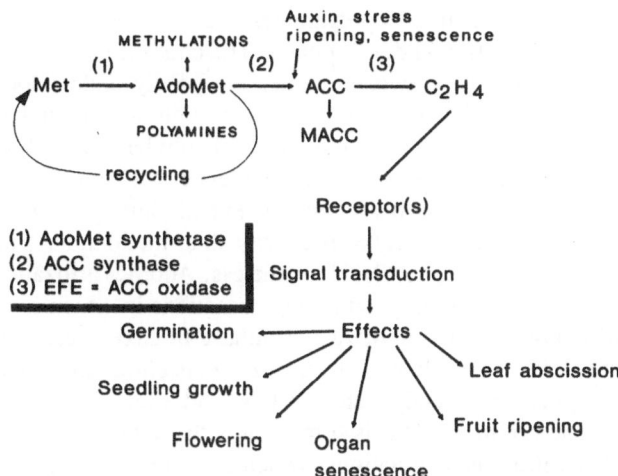

FIGURE 1. Schematic overview of ethylene biosynthesis, signal transduction, and effects.

2.   MOLECULAR ASPECTS OF ETHYLENE BIOSYNTHESIS

2.1.   Biosynthetic Cycle

Biochemical aspects of ethylene biosynthesis have been extensively re-viewed by Yang and Hoffman (1984) and recently by Kende (1989). Figure 1 schematically presents ethylene biosynthesis and effects in higher plants. The effective precursor of ethylene is L-methionine. S-adenosyl-L-methionine (AdoMet) and 1-aminocyclopropane-1-carboxylic acid (ACC) were identified as intermediates between methionine and ethylene. The methylthioadenosine moiety of AdoMet is recycled back to methionine, in the same way as was established for bacteria and mammals. As mentioned above, three enzymes of the ethylene pathway have been cloned: they are the first—and to date the only—plant-encoded hormone biosynthetic enzymes to be cloned. In the following sections we describe the current molecular data available on these three important enzymes involved in ethylene formation.

2.2.   From L-Methionine to S-Adenosyl-L-Methionine: A Branch Point between the Ethylene and Polyamine Pathways

ATP:L-methionine S-adenosyltransferase (AdoMet synthetase, EC 2.5.1.6) catalyzes the formation of AdoMet, which serves in a variety of reactions in all living organisms (Tabor and Tabor, 1984). First, AdoMet acts as a methyl donor in transmethylations of nucleic acids, lipids, polysaccharides, and proteins. In addition, it is the precursor of both polyamines and ethylene, which play antagonistic roles in plant growth and development. The controlling elements regulating the balance between these two hormones are poorly understood. There is some indication that the concentration of AdoMet might affect ethylene production, as deduced from studies with inhibitors of ethylene or polyamine biosynthesis (Roberts *et al.*, 1984). Additional evidence that AdoMet levels might regulate ethylene production is based on selenomethionine-enhanced ethylene formation, probably due to its higher reactivity than methionine with AdoMet synthetase (Konze and Kende, 1979). Nevertheless, AdoMet utilization for ACC synthesis is probably minor compared to other reactions in which it is involved. Neither inhibition nor stimulation of ACC synthase in auxin-treated mung bean hypocotyl tissue leads to a change in AdoMet concentration (Yu and Yang, 1979), which implies that steady-state levels of AdoMet are not significantly affected by ethylene synthesis. This is consistent with the fact that the majority of AdoMet is utilized in transmethylations, which is a housekeeping function, requiring constant supplies of substrate. This idea is supported by analysis of the expression pattern of AdoMet synthetase genes in *Arabidopsis thaliana* (Peleman *et al.*, 1989a,b). *Arabidopsis* possesses two genes encoding AdoMet synthetase, designated *sam-1* and *sam-2*, highly conserved, both on the nucleotide and on the deduced amino acid level (89% and 97%, respectively). Northern analysis revealed that these genes are specifically expressed in stem, root, and callus tissue and undetectable in leaf, inflorescence, and seed pods. Histochemical analysis of *Arabidopsis* and tobacco plants transformed with a *sam-1* promoter GUS construct demonstrated that the expression of the *sam-1* gene is primarily confined to the vascular bundles (xylem and phloem) and to the surrounding sclerenchyma tissue. These data were confirmed by transient expression in rice (Dekeyser *et al.*, 1990). The observed expression pattern can be associated with the degree of lignification, involving transmethylations. Elevated levels of *sam-1* expression were also remarked in the parenchyma cells of the root cortex in *Arabidopsis*, which might be correlated with increased ethylene or polyamine biosynthesis. Further investigations will be necessary to reveal specific roles, if any, for each of the AdoMet synthetase genes and might offer a more profound understanding of the interrelationship of ethylene and polyamines.

2.3. From AdoMet to ACC: ACC Synthase, a Key Regulatory Enzyme in Ethylene Formation

The three-membered ring nonprotein amino acid ACC is the immediate precursor of ethylene in higher plants and is formed by an α,γ-elimination reaction to the methionyl side chain of AdoMet with pyridoxal-5'-phosphate (PLP) as a cofactor (Ramalingam *et al.*, 1985). There is ample experimental evidence that the rate of ethylene synthesis is limited by the availability of ACC in the tissue (Yang and Hoffman, 1984). This implies that those factors which enhance ethylene production are in fact inducers or activators of ACC synthase. Pulse labeling with ^{35}S-methionine demonstrated that the wound-induced increase in ACC synthase activity involves *de novo* synthesis of a rapidly turning-over polypeptide (Bleecker *et al.*, 1988b).

Because of its key regulatory role in ethylene synthesis, there has been considerable interest in understanding the biochemistry of ACC synthase (*S*-adenosyl-L-methionine methylthioadenosinelyase; EC 4.4.1.14), and more recently, a number of laboratories have attempted to clone the corresponding gene(s). Both an antibody-based detection system (Sato and Theologis, 1989; Nakajima *et al.*, 1990) and a peptide-sequencing approach (Van Der Straeten *et al.*, 1990) have been followed.

2.3.1. Biochemistry of ACC Synthase

The ACC synthase protein has been purified and characterized from the following sources: tomato pericarp, winter squash mesocarp, zucchini fruit, mung bean hypocotyl, and apple. Until recently, there was considerable confusion about the molecular weight of the enzyme. Cloning of the corresponding mRNA indicated that ACC synthase is synthesized as a precursor of 56 kDa (Sato and Theologis, 1989; Van Der Straeten *et al.*, 1990; Nakajima *et al.*, 1990), confirming estimates obtained on gel filtration columns (Boller *et al.*, 1979; Yu *et al.*, 1979; Acaster and Kende, 1983; Bleecker *et al.*, 1986). This precursor is naturally or artificially processed to a polypeptide of 45–50 kDa, as found after enrichment of the enzyme from different origin (Bleecker *et al.*, 1986; Nakajima *et al.*, 1988; Satoh and Yang, 1988; Sato and Theologis, 1989; Van Der Straeten *et al.*, 1989b). Privalle and Graham (1987) found a 50-kDa labeled protein after reduction with sodium borotritide of a 400-fold enriched ACC synthase preparation of wounded tomato pericarp. Although this reducing agent can attack the double bond between ACC synthase and its cofactor, it is difficult to interpret their result since the specificity of the reaction has been questioned by both Satoh and Yang (1988) and ourselves (D. Van Der Straeten, unpublished results).

Biochemical studies demonstrated the highly labile nature of ACC synthase as well as its extremely low abundance, estimated to be in the range of 0.0001% of total protein in naturally ripened fruit (Bleecker *et al.*, 1986; Van Der Straeten *et al.*, 1989b). Most efforts were invested in the purification of the tomato enzyme, and monoclonal antibodies have been made against it (Bleecker *et al.*, 1986; Privalle and Graham, 1987; Mehta *et al.*, 1988; Satoh and Yang, 1988; Van Der Straeten *et al.*, 1989b,c). A major advantage of using climacteric fruit tissue is the high inducibility of ACC synthase by a combination of LiCl treatment with wounding (Kende *et al.*, 1986). A kinetic analysis of the induction process revealed that ACC synthase activity peaks after 16 hr of LiCl incubation and 3 hr of wounding (Van Der Straeten and Van Montagu, 1990). Optimal inductions yielded specific activities of 15–20 U/g fresh weight, corresponding to a 50- to 100-fold increase over the control values. After almost 5000-fold enrichment, ACC synthase still did not appear to be homogeneously pure. However, two independent approaches allowed us to assign ACC synthase activity to a 45-kDa polypeptide (Van Der Straeten et al., 1989b). Evidence was based on two-dimensional gels and on a (3,4-^{14}C)AdoMet binding experiment. Comparison of amino acid sequence data between this 45-kDa protein and the 50-kDa species identified by Bleecker et al. (1986) and Satoh and Yang (1988) confirmed that these polypeptides are identical (Yip et al., 1990; H. Kende, personal communication). Peptide sequences were obtained both from the N-terminus after electroblotting and from tryptic peptides separated by reversed-phase chromatography (Van Der Straeten et al., 1990).

ACC synthase has also been purified from two different *Cucurbita* species. Nakajima et al. (1988) were able to homogeneously purify ACC synthase from wound-induced winter squash (*Cucurbita maxima* Duch.) by including an immunoaffinity gel. The apparent subunit molecular weight as estimated by SDS-PAGE was about 50 kDa. However, the native enzyme seemed to be a multimer (Nakajima and Imaseki, 1986), in contrast to its tomato counterpart (Bleecker et al., 1986; D. Van Der Straeten, unpublished results), which is a monomer.

Sato and Theologis (1989) reported purification of zucchini (*Cucurbita pepo*) ACC synthase, induced by a mixture of LiCl, IAA, BA, and AOA. A crude antiserum was obtained after a 6000-fold enrichment of the enzyme and further purified by affinity chromatography with total proteins from uninduced tissue. This antibody preparation immunoprecipitated ACC synthase activity and recognized a 46-kDa polypeptide on Western blots in crude and highly purified preparations.

Recently, Yip et al. (1990) have partially purified ACC synthase from apple fruit. Both in the case of radiolabeling with AdoMet and sodium borohydride, a 48-kDa polypeptide was recovered using a monoclonal antibody directed against apple fruit ACC synthase. The monoclonal, however, fails to recognize the tomato enzyme, indicating that there are probably structural differences between ACC synthases from distant origin.

Although it is clear from the data described above that ACC synthase has a consensus molecular mass of approximately 47–48 kDa, there are two reports of isoforms with a molecular mass of about 65 kDa. Both the wound–induced tomato isoform of 67 kDa (Mehta et al., 1988) and the auxin-induced mung bean ACC synthase of 65 kDa (Tsai et al., 1988) were identified with a monoclonal antibody on Western blots. The existence of different isoforms of ACC synthase arising from different genes cannot be excluded until cloning of the complete ACC synthase gene family. Alternatively, these isoforms might occur after glycosylation or other posttranslational modifications. However, it is not impossible that, because of the extreme low abundance of the enzyme, an abundant contaminant was mistaken for ACC synthase.

2.3.2. Cloning the Genes Encoding ACC Synthase

To date, cloning of ACC synthase has been reported from three different sources: zucchini (Sato and Theologis, 1989), tomato (Van Der Straeten et al., 1990), and winter squash (Nakajima et al., 1990). The authenticity of the clones was proved by an active recombinant protein in *Escherichia coli* and yeast in the case of zucchini and winter squash. In tomato, the identity was confirmed by immunoinhibition and -precipitation of activity using an antibody against the denatured recombinant polypeptide.

Sequence data were published for both the tomato (Van Der Straeten et al., 1990) and the winter squash ACC synthase (Nakajima et al., 1990), allowing a comparison of structural features between these two species. The tomato pcVV4A cDNA (renamed tACC1) is 1846 bp long and contains an open reading frame encoding a polypeptide of 55 kDa (485 amino acids) (Van Der Straeten et al., 1990). The wound-induced cDNA clone from winter squash is 1748 bp long, with an open reading frame encoding 493 amino acids. Figure 2 shows an alignment of both amino acid sequences. The overall identity is 66%. This significant variation between ACC synthases explains the absence of interspecies cross-reactivity of antibodies against ACC synthase, at least in certain cases (Yip et al., 1990). The tomato and squash enzyme have seven stretches of at least 10 identical amino acids, six of which are located in the first three-fifths of the enzyme. As a consequence, the carboxyl terminal part of the protein is highly divergent. In both cases it possesses the most hydrophilic regions of the protein. In the case of squash, the last 60 amino acids form three distinct, highly hydrophilic clusters (Nakajima et al., 1990), which are not so pronounced in its tomato analog (Figure 3). The region covering amino acids 274–292 in the tomato enzyme and 275–293 in that of squash is 100% conserved and has been proposed to be the active site region upon comparison to the pyridoxyl peptides of other PLP-dependent enzymes (Van Der Straeten and Van Montagu, 1990; Nakajima et al., 1990). Lysine-278 (279 in squash) is the active site lysine, which forms an aldimine with the pyridoxal residue, as was confirmed by radiolabeling of the tomato enzyme and subsequent sequence analysis of the labeled pyridoxyl peptide (Yip et al., 1990). The active site region resides in a possible random coil structure (Chou and Fasman, 1978). In addition, neither of these two ACC synthases shows any homology with consensus organelle target sequences (Van Der Straeten et al., 1990), implying cytoplasmic localization. Some evidence was already provided by Boller et al. (1979), but confirmation by immunolocalization is needed.

On the nucleotide level, there is 67% identity between the coding regions of

```
1  -  MEFHQIDERNQALLSKIAVDDGHGENSPYFDGWKAYDNDPFHPEDNPLGV  -50
          :  :    :      :      ::: :      :::::::::::::::: ::::: :: ::
2  -  MGF-EIAKTN-SILSKLATNEEHGENSPYFDGWKAYDSDPFHPLKNPNGV  -48

1  -  IQMGLAENQLSFDMIVDWIRKHPEASICTPKGLERFKSIANFQDYHGLPE  -100
          :::::::::::  : : :::    :  ::: :    :: ::::::::::::
2  -  IQMGLAENQLCLDLIEDWIKRNPKGSICS-EGIKSFKAIANFQDYHGLPE  -97

1  -  FRNGIASFMGKVRGGRVQFDPSRIVMGGGATGASETVIFCLADPGDAFLV  -150
          ::   :: :: :  ::::: ::: : :: :::::: :: :::::::::::::
2  -  FRKAIAKFMEKTRGGRVRFDPERVVMAGGATGANETIIFCLADPGDAFLV  -147

1  -  PSPYYAAFDRDLKWRTRAQIIRVHCNSSNNFQVTKAALEIAYKKAQEANI  -200
          :::::: ::  ::: ::: : : :: ::::: : : :: :: ::
2  -  PSPYYPAFNRDLRWRTGVQLIPIHCESSNNFKITSKAVKEAYENAQKSNI  -197

1  -  KVKGVIITNPSNPLGTTYDRDTLKTLVTFVNQHDIHLICDEIYSATVFKA  -250
          :::: ::::::::::: :::::::   : :: ::: :::: ::::
2  -  KVKGLILTNPSNPLGTTLDKDTLKSVLSFTNQHNIHLVCDEIYAATVFDT  -247

1  -  PTFISIAQIVEEMEH--CKKELIHILYSLSKDMGLPGFRVGIIYSYNDVV  -298
          :  : ::: :   : : :    :   : :: :::::::::::::::::::: :: :
2  -  PQFVSIAEILDEQEMTYCNKDLVHIVYSLSKDMGLPGFRVGIIYSFNDDV  -297

1  -  VRRARQMSSFGLVSSQTQHLLAAMLSDEDFVDKFLAENSKRLAERHARFT  -348
          :  :: :::::::: :::  :::: ::: ::: : : :: :: ::
2  -  VNCARKMSSFGLVSTQTQYFLAAMPSDEKFVDNFLRESAMRLGKRHKHFT  -347

1  -  KELDKMGITCLNSNAGVFVWMDLRRLLKDQTFKAEMELWRVIINEVKLNV  -398
          :   :: :: :::  : ::::: ::    :: :: :::::::: :::::
2  -  NGLEVVGIKCLKNNAGLFCWMDLRPLLRESTFDSEMSLWRVIINDVKLNV  -397

1  -  SPGSSFHVTEPGWFRVCFANMDDNTVDVALNRIHSFVENIDKKEDNTVAM  -448
          : :::: :::::::::::: ::: :: :: ::    :   : :
2  -  SLGSSFECQEPGWFRVCFANMDDGTVDIALARIRRFVG-VEKSGDKSSSM  -446

1  -  PSKTRRRENKLRLSFSFSGRRYDEGNVLNSPHTMSPH--SPLVIAKN    -493
          : :::::::  : ::: :: :: :   :::: :
2  -  EKKQQWKKNNLRLSFS--KRMYDES-VL-SPLS-SPIPPSPLVR    -485
```

FIGURE 2. Comparison of the amino acid sequences of winter squash (1) and tomato (2) ACC synthase.

the tomato and the squash clone. There is a striking difference in codon usage, resulting in a relatively low percent of GC in tomato (38.7% versus 46.2% in squash). It is notable from Table I that tomato has a pronounced codon bias, at least for the amino acids Arg, Asp, Cys, Glu, Gly, and Pro. In squash there is a much more uniform distribution among possible codons. In both cases, TAA was used as a stop codon. 5'- and 3'-untranslated regions are divergent and very AT rich.

Expression of the 1.9-kb ACC synthase messenger was studied under various conditions. In zucchini, ACC synthase messenger was induced by IAA,

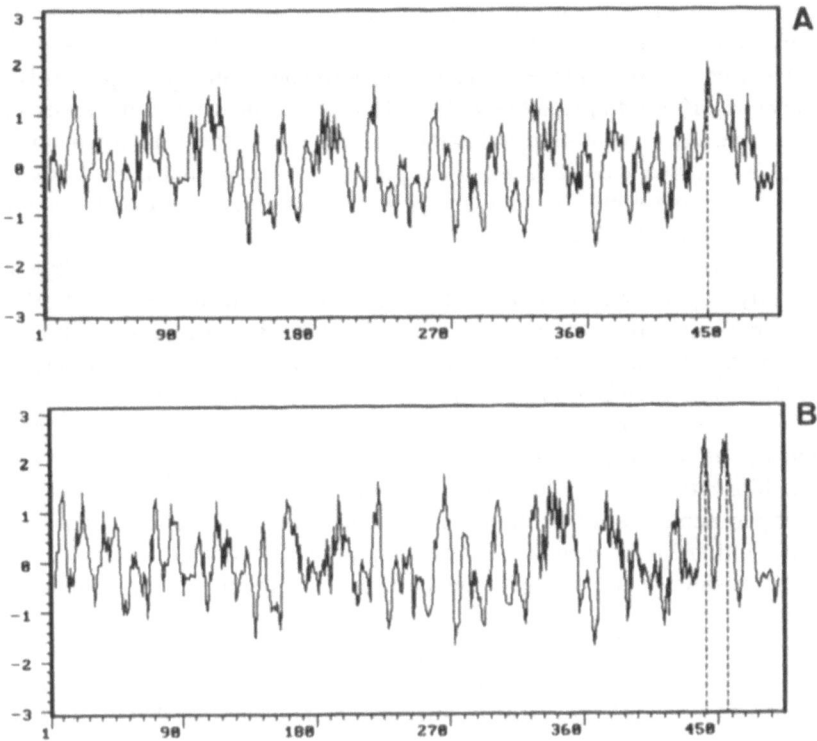

FIGURE 3. Hydrophilicity plots of tomato (A) and winter squash (B) ACC synthase.

Table I
Comparison of Codon Usage of Tomato and
Winter Squash ACC Synthase

Amino acid	Codon	Usage in tomato (%)	Usage in squash (%)
Ala	GCC	14 (4/28)	29 (10/34)
Arg	CGA	4 (1/24)	27 (8/30)
Arg	AGA	63 (15/24)	13 (4/30)
Asp	GAT	72 (21/29)	42 (14/33)
Cys	TGT	82 (9/11)	43 (3/7)
Glu	GAA	75 (21/28)	46 (13/28)
Gly	GGA	47 (14/30)	18 (5/28)
His	CAT	33 (3/9)	62 (10/16)
Pro	CCA	58 (14/24)	29 (6/21)

Note: The numbers in parentheses are the number of times that the codon appears over the total number of amino acid mentioned.

LiCl, and wounding (Sato and Theologis, 1989). An enhancement of the auxin effect was observed upon addition of BA, although BA did not have any effect alone. The tomato tACC1 clone seemed to be correlated with ripening (Van Der Straeten et al., 1990). In green tissue the mRNA was undetectable, but appeared in pink tissue. At least 50-fold induction was observed after a LiCl-wounding treatment, combined with IAA and BA. Kinetics of the wound inducibility of squash ACC synthase mRNA were analyzed and indicated a very high level of expression after 6–8 hr (Nakajima et al., 1990). The signal was detectable after 2 hr, but increased at least 50- to 100-fold after 6 hr. Treatment with ethylene significantly suppressed the induction. This reduction in abundance of ACC synthase messenger by ethylene explains, at least in part, the autoinhibition of ethylene as reported previously (Riov and Yang, 1982a,b). On the other hand, a stimulatory effect occurred upon administration of 2,5-norbornadiene, an inhibitor of ethylene action, after wounding. One explanation could be that as a consequence of ethylene action, a trans-acting factor is made that represses ethylene biosynthesis at the level of ACC synthase. However, the above-mentioned experiments do not allow us to distinguish between transcriptional and posttranscriptional control. Nuclear runoff experiments and inhibitor studies will determine at what level the ACC synthase gene expression is predominantly regulated.

The fact that, in all systems investigated, a polypeptide of 55–56 kDa is encoded by the entire reading frame, which is about 6–10 kDa more than the ACC synthase protein recovered after purification, indicates possible in vivo or in vitro degradation or processing. Heterologous expression of the precursor, however, proves that it is a functional enzyme. Both Van Der Straeten et al. (1990) and Nakajima et al. (1990) suggested that this degradation occurred at the carboxyl terminus. It was proposed that functionless proteolytic degradation rather than a biologically significant processing would be involved on the hydrophilic C-terminal cluster (Nakajima et al. 1988, 1990). Additional experiments are required to sustain this hypothesis.

In all described species, there are at least two genes for ACC synthase. In tomato, two different cDNAs were isolated, with 82% homology. On genomic Southern blots, it appears that the two corresponding genes have identical restriction patterns or that they are tandemly arranged in the tomato genome (Van Der Straeten et al., 1990). Tandem organization is observed in zucchini (Theologis et al., 1990), where two ACC synthase genes are transcribed toward each other, separated by a 3-kb intergenic region. DNA sequencing revealed a very high degree of conservation (95%), in both exons and introns. Their promoters, however, show no similarity. The presence of at least two different genes was also confirmed in winter squash (Nakajima et al., 1990). As stated in the Introduction, endogenous activity of ACC synthase can be stimulated by very different factors. How these diverse inducers can trigger the formation of the same enzyme remains an intriguing problem. It is conceivable that one of the ACC synthase genes would respond to internal signals (ripening, auxin), whereas the other one would be triggered by external factors (wounding, pathogens). At least in winter squash, it was observed that different forms of the enzyme can be very distinct. There is no immunochemical relation between the wound-induced and the auxin-induced isoforms (Nakagawa *et al.*, 1988; Imaseki *et al.*, 1990), and similarly, the mRNA for the auxin-induced form is not recognized by the wound-induced cDNA on Northern blot (Nakajima *et al.*, 1990). It might turn out that the amino acid sequence of the active site peptide gives an indication for the type of induction of the corresponding gene. So far, there is 100% homology between the 12-amino-acid pyridoxyl peptides of the wound-induced tomato and winter squash enzymes (Van Der Straeten *et al.*, 1990; Yip *et al.*, 1990; H. Kende, personal communication; Nakajima *et al.*, 1990). The ripening-induced form of apple ACC synthase, however, has one amino acid difference in position 6, where leucine was found instead of methionine. From the observations mentioned above, it is clear that many more investigations will be necessary to clarify the functional significance of these structural variations.

2.4. From ACC to Ethylene: Ethylene-Forming Enzyme or ACC Oxidase

The final step in ethylene formation is the conversion of ACC to ethylene. Since a marked increase of ethylene production was observed upon application of ACC to plant organs with the exception of preclimacteric tissues (Cameron *et al.*, 1979; Lürssen *et al.*, 1979), ethylene-forming enzyme (EFE) was proposed to be largely constitutive (Yang and Hoffman, 1984). Although never rate limiting for ethylene synthesis, increased EFE activity has been reported in both wounded (Hoffman and Yang, 1982) and ripening fruit (Liu *et al.*, 1985b) treated with exogenous ethylene.

Yang and Hoffman (1984) proposed a mechanism of ethylene formation by which ACC would first be hydroxylated, to *N*-hydroxy-ACC, and then broken down into ethylene (from C3 and C4 of methionine) and cyanoformic acid. The last compound is very labile and would fragment spontaneously into CO_2 and cyanide. Supporting evidence came from observations by Peiser *et al.* (1984) demonstrating that the α-carbon of ACC was incorporated into asparagine or β-cyanoalanine, expected if the amino group of ACC would give rise to cyanide. It is important to mention that the biological reaction does not proceed in a concerted way, but rather by a stepwise mechanism. This was concluded from feeding experiments with 2,3-dideuterio-ACC isomers, resulting in scrambling of the hydrogen configuration (Adlington *et al.*, 1983; Pirrung, 1983; Pirrung and McGeehan, 1983). This is in contrast to the chemical oxidation proceeding with complete retention of configuration (Adlington *et al.*, 1983).

A major problem in identification of the EFE was the possibility of non-enzymatic conversion of ACC by oxidants, so that enzyme systems catalyzing the activation of oxygen to hydroxyl or superoxide radicals or those generating hydroperoxides might lead to nonenzymatic ethylene formation from ACC (Lizada and Yang, 1979; Legge *et al.*, 1982; McRae *et al.*, 1982; Bousquet and Thimann, 1984; Lynch *et al.*, 1985; Pirrung, 1986). However, the discovery by Hoffman *et al.* (1982a) that EFE stereodifferentiates between different isomers of ethyl-ACC (AEC) provided an excellent tool to discriminate nonphysiological systems from the one operating *in vivo* (McKeon and Yang, 1984; Guy and Kende, 1984; Venis, 1984). Only the (1R,2S) diastereoisomer was converted to 1-butene, whereas artifactual systems did not show this preference. Based on the facts that isolated vacuoles exhibited the same stereospecificity as the *in vivo* enzyme and that activity was destroyed by lysis of the vacuole (Guy and Kende, 1984; Mayne and Kende, 1986), it was concluded that EFE is associated with the tonoplast membrane. However, this is probably not the only site of ethylene formation, since it was demonstrated that vacuoles do not account for more than 80% of ethylene formation of protoplasts (Guy and Kende, 1984). It is therefore not unlikely that EFE is located both in the tonoplast and in the plasmamembrane. John (1983) hypothesized the existence of a transmembrane proton gradient, confirmed in part, at the vacuolar level, by inhibition of ethylene synthesis by ionophores and protonophores (Mayne and Kende, 1986). Mayne and Kende also provided evidence that EFE was located at the inner face of the tonoplast. The observed dependence of EFE activity on membrane integrity (Odawara *et al.*, 1977; Apelbaum *et al.*, 1981) probably reflects the requirement for the transmembrane gradient.

Despite the important insights gained into the mechanism of ACC oxidation and its intracellular location, isolation of the ethylene-forming system has been seriously hampered by dependence on membrane integrity. However, genes encoding proteins that are difficult to purify can be identified by a combination of cDNA and antisense technology (Cabrera *et al.*, 1987). By this approach, Grierson and co-workers identified one of their ripening- and wounding-specific clones in tomato, designated pTOM13, as a candidate for the ACC oxidase or EFE (Hamilton *et al.*, 1990). They found that ethylene production in wounded leaves of pTOM13 antisense transformants was inhibited by 68% and in ripening fruit by 87%. Progeny of self-fertilized transformants inheriting two antisense genes retained only 3% of the control production in ripening fruit. Moreover, when EFE activity was assayed in wounded leaves of homozygous pTOM13 antisense plants, according to Hoffman and Yang (1982), 93% was inhibited. Smith *et al.* (1986) and Holdsworth *et al.* (1987a) characterized the messenger homologous to the pTOM13 cDNA clone, isolated from a ripe tomato library (Slater *et al.*, 1985). Expression of pTOM13 was confined to tomato fruit ripening and wounded leaf or green fruit. Accumulation of the pTOM13 mRNA during fruit ripening reached its maximum in orange fruit. In wounded tomato leaf the induction was very rapid, within 30 min. As both these responses require ethylene synthesis, which increased parallel with pTOM13 mRNA accumulation, it was suggested (Smith *et al.*, 1986; Holdsworth *et al.*, 1987a) that pTOM13 might be involved in the ethylene pathway. Holdsworth *et al.* (1988) examined the expression and organization of the pTOM13 gene family. The tomato genome contains at least three classes of pTOM13-related sequences. One of these is the gene corresponding to the pTOM13 cDNA. A second gene is preferentially expressed in wounded leaves and the encoded protein is 90% homologous to the pTOM13-derived protein and 20 amino acids longer at the N-terminus (Holdsworth *et al.*, 1987b). It is possible that the third copy is a pseudogene or is involved in other ethylene-related responses, although this remains to be proven (Holdsworth *et al.*, 1988). Homologies within the three genes are confined to the coding regions.

The deduced polypeptide encoded by pTOM13 (Holdsworth *et al.*, 1987a) shows 33% identity and 58% similarity to the flavone-3-hydroxylase encoded by the *incolorata* gene of *Antirrhinum majus* (Hamilton *et al.*, 1990). Hydroxylase activity would fit according to the aforementioned mechanistic hypothesis of ACC oxidation. Homology was also found with the ethylene-responsive, fruit-ripening gene product E8 (Deikman and Fischer, 1988). Alignment of amino acids showed three domains of 29–85 amino acids with 52–59% identity. The function of the E8 protein remains as yet unknown. Nevertheless, further evidence will be needed to establish the role of the pTOM13 polypeptide unequivocally. Immunolocalization of the different pTOM13-related proteins is one of the experiments on this line. Since the 33.5-kDa polypeptide encoded by pTOM13 does not contain the consensus ER target signal (von Heijne, 1983) at the N-terminus, and since the literature to date indicates initial ER routing for vacuolar proteins (Voelker *et al.*, 1989; Wilkins *et al.*, 1990), it remains to be proven whether at least one of the pTOM13-related polypeptides is targeted to the vacuole. Moreover, tonoplast-associated proteins have been shown to possess several membrane-spanning domains, connected by short loops (Nelson and Taiz, 1989; Johnson *et al.*, 1990). This structural prediction cannot be derived from a hydropathy plot for the pTOM13 protein (Holdsworth *et al.*, 1987a).

2.5. Mutants in Ethylene Biosynthesis and Its Regulation

The isolation of genes encoding enzymes in ethylene biosynthesis has been achieved through classical molecular biology techniques such as the use of oligonucleotide probes to screen libraries, differential hybridization for cDNAs corresponding to developmentally regulated genes, and antibody screening of expression libraries. However, recent progress in molecular genetics allows identification of genes for which only a mutant phenotype is known. This includes the use of integrated physical/RFLP maps (Chang *et al.*, 1988; Nam *et al.*, 1989; Hauge *et al.*, 1990) and yeast artificial chromosome libraries (Guzmán and Ecker, 1988; Grill and Somerville, 1989), or insertional mutagenesis with T-DNA (Feldmann *et al.*, 1989; Marks and Feldmann, 1989; Koncz *et al.*, 1990; Yanofsky *et al.*, 1990) and transposons (Dean *et al.*, 1990; Honma *et al.*, 1990). Efforts are concentrated on the small crucifer *Arabidopsis thaliana*, which is an excellent system for both classical and molecular genetics (Meyerowitz, 1987). More than 75 loci were assembled in a genetic linkage map (Koornneef *et al*, 1983), which to date has been extended to approximately 100 markers (Koornneef, 1990). The genome of *Arabidopsis* is the smallest known among Angiosperms (estimated 100,000 kb; H. M. Goodman, personal communication) and is nearly devoid of interspersed repeated elements. These features have allowed successful application of the above-mentioned techniques in *Arabidopsis*. However, tomato also offers a suitable system for some of these approaches. An extensive linkage map based on isozymes and RFLPs has been published (Bernatzky and Tanksley, 1986; Tanksley and Mutschler, 1990), and transposon-tagging systems are being developed (Rommens *et al.*, 1990). Moreover, the first ethylene-related mutants were isolated in tomato. The single gene mutant diageotropica (*dgt*), isolated by Zobel (1972), exhibits diageotropic growth characteristics. Although its phenotype can be partially reverted by exposure to low ethylene concentrations (Zobel, 1973), it was demonstrated that the fundamental lesion of *dgt* is an insensitivity to auxin (Kelly and Bradford, 1986). This was confirmed by the results of Hicks *et al.* (1989), which indicate that mutant plants have reduced levels of auxin-binding sites in the stem. In contrast, the epinastic (*epi*) mutation results in an overproduction of ethylene, as a consequence of higher intracellular ACC levels (Fujino *et al.*, 1988). However, ethylene biosynthetic and action inhibitors were ineffective in phenotypical reversion of the *epi* phenotype (Fujino *et al.*, 1989). It was concluded that the primary physiological lesion resulting from the *epi* mutation does not result from ethylene overproduction or increased sensitivity, but rather from an altered target cell specificity. Target cells seem to constitutively exhibit the phenotype normally induced by ethylene, regardless of the presence or absence of the hormone (Fujino *et al.*, 1989). Additional evidence for this hypothesis is given by Ursin and Bradford (1989), who demonstrate that the target cells for epinastic growth in *epi* have the ability to respond to auxin alone, in contrast to wild type.

Constitutive ethylene response mutants were also isolated in *A. thaliana* (Guzmán and Ecker, 1990; Van Der Straeten and Van Montagu, 1990). In both cases, screening by simulation of "triple response" in the absence of exogenously supplied hormone was adopted. Triple response of etiolated seedlings was first described by Neljubow on pea (1901) and exhibits three distinct morphological changes: inhibition of stem and root elongation, radial stem expansion, and loss of geotropism. The latter may be the cause of an exaggerated apical hook in *Arabidopsis*, as suggested by Guzmán and Ecker (1990). Studies by Goeschl *et al.* (1966) indicate that ethylene determines the shape of dark-grown seedlings and that ethylene production is localized in the apical hook region (Goeschl *et al.*, 1967; Taylor *et al.*, 1988). Constitutive triple-response mutants were isolated from a population of chemically mutagenized M2 seeds and designated ethylene overproducer (*eto*) by Guzmán and Ecker (1990). Segregation analysis indicated that the phenotype was caused by a single recessive mutation, located on chromosome 5. The recessive nature suggests that a negative regulatory mechanism is affected by the mutation. There was at least 40-fold more ethylene production in etiolated seedlings, and two- to fivefold more in light-grown seedlings and adult plant tissues, than normally found. This seemed to be due to stimulation of a step before ACC oxidation, most likely ACC synthase, since it is the rate-limiting conversion in the ethylene pathway. Ethylene biosynthesis inhibitors reverted the constitutive triple-response morphology to wild type. It was proposed that the *ET01* gene product could play a role in ACC synthase gene regulation. Application of ethylene action inhibitors resulted in a two- to threefold increased production of the hormone, most likely by affecting the autoregulation of ethylene formation.

Our laboratory has isolated some constitutive triple-response mutants named *epi*, for epinastic hypocotyl (Van Der Straeten and Van Montagu, 1990). The phenotype was confirmed on the level of M3 seeds, and plants are rescued *in vitro* in the presence of silver thiosulfate. The phenotype seemed to be caused by a recessive mutation. It is not yet clear whether this mutation affects the ethylene pathway directly or triggers auxin overproduction or increased sensitivity to either of the hormones. There is no doubt that the molecular analysis of the above-mentioned mutations will be instructive for our understanding of regulation of the ethylene pathway.

2.6. Gene Engineering for Manipulation of Endogenous Ethylene Formation: Perspectives

The control of endogenous ethylene synthesis by gene engineering opens perspectives of great economic importance to agriculture and horticulture. For example, reduction of ethylene levels might overcome substantial postharvest losses that result from overripening of fruits and vegetables during transportation. With the data presented above, and from the schematic overview of ethylene biosynthesis in Figure 1, one can postulate several ways to manipulate endogenous ethylene production. The obvious first choices would be to reduce ACC synthase or ACC oxidase levels, for instance by means of antisense RNA production. Successful results were obtained in the latter case (Hamilton *et al.*, 1990) where, in the progeny containing two antisense copies, the ethylene production was inhibited by 97%. These fruits could be kept for several weeks at room temperature and were less prone to overripening and shriveling than controls. This proves that manipulation of ethylene may offer a possibility to control ripening, and a similar approach can be followed for ACC synthase. In the future, gene disruption by homologous recombination (Scherer and Davis, 1979) or expression of antibody genes (Carlson, 1988; Hiatt et al., 1989; Düring et al., 1990) might become feasible in plants. Conversely, controlled overexpression could also offer several possibilities for applications in agriculture and horticulture, from synchronizing harvest in cotton, grape, and tobacco, to induction of flowers in Bromeliaceae and Cucurbitaceae (Lürssen and Konze, 1985; Bangerth, 1986; Morgan, 1986). In either case, strong and inducible promoters will be required. A more tight control might result from combination of the natural ACC synthase or ACC oxidase promoter with enhancers from other promoters. It is also worth mentioning that the role of each of the gene family members will first need to be evaluated, requiring much more fundamental research. It is not yet proven that overproducing or antisense ACC synthase or EFE plants can be propagated without any harmful effect on their normal physiology.

Finally, one could also propose to manipulate the ACC content in a tissue by influencing ACC malonyl transferase, catalyzing the conversion of ACC to N-malonyl-ACC. This process is known to participate in regulation of ethylene production (Amrhein et al., 1981; Hoffman et al., 1982b; Liu et al., 1985a). The ACC N-malonyltransferase has been partially purified from mung bean hypocotyls (Kionka and Amrhein, 1984). Moreover, it could even be possible to modulate ACC levels by introduction of ACC deaminase (Walsh et al., 1981) from *Pseudomonas* species into plants, under control of an inducible plant promoter. All these approaches are dreams, with substantial economic impact, to be realized in the near future.

3.   ETHYLENE MODE OF ACTION: ON THE WAY TO UNRAVEL HOW A PLANT PERCEIVES A TWO-CARBON ATOM MOLECULE

By analogy to animal cell hormones, it was proposed that plant hormone action is mediated by a receptor, a specific cellular recognition site that binds the hormone and consequently instructs the cell to respond in the appropriate manner. Although a lot of effort was concentrated on the biochemistry and purification of plant hormone receptors (Venis, 1985; Napier and Venis, 1990; Sisler, 1991), only one putative receptor—an auxin-binding protein—has been cloned to date (Hesse et al., 1989; Inohara et al., 1989). Obvious obstacles such as membrane association and the difficulty to prove that a given hormone binding–site is a functional receptor have hampered progress in this aspect of plant biology. However, recent advances in molecular genetics (see section 2.5) allow identification of genes for which only a mutant phenotype exists and will certainly enable scientists to get their hands on plant hormone receptor genes in the near future. Using the approach of combined physical/RFLP maps and YAC libraries, it has been possible to localize genes encoding a putative ABA and ethylene receptor to specific cosmids (Giraudat et al., 1990; Bleecker et al., 1990). It is clear that these revolutionary techniques will rapidly expand our knowledge of the principles of recognition of plant hormones.

3.1. Biochemistry and Physiology of the Ethylene Receptor

It has been recognized that the correlation between ligand structure and biological activity has important implications for receptor topography. Initial observations about chemical requirements for ethylene action came from Burg and Burg (1967). Using the pea epicotyl straight growth test, they presented evidence that biological activity of ethylene analogs (1) requires an unsaturated bond adjacent to a terminal carbon atom, (2) is inversely related to molecular size, and (3) decreases by substitutions that lower the electron density in the unsaturated position. It was also shown that the biological activity of different unsaturated hydrocarbons is related to the stability of their respective complexes with silver. Similarly, Sisler and Yang (1984) reported that cyclic olefins bind to silver in the same order that they inhibit ethylene action. In addition, Sisler (1982a) demonstrated that ethylene binding is inhibited by typical inhibitors of metalloenzymes. These data together implied the existence of a metal-containing receptor site, an idea now generally accepted. The most likely candidate would be Cu^+ (Beyer, 1976; Sisler, 1977; Thompson et al., 1983). Studies by Sisler (1977) with π acceptor compounds as isocyanides, resulting in ethylene-like responses, led to the proposal that π complex formation is the actual basis for ethylene action in plants. Silver ion, a known noncompetitive inhibitor of ethylene action in plants (Beyer, 1976; Sisler, 1982a), probably interferes with the ethylene receptor complex by removing an essential ligand, resulting in interference with ethylene binding.

One of the problems specific for ethylene receptor studies is the fact that many effects of ethylene on plants show short response times, indicating the presence of receptors with high rates of association and dissociation. Such sites would be very hard to detect, since ethylene can rapidly diffuse away from the binding site. However, the discovery of binding sites with low rate constant of association/dissociation has facilitated the work quite considerably. At least three different binding sites have been distinguished by their dissociation rates (Sisler, 1990). Ethylene–binding systems have been characterized in several plant tissues, including tobacco leaves (Sisler, 1979), *Phaseolus* cotyledons (Bengochea et al., 1980), mung bean shoots (Sisler, 1982a), tomato leaves and fruit (Sisler, 1982b), carnation petals and leaves (Sisler, *et al.*, 1986), pineapple (Goren and Sisler, 1986) and pea epicotyls (Sanders *et al.*, 1989b). Light microscopic and high-resolution electron microscopic autoradiography in *Phaseolus* cotyledons indicated that ethylene-binding sites are primarily localized on the rough endoplasmic reticulum and on protein body membranes (Evans *et al.*, 1982). Recently, Hall *et al.* (1990) succeeded in purifying the ethylene-binding protein from this system. The polypeptide seems to be an extremely hydrophobic glycoprotein, and amino acid sequence data were obtained. Atomic spectroscopy indicated the presence of a copper atom in the molecule. An antibody was raised against the purified ethylene-binding protein, which was found to be more abundant in the abscission zone than in the adjacent petiolar tissue (Connern *et al.*, 1989; Hall *et al.*, 1990). Cross-reactivity in pea, tomato, and *Arabidopsis* was demonstrated. The question remains whether all these ethylene-binding sites are functional. In the case of the pea-binding protein, a strong correlation was observed between the binding properties of the fast associating site with ethylene analogs and the growth responses (Hall *et al.*, 1990), suggesting its functionality as a true receptor. Similarly, Sisler presented evidence that ethylene-binding *in vivo* as well as *in vitro* is to a physiological receptor (Sisler, 1987; Sisler and Wood, 1987). The conclusion was based on the comparison of binding characteristics and activity of ethylene analogs, and ethylene diffusion rates to and from the binding site, in extracts from mung bean sprouts and tobacco leaves.

Finally, it needs to be mentioned that ethylene is metabolized to CO_2, or ethylene oxide and ethylene glycol (Beyer, 1975, 1985; Blomstrom and Beyer, 1980; Sanders *et al.*, 1989a,b). At least 80% of ethylene in the plant is converted to ethylene oxide (Sanders *et al.*, 1989b). Abeles (1984a,b) proposed several roles for ethylene metabolism in plants, including that ethylene metabolism is involved in its mode of action, as first suggested by Beyer (1972, 1979). Recently, Sanders *et al.* (1989b) presented strong evidence that ethylene action and metabolism are two independent processes. Future characterization of enzymes involved in ethylene catabolism might reveal potential regulatory aspects on its biosynthesis or action.

3.2. Toward the Isolation of Genes from *Arabidopsis thaliana* involved in Ethylene Signal Transduction

As previously mentioned, molecular genetics offers a powerful tool to answer the question of how a plant recognizes ethylene and subsequently transduces that signal to lead to the well-documented biochemical and physiological changes. Three laboratories have been concentrating on ethylene response mutants in *Arabidopsis* and have applied either ethylene (Bleecker *et al.*, 1988a; Guzmán and Ecker, 1990) or ACC (Van Der Straeten *et al.*, 1989a) as a screening agent. In all cases, the absence of triple response upon hormone treatment

Table II
Classes of *Arabidopsis* Mutants Affected in C_2H_4 Biosynthesis and Response[a]

Type	Mutation	Screening	Type	Chromosome no.	Possible target gene(s)	Reference
Biosynthesis	etol		Recessive	5	ACC synthase	1
	epi					2
Response	etr1	C_2H_4	Dominant	1	C_2H_4 receptor	3
	ein1	C_2H_4	Dominant	1		1
	ein2	C_2H_4	Recessive	4		1
	hin	C_2H_4				4
	hls	C_2H_4	Recessive			1
	sin1	*	Dominant	4		4
	ran	2,5-NBD	Dominant		C_2H_4 receptor	4
	etr2	ACC	Recessive		C_2H_4 signal transduction	5
	ain1	ACC	Recessive	1	C_2H_4 signal transduction	2

[a]ain, ACC insensitive; ein, ethylene insensitive; eto, ethylene overproducer; epi, epinastic hypocotyl; etr, ethylene resistant; hin, hypocotyl insensitive; hls, hookless; ran, responsive to antagonist; sin, suppressor of ethylene insensitivity; *, pseudorevertant of ein2; 2,5-NBD, 2,5-norbornadiene. (1) Guzmán and Ecker (1990); (2) Van Der Straeten and Van Montagu (1990); (3) Bleecker et al. (1988a); (4) Guzmán and Ecker (1989); (5) Pickett et al. (1989).

was searched (normal etiolated phenotype). Triple-response screening has allowed identification of at least seven different loci involved in ethylene signal transduction, based on the phenotypic differences and allelism tests of the mutants (Table II). It is hard to estimate the number of additional genes involved in the pathway at this point, since the distinct biochemical processes are totally unknown and the mutational analysis is probably incomplete.

3.2.1. Screening on Ethylene: The *etr, ein, hls, hin, sin,* and *ran* Loci

The first well-characterized ethylene-resistant *Arabidopsis* mutant was described by Bleecker et al. (1988a). Chemical mutants resistant to 5 ppm C_2H_4 were isolated in a Columbia background. Segregation of ethylene insensitivity indicated that the *etr-1* mutation is dominant. The mutant plant was lacking all the ethylene responses examined, including inhibition of cell elongation, promotion of germination and leaf senescence, feedback suppression of ethylene synthesis, and stimulation of peroxidase activity. Since saturable ethylene binding in the mutant was only one-fifth of the wild-type value, it was concluded that the *etr-1* mutation probably affects the ethylene receptor. The mutation was mapped to chromosome 1 in *Arabidopsis*. Progress on the cloning of the *etr* gene was reported recently (Bleecker et al., 1990). The *etr* gene was confined to the 23-kb overlap of two cosmids, which are conferring ethylene insensitivity to wild type. Sequence analysis of this region might reveal some of the physical properties of the protein and thereby offer some clues for understanding the function of the receptor.

Genetic analysis by Guzmán and Ecker (1990) indicated that at least one additional locus is involved in insensitivity to ethylene. Of 25 independent isolates, obtained by chemical mutagenesis, six were further characterized. Mendelian analysis indicated that one of them was dominant (*ein1-1*), and five others were recessive alleles of a second locus (*ein2-1,5*). In both cases, rosettes were larger than wild type, and a delay in bolting was observed. *Ein1* was mapped to chromosome 1, close to the *ap1* locus, and therefore possibly allelic with *etr1*, which also exhibits these characteristics. *Ein2* was positioned on chromosome 4 by RFLP linkage analysis. In both mutants a two- to threefold increase in ethylene production was demonstrated compared to wild type. This may indicate a defect in feedback control, as was concluded for *etr1*. Leaves from these plants exposed to ethylene did not display any significant reduction in ethylene synthesis, as opposed to wild type.

In addition, Guzmán and Ecker (1990) have described the isolation of the hookless (*hls*) phenotype. As mentioned above, etiolated *Arabidopsis* seedlings display an arch-shaped apical region. Ethylene was proved to play an important role in the morphology of dark-grown seedlings (Goeschl *et al.*, 1966). It was demonstrated that the localized production of the hormone is responsible for the formation and maintenance of the hook (Goeschl *et al.*, 1967; Taylor *et al.*, 1988). Upon screening etiolated seedlings in the absence of ethylene, hookless mutants were observed. Two mutations were further investigated and segregated as single, recessive, nonallelic traits. *Hls1-1* plants had small, narrow rosette leaves; moreover, they bolted and senesced much earlier than wild type. In addition, these plants produced reduced amounts of ethylene. Nevertheless, this is unlikely to be the cause for the profound morphological change in the hook region. When treated with exogenous ethylene, *hls1-1* seedlings did not evidence any hook curvature, although hypocotyl and root responses were normal. This indicates that the latter are independent from the hook response. Guzmán and Ecker (1990) speculate that the primary action of the *hls* mutation affects an antagonist of ethylene, based on the speculation that hook formation results from the production of such antagonist in the outer region of the apical hook. Further analysis will have to prove this role as a modulator in ethylene action.

Three additional classes of mutations were isolated. Hypocotyl-insensitive (*hin;* Guzmán and Ecker, 1989) mutants have a long hypocotyl in the presence of ethylene, but display an apical hook, as opposed to *hls*. Another class of mutants was identified as suppressors of ethylene sensitivity (*sin;* Guzmán and Ecker, 1989). *Sin1* is a dominant mutation closely linked to *ein2* on chromosome 4. Finally, mutants responsive to antagonist (*ran;* Guzmán and Ecker, 1989) show a typical ethylene response in the presence of 2,5-norbornadiene, a competitive inhibitor of ethylene binding (Sisler and Yang, 1984). *ran* is a dominant mutation, probably affecting the ethylene receptor(s).

3.2.2. Screening on ACC: The *ain* Phenotype

Screening with the ethylene precursor ACC offers the possibility of recovering mutations both in EFE and in the ethylene signal transduction pathway. ACC-insensitive mutants were isolated independently by Pickett *et al.* (1989) and Van Der Straeten *et al.* (1989a). The *etr2* mutant (Pickett *et al.*, 1989) was identified on the basis of its ability to elongate roots in the presence of high ACC concentrations. Genetic analysis indicated that *etr2* is recessive and confers resistance to some, but not all, effects of ethylene exposure.

ain mutants were identified by their lack of triple response when germinated on 0.5 mM ACC in the dark (Van Der Straeten *et al.*, 1989a; Van Der Straeten and Van Montagu, 1990). Two mutants were subjected to further investigation and appeared to be allelic. Genetic analysis indicated that the phenotype was caused by a single recessive mutation. Adult plants are indistinguishable from wild type above ground, but their abundant root system grows ageotropically. Bolting was delayed 1–3 weeks; however, none of the late-flowering mutants described by Koornneef *et al.* (1990) are ACC insensitive, and thus not allelic to *ain1* (D. Van Der Straeten and A. Djudzman, unpublished results). Moreover, the *ain* mutants were insensitive to 10 ppm ethylene, which means that none of them appears to be a candidate EFE mutant, and which also excludes an ACC uptake mutation. *Ain1* was mapped on chromosome 1, 30 cM from the *ap1* locus. Work is underway to isolate the corresponding gene, by an integrated physical/RFLP map and cosmid library (Hauge *et al.*, 1990).

From the overview presented above, it is clear that at least seven loci are involved in ethylene insensitivity (Table II). Nucleotide sequences of the corresponding clones might provide more insight into the nature of polypeptides involved in ethylene signal transduction and expand our view on plant hormone perception in general.

4. MOLECULAR ASPECTS OF ETHYLENE INDUCIBILITY

It is generally accepted that hormones exert their effects by activating certain genes transcriptionally or posttranscriptionally. The assessment of the primary role for ethylene in activation of gene expression is relatively straightforward by analysis of the effect of biosynthetic and action inhibitors. This approach allows us to obtain a critical view on this matter. In the next section, we will elaborate on a number of genes induced by ethylene under specific developmental circumstances, or under stress conditions. We will focus on those for which a substantial amount of information has been obtained. As for developmental situations, ripening, senescence, and abscission have been studied in most detail. On the other hand, the role of ethylene in triggering the plant's defense against pathogens will be discussed. In addition, the importance of ethylene for the reaction of plants toward some abiotic factors will be presented.

4.1. Ethylene-Induced Genes in Plant Development: Ripening, Senescence, and Abscission-Specific Genes

4.1.1. Ethylene-Responsive Fruit-Ripening Genes

Ripening is the final phase in fruit development, in most cases ultimately leading to a soft, edible fruit. The process involves a series of highly coordinated physiological and biochemical changes, affecting color, flavor, and texture of the fruit (Sacher, 1973). These changes involve alterations in several enzyme activities, including invertase (Iki et al., 1978) and polygalacturonase (PG; Brady et al., 1982; Tucker and Grierson, 1982) during tomato fruit ripening, and endo-1,4-β-glucanase (cellulase) and later also PG in avocado ripening (Pesis et al., 1978; Awad and Young, 1979). All these events occur in conjunction with an increase in ethylene production and a consequent raise in CO_2, the respiratory climacteric (Brady, 1987). A direct role for ethylene in ripening of climacteric fruit has been unequivocally established (Biale and Young, 1981; Yang, 1987). This is based on three different facts: First, the onset of ripening is accompanied by a burst of autocatalytic ethylene production, named "system 2" ethylene (McMurchie et al., 1972) in contrast to "system 1" ethylene operating until ripening starts. Second, exposure of unripe fruit to exogenous ethylene will hasten ripening (McGlasson and Pratt, 1964; McGlasson et al., 1975a; Rai et al., 1983). Third, exposure to ethylene biosynthetic or action inhibitors will delay ripening (Ness and Romani, 1980; Child et al., 1984; Saltveit et al., 1978; Hobson et al., 1984). It has been proposed that ethylene is inducing a set of genes whose corresponding proteins are responsible for the different physiological and biochemical events during ripening. This hypothesis was pursued by the cloning of cDNAs and genes encoding cell wall softening enzymes as PG (Della-Penna et al., 1986; Grierson et al., 1986; Sheehy et al., 1987; Bird et al., 1988; Rose et al., 1988), pectin esterase (Ray et al., 1988), and cellulase (Christoffersen et al., 1984; Tucker et al., 1987; Cass et al., 1990). In addition, a proteinase inhibitor I gene (Margossian et al., 1988) and several ripening-associated genes of unknown function (Slater et al., 1985; Mansson et al., 1985; Lincoln et al., 1987; Pear et al., 1989) have been cloned. In many cases, the primary role of ethylene on mRNA induction was demonstrated by application of action inhibitors (Lincoln et al., 1987; Davies et al., 1988; Davies and Grierson,

1989) or exogenous ethylene treatment of unripe fruit (Lincoln *et al.*, 1987; Maunders *et al.*, 1987), respectively inhibiting and stimulating accumulation of specific messengers. Inhibition by silver affected mRNAs to different extents, due either to variations in stability, or to a secondary rather than a primary ethylene requirement (Davies *et al.*, 1988).

Three aspects of regulation of gene expression by ethylene can be considered: spatial and temporal aspects of expression (kinetics of induction); response to concentration or sensitivity changes; finally, transcriptional and posttranscriptional regulation.

Studies of activation of ripening–related genes have been performed in two different ways. On the one hand, kinetics of expression were investigated in unripe fruit (tissue) exposed to elevated ethylene levels (Nichols and Laties, 1984; Lincoln *et al.*, 1987; Maunders *et al.*, 1987); on the other hand, timing of expression was analyzed during development of the fruit (Mansson *et al.*, 1985; DellaPenna *et al.*, 1986; Lincoln *et al.*, 1987; Maunders *et al.*, 1987; Davies and Grierson, 1989; Pear *et al.*, 1989). The first approach allows us to divide ethylene-responsive mRNAs in two classes: (1) genes that are activated within the first 2 hr of exogenous ethylene exposure, likely to be the result of a primary ethylene effect (Lincoln *et al.*, 1987); (2) genes for which the first detectable messenger appears only after a few hours of treatment, where chances are that these are the result of a series of metabolic events initiated by ethylene, rather than a direct induction (Nichols and Laties, 1984; Tucker and Laties, 1984; Maunders *et al.*, 1987). The ripening-specific proteinase inhibitor I (Margossian *et al.*, 1988) and the E8 gene of unknown function belong to the first group. Polygalacturonase and cellulase genes picture in the second one. Apart from ethylene responsiveness, an additional level of regulation is involved. In some cases, the message appears before full maturity of the fruit has been reached (Lincoln *et al.*, 1987; Pear *et al.*, 1989); in other cases, the message is enhanced at mature green stage or in ripening fruit (Mansson *et al.*, 1985; Lincoln *et al.*, 1987; Maunders *et al.*, 1987; Davies and Grierson, 1989). For the E8 gene of tomato, this developmental regulation was correlated with the increase of a DNA-binding activity interacting with the region -1088 to -682 from the transcriptional start (Deikman and Fischer, 1988). The large majority of the ripening-specific genes described are essentially fruit specific, although some are also transcribed in leaf or root (Mansson *et al.*, 1985; Maunders *et al.*, 1987; Lincoln and Fischer, 1988a; Davies and Grierson, 1989).

A second aspect of ethylene regulation of expression is the existence of diverse mechanisms of ethylene inducibility. Fischer and his co-workers (Lincoln *et al.*, 1987; Lincoln and Fischer, 1988a) presented evidence supporting the idea that changes both in ethylene concentration and in sensitivity may regulate gene activation. In some cases, the onset of gene transcription and mRNA accumulation coincides with ethylene evolution. In others, it precedes the ethylene increase and therefore is probably due to a higher sensitivity to basal ethylene levels. These molecular data elegantly support earlier physiological experiments suggesting that ethylene-mediated processes involve perception of changes in ethylene levels and sensitivity to the hormone (Abeles, 1967; Kende and Hanson, 1976; McGlasson, 1978).

Finally, there is experimental evidence that the ethylene-responsive proteinase inhibitor I gene is regulated not only on the transcriptional, but also on the posttranscriptional level (Lincoln and Fischer, 1988a). Posttranscriptional processes were also suggested to contribute to the large accumulation of PG mRNA during ripening (DellaPenna et al., 1989).

In conclusion, we would like to mention the existence of some ripening-impaired mutants of tomato, which offer a tremendous advantage in the study of fruit ripening in general and, in particular, softening. The ripening inhibitor (rin) and nonripening (nor) mutants are both recessive and produce a nonclimacteric fruit, with very low levels of ethylene, lycopene, and PG. These fruits can be stored over very long periods with almost no softening of the cell walls (Tigchelaar et al., 1978; Grierson et al., 1987). In contrast, the never-ripe (Nr) mutation is a dominant locus, conferring a slow-ripening phenotype. Studies on expression of the ethylene-related genes E4, E8, E17, and J49 during normal rin development and under exogenous ethylene imply different mechanisms for ethylene inducibility, as mentioned previously (Lincoln and Fischer, 1988b). The above-mentioned mutations repressed PG transcription. In contrast, transcription of the ripening-associated genes E4, E8, E17, and J49 was partly inhibited or enhanced, depending on the gene in question and on the genetic background (DellaPenna et al., 1989). Moreover, it has been possible to assess the complexity of the process of fruit softening using rin-type tomato fruit. PG had been previously regarded as the primary agent regulating tomato fruit softening (Hobson, 1964; Brady et al., 1982). However, expression of the PG gene under control of the E8 promoter, which is inducible in rin fruit (Lincoln and Fischer, 1988b), leads to transgenic rin tomato with significant cell wall polyuronide degradation, but not to softening (Giovannoni et al., 1989). A similar conclusion could be drawn from antisense experiments. Overexpression of antisense PG under control of the cauliflower mosaic virus 35S promoter (Smith et al., 1988; Sheehy et al., 1988) leads to 90% reduction in PG mRNA and activity, but no difference in softening was observed. In this case, however, owing to the abundance of PG in normal fruit, it is possible that the remaining 10% of PG activity is responsible for the ripening effect seen.

4.1.2. Regulation of Senescence-Related Gene Expression by Ethylene

Senescence was defined as the sequence of metabolic events culminating in cell death (Roberts et al., 1985). Senescence and ripening share some common features at the physiological and biochemical level (Borochov and Woodson, 1989). These include an increase in hydrolytic enzymes (Baumgartner et al., 1975; Hobson and Nichols, 1977), degradation of starch and chlorophyll (Roberts et al., 1985), loss of cellular compartmentalization (Suttle and Kende, 1980), and finally, a climacteric surge in respiration (Nichols, 1973; McGlasson et al., 1975b). As for ripening, there is clear-cut evidence that ethylene plays a key regulatory role in senescence. A climacteric rise in ethylene production, coincident with the first visible signs of senescence and primarily due to an increase in ACC synthase activity, was observed on both leaf and flower senescence (McGlasson et al., 1975b; Nichols, 1966; Bufler et al., 1980; Peiser, 1986). It was suggested that ACC would be transported to the petals from other floral parts, thereby inducing flower senescence (Bufler et al., 1980; Hsieh and Sacalis, 1986), but more recent evidence weighs against this hypothesis (Hanley et al., 1989). A second factor supporting the key role for ethylene in senescence is that exposure to exogenous ethylene hastens the onset of the senescence program in leaves (Gepstein and Thimann, 1981) and flowers (Mayak et al., 1977). Finally, both processes have been shown to be delayed by inhibitors of ethylene biosynthesis (Bufler et al., 1980) or action (Veen, 1979; Goren et al., 1984). Recently, it was even demonstrated that petal senescence of carnation flowers, which already have entered the ethylene climacteric, can be reversed by norbornadiene treatment (Wang and Woodson, 1989).

For reasons mentioned above, it is not surprising that similarities were also found in gene expression during fruit ripening and leaf senescence, including an increase in the putative ACC oxidase message (Davies and Grierson, 1989). At present it is not yet clear whether mRNAs in either situation are transcribed from the same gene, rather than from different members of a multigene family. Moreover, the majority (five of seven) of these mRNAs showed a maximal prevalence at peak ethylene production (at the onset of color loss) and an inhibition of accumulation by silver treatment, providing direct evidence for ethylene-regulated gene expression in leaf senescence. Loss of other mRNAs was prevented by silver, indicating that ethylene regulates the decrease of specific messages upon senescence, too. Although the function of the proteins encoded by the majority of senescence-induced mRNAs is unknown, possible candidates include proteases, ribonucleases, peroxidases, and photorespiratory enzymes. An increase in both acidic (60 kDa) and basic (33 kDa) peroxidases during ethylene-enhanced senescence of cucumber cotyledons has been demonstrated (Abeles et al., 1988). cDNA clones encoding ethylene-induced peroxidases from cucumber (Morgens et al., 1990) and a tobacco peroxidase (Lagrimini et al., 1987) have been cloned. The former exhibits an evident increase in mRNA level after 3 hr of induction and has a maximal prevalence after 15 hr. In addition, it was shown that overexpression of tobacco acidic peroxidase (36 kDa) resulted in severe wilting initiated at the time of flowering (Lagrimini et al., 1990).

Gene expression during senescence of carnation (*Dianthus caryophyllus* L.) flowers and possible mechanisms of ethylene regulation have been studied in detail. *In vitro* translation of polyadenylated RNA from presenescent flowers revealed significant increases in specific polypeptides after 3 hr of ethylene exposure and maximal accumulation after 6 hr (Woodson and Lawton, 1988), whereas other products decreased. In four of six cases an almost complete inhibition by silver ion and 2,5-norbornadiene was noted. Taken together, these facts implicate a primary role for ethylene in regulation of gene expression in senescent carnation flowers. Furthermore, the ethylene-enhanced mRNAs are similar to those shown to accumulate during natural senescence at the ethylene climacteric (Woodson, 1987). One of these, encoding an 81-kDa protein, occurs prior to the increase in ethylene production and was postulated to result from an increased sensitivity to the hormone. The same group of researchers subsequently isolated three senescence-related cDNA clones from carnation (Lawton *et al.*, 1989) and analyzed their regulation by ethylene (Lawton *et al.*, 1990). Their respective mRNAs substantially accumulate 4–5 days after harvest of the flowers, or between 0.5 and 6 hr of exogenous ethylene treatment to presenescent flowers (Lawton *et al.*, 1989). It can be concluded that ethylene is able to modulate the expression of senescence-related RNAs before any visible symptom occurs.

With respect to ethylene regulation, several important points have to be mentioned. First, the cDNAs described fall into two classes. The first class is represented by the pSR5 clone, detectable after 0.5 hr of ethylene exposure, encoding the 81-kDa polypeptide of unknown function previously mentioned. Its mRNA accumulated prior to increased ethylene evolution during natural senescence and showed only a minor reduction upon treatment of presenescent carnation with aminooxyacetic acid (inhibitor of ACC synthase) or silver thiosulfate, or exposure of climacteric flower petals to norbornadiene (Lawton *et al.*, 1989, 1990). This confirmsthat the induction of some RNAs is regulated by a change in ethylene sensitivity, but also suggests that factors other than ethylene may modulate their expression. Class 1 is noticed in petals and leaves upon ethylene exposure. A second class is clearly regulated by a change in ethylene concentration, increasing in parallel with the ethylene climacteric. The two clones representing this class are both significantly repressed upon treatment with AOA, silver thiosulfate, and norbornadiene (Lawton *et al.*, 1989, 1990). Expression was confined to floral organs exposed to ethylene and was most abundant in petals.

An increased transcription rate in response to ethylene was noted for both classes (Lawton *et al.*, 1990). Moreover, it was apparent that ethylene is needed not only for the induction of senescence-related gene expression, but also for the maintenance of elevated RNA levels of these genes. Transient exposure to ethylene leads to a decrease in abundance of mRNA, unless it was sufficient to trigger autocatalytic enhancement of ethylene (Lawton *et al.*,, 1990). Similarly, it was proposed that continuous perception of ethylene is required in order to sustain high levels of ethylene production in climacteric tissue (Tucker and Brady, 1987; Wang and Woodson, 1989).

Trewavas (1982) proposed that the response to a hormone is dependent on two different factors: tissue sensitivity to the hormone and hormone concentration. Lawton *et al.* (1990) present evidence that flower maturity can enhance the capacity of ethylene to induce a particular RNA species. It was demonstrated earlier that responsiveness of petal tissue to ethylene increases with age (Nichols, 1968; Mayak *et al.*, 1977). However, since this is not accompanied by an increased ethylene-binding capacity (Brown *et al.*, 1986), it is likely that this developmental increase in tissue responsiveness is the result of a regulation beyond the receptor, at a specific point in signal transduction.

The cloning of the genes encoding the senescence-related cDNAs and analysis of their *cis*- and *trans*-acting regulatory factors will allow a more profound understanding of this developmental program and its control by ethylene.

4.1.3. Ethylene Regulation of Gene Expression upon Abscission

Abscission is the developmental process by which a plant sheds organs, as leaves, fruits, petals, or other floral structures. It is accompanied by a series of complex morphological and biochemical events, which have recently been reviewed (Sexton and Roberts, 1982; Sexton *et al.*, 1985). The term "abscission zone" refers to the predictable places in the plant where, in a discrete one- to three-cell-wide "separation layer," wall-to-wall adhesion is lost by secretion of hydrolytic enzymes that dissolve the middle-lamella polysaccharides. These separation-layer cells are different from the surrounding cells in that they show an expansion induced by ethylene and repressed by auxin (type II cells; Osborne, 1982; Osborne *et al.*, 1985). This negative regulation is apparent throughout the entire abscission process. It is likely that ethylene acts by altering the concentration in auxin through enhanced degradation, inhibition of transport, or increased conjugation (Beyer and Morgan, 1971; Riov and Goren, 1979; Wood, 1985). In addition, ethylene might be a second messenger for abscisic acid, although this hypothesis is contentious (Sexton *et al.*, 1985). There is biochemical evidence that abscission involves cell wall solubilization, similar to the ripening process. Increases in cellulase (Horton and Osborne, 1967; Lewis and Varner, 1970; Pollard and Biggs, 1970), polygalacturonase (Riov, 1974; Greenberg *et al.*, 1975), and chitinase (Gomez *et al.*, 1987) activities in the abscission zone have been reported. Cellulase was shown to be associated to the separation layer (Sexton *et al.*, 1981). Recently it was proved to be a basic cellulase accumulating upon abscission, in contrast to the acidic form present before onset of the process, most probably derived from a different gene (del Campillo *et al.*, 1988).

The function of chitinase in abscission is unknown, but is probably not direct (Mauch and Staehelin, 1989).

A great number of papers support the idea that endogenous ethylene production correlates with the onset of abscission and therefore probably coordinates and accelerates the whole process (Jackson and Osborne, 1970; Sexton *et al.*, 1985). Several authors also propose ethylene as the actual inducer of natural abscission. This idea is based on a delay in abscission obtained after application of ethylene biosynthesis and action inhibitors (Baird *et al.*, 1984; Kushad and Poovaiah, 1984; Sisler *et al.*, 1985). Recent molecular data on the accumulation of abscission cellulase mRNA sustain this hypothesis (Tucker *et al.*, 1988). Expression of abscission cellulase (9.5 cellulase) was entirely blocked after a 24-hr exposure to 2,5-norbornadiene, following a 31-hr ethylene treatment to initiate abscission, accompanied by moderate cellulase mRNA accumulation. In addition, this indicates that ethylene is necessary for continued expression of cellulase mRNA. The same authors demonstrate the existence of a very long lag phase (24 hr) preceding cellulase messenger appearance after ethylene exposure (maximal accumulation occurs after 60 hr), implying that ethylene might not be the prime factor inducing cellulase but rather triggering a second messenger. Expression was at least 10-fold higher in abscission zones than in stems or petioles. Interestingly, they also observed that application of auxin to the distal end of explant petioles, prior to ethylene exposure, inhibits abscission and cellulase messenger accumulation. This treatment, however, did not significantly affect the expression of chitinase in the explant. In conclusion, the molecular data seem to imply that ethylene possibly acts by first decreasing auxin concentration to noninhibitory levels and subsequently triggering the activation of the abscission cellulase gene.

Finally, we would like to mention a study on the localization of 9.5 cellulase in abscising and nonabscising tissues of *Phaseolus* by nitrocellulose tissue printing (del Campillo *et al.*, 1990). In both proximal and distal abscission zones exposed to ethylene, cellulase was found in the cortical cells of the separation layer and associated with vascular traces of the adjacent tissue. Surprisingly, it was also detected in the vascular traces of the stem and pulvinus without developing a separation layer. Its role at those sites remains to be elucidated.

4.2. Gene Activation and Stress Ethylene

4.2.1. Ethylene Formation in Response to Stress

One of the physiological consequences of stress is a boost of ethylene production (Yang and Hoffman, 1984). Stress ethylene formation has been recognized upon infection by pathogens—viral, bacterial, or fungal—or application of biotic elicitors (cell wall fragments, oligosaccharides) and upon various forms of abiotic stress. These can be mechanical wounding, radiation, excessive temperatures, drought, flooding, or exposure to certain chemicals, including abiotic elicitors (polyacrylic acid, α-amino butyric acid), herbicides, and pollutants such as heavy metals, ozone, and SO_2. It is generally accepted that the conversion of AdoMet to ACC is the key reaction controlling stress ethylene formation (Boller and Kende, 1980; Kende and Boller, 1981; Konze and Kwiatkowski, 1981; de Laat and van Loon, 1982; Fuhrer, 1982; Hoffman and Yang, 1982; Wang and Adams, 1982; Chappell et al., 1984; Mauch et al., 1984; Cohen and Kende, 1987; Spanu and Boller, 1989; Larrigaudière et al., 1990). As mentioned above, ethylene may be formed from ACC by reaction with certain oxygenated molecules. Moreover, there is evidence that the formation of toxic oxygen species is also enhanced by some types of stress (Finlayson and Pitts, 1986; Asada and Takahashi, 1987; Dixon and Lamb, 1990). It was proposed that ethylene production in injured tissue could be the result of lipoxygenase activity (Kacperska and Kubacka-Zebalska, 1985), but recent data from Spanu and Boller (1989) indicate that this is not the case, as demonstrated in Phytophthora-infected tomato plants. Their conclusion is based on low affinity for the ACC to ethylene conversion and inhibition by Co^{2+}, which are both characteristic for EFE and not expected for the lipoxygenase system (Wang and Yang, 1987). Nevertheless, this does not exclude the possibility of nonbiological or lipoxygenase-mediated ethylene formation when a plant is subjected to ozone or other air pollutants that might react with ozone and lead to highly reactive free radicals (Finlayson and Pitts, 1986).

Stress conditions trigger the induction of a wide series of genes, with a central role for enzymes of the phenylpropanoid pathway (PAL, 4-CL, CHS), leading to biosynthesis of lignin, phytoalexins, and anthocyanins. Newly formed polypeptides may play a role in cell wall rigidity (extensins, glycine–rich proteins, peroxidases, cinnamylalcohol dehydrogenase, or callose synthase) or be associated with an antimicrobial activity (endohydrolases, thionins, thaumatin-like proteins, proteinase inhibitors). For others, the function remains unknown. Since the molecular basis of the plant defense response has been reviewed extensively (Bowles, 1990; Carr and Klessig, 1990; Dixon and Lamb, 1990), we will limit the discussion to a very fundamental question: Is ethylene a signal in triggering the molecular events of defense? Or, in other words, is the ethylene signal transduction pathway of primary importance in the induction of the defense response?

4.2.2. Both Ethylene- and Nonethylene-Dependent Pathways Operate in Response to Stress

The observations that stress enhances endogenous ethylene formation and that certain biochemical defense mechanisms are triggered by exogenous ethylene led to the proposal that it may function as the factor activating the plant's defense (Pegg, 1976; Boller, 1982). Ethylene induction has been reported for several defense-related genes (Broglie *et al.*, 1986; Ecker and Davis, 1987; Margossian *et al.*, 1988; Vögeli *et al.*, 1988; Morgens *et al.*, 1990). In all cases of stress examined, ethylene increase preceded the respective biochemical response, a kinetic property expected if ethylene played a primary role. For instance, ACC synthase induction and ethylene production increased within 15 min after wounding of green tomato fruit or fungal elicitor application to cultured parsley cells (Boller and Kende, 1980; Konze and Kwiatkowski, 1981; Chappell *et al.*, 1984), or reaching a maximum 6 hr after fungal infection or chitosan application in immature pea pods, preceding substantial accumulation of endohydrolases by several hours (Mauch *et al.*, 1984).

The issue of ethylene's primary role in induction of defense has been investigated by several groups using inhibitors of ethylene biosynthesis or action. These studies allow us to draw the conclusion that some inducers, including ozone (Mehlhorn and Wellburn, 1987), TMV, and α-aminobutyric acid (Lotan and Fluhr, 1990), might use an ethylene pathway to trigger the response. In the case of ozone, this response does not seem to be defense but rather cell death that is induced by an ethylene pathway (Mehlhorn and Wellburn, 1987). It was shown that AVG offers almost complete protection against visible ozone injury of the leaf. On the other hand, it was demonstrated that both AVG and STS inhibit the accumulation of chitinase in tobacco leaves (Lotan and Fluhr, 1990). AVG treatment blocked more than 90% of the endogenous ethylene production and, therefore, must be acting specifically. Moreover, it was shown that this ethylene-dependent pathway is also light regulated, although not phytochrome mediated. A third positive correlation was found between ethylene formation and chitinase induction in melon leaves or seedlings treated with elicitors from *Colletotrichum lagenarium* (Roby *et al.*, 1986). Unfortunately, their results are difficult to interpret since they were able to demonstrate only a very limited inhibition of ethylene production by AVG and even obtained some values below controls.

In contrast, Lotan and Fluhr (1990) suggested the existence of a non-ethylene pathway for induction of chitinase by *Pseudomonas syringae* pv *tabaci* and xylanase, a fungal elicitor from *Trichoderma viride*. In this case, levels of chitinase remained unaffected by the inhibitors. A similar situation was discovered earlier for chitinase and β-1,3-glucanase in pea pods infected with compatible and incompatible *Fusarium solani* species (live or autoclaved) or treated with chitosan or heavy metals such as Cd^{2+} (Mauch *et al.*, 1984). Although application of AVG inhibited ethylene production by about 80%, it did not significantly affect either endohydrolase activity in each of these cases. Finally, recent data from Henstrand and Handa (1989) indicate that upon wounding of unripe tomato pericarp both pathways may be involved, implying existence of more than one signal regulating gene expression. At least 50 mRNA species are affected and mostly up-regulated after wounding, but less than 15% is influenced by treatment with ethylene action inhibitors STS and norbornadiene.

That defense-related genes may be induced by several independent factors is indirectly supported by the following studies on endohydrolase expression and localization. It has been proposed that ethylene systemically induces basic iso-forms of chitinase and β-1,3-glucanase which are confined to the vacuole (Boller and Vögeli, 1984; Mauch and Staehelin, 1989; Keefe *et al.*, 1990) and probably play a role as a second and drastic line of defense against pathogens (Mauch and Staehelin, 1989). On the other hand, both genes seem to be developmentally regulated and enhanced in older leaves, roots (Felix and Meins, 1986), and flowers of tobacco (Lotan and Fluhr, 1989). These different regulations might be a reflection of the existence of several genes encoding basic isoforms, each regulated in a different manner. However, Samac *et al.* (1990) found a single gene encoding basic chitinase in *A. thaliana* which was homologous to a maize basic chitinase, and which shows an age- and organ-specific expression. The *etr1* mutant also accumulated basic chitinase mRNA in the roots, indicating that the organ-specific expression was absolutely ethylene independent.

In conclusion, the general picture supports the existence of at least two different pathways for induction of stress-related genes. Possible signal transduction mechanisms have been proposed by Dixon and Lamb (1990). However, since endogenous ethylene formation is enhanced by every factor triggering the defense response with the exception of salicylic acid (Leslie and Romani, 1986), it is not unlikely that ethylene is a general enhancer of biochemical defense, possibly acting at a distance, even in those cases where it is not a real second messenger. That the regulation is complex and also involves hormone balances is suggested by the investigation of β-1,3-glucanase induction in tobacco callus treated with or starved for auxin and cytokinin (Felix and Meins, 1987). In the presence of hormones, ethylene levels are high and very little glucanase activity is detected, whereas in hormone-free medium ethylene induction of the enzyme seems possible.

Future research on ACC- or ethylene-insensitive mutants and plants containing antisense ACC synthase or oxidase constructs will allow further dissection of ethylene- and nonethylene-dependent pathways in a plant's response to stress.

4.3. Molecular Requirements for Ethylene Inducibility of a Gene

Genomic sequences including 5'-upstream regions of ethylene-inducible genes have been published in six different cases. This includes the sequence of one of the tomato pTOM13 (putative ACC oxidase) genes (Holdsworth *et al.*, 1987b), the unique tomato PG gene (Bird *et al.*, 1988; Rose *et al.*, 1988), one of the tomato E8 fruit-ripening genes (Deikman and Fischer, 1988), the tomato E4 gene (Cordes *et al.*, 1989), one of the ethylene-responsive bean chitinase genes (Broglie *et al.*, 1989), and finally, one of the avocado cellulase genes (Cass *et al.*, 1990). Deikman and Fischer (1988) identified a DNA-binding factor that interacts with the E8-flanking sequences (-1088 to -682) and with the 5'-

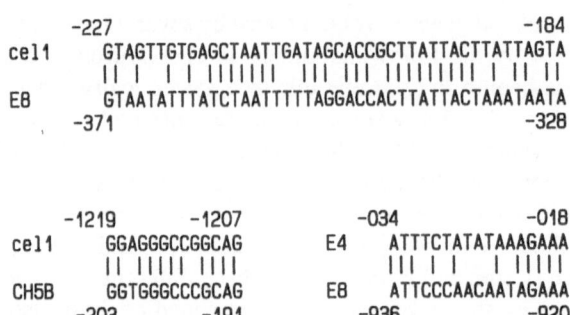

FIGURE 4. Regions of similarity between the avocado cell (cellulase) upstream sequences and upstream sequences of the tomato E8 gene and the bean CH5B (chitinase) gene, and between the upstream sequences of the tomato E4 and E8 genes. Nucleotide positions are relative to the start of transcription.

flanking region of the E4 gene (−403 to +65), coordinately expressed during fruit ripening. A more detailed analysis by gel retardation and methylation interference revealed conserved features of the binding sites (Cordes *et al.*, 1989). On a stretch of 17 bp, 11 bp are identical, flanking a nonconserved AT-rich core (Figure 4). Furthermore, Broglie *et al.* (1989) were able to delineate the sequences responsible for ethylene induction in the bean chitinase promoter to a 228-bp region (−422 to −195). The results obtained with stably transformed tobacco plants were confirmed by transient expression in protoplasts (Roby *et al.*, 1991). In addition, some interesting similarities were found between the 5′-flanking regions of the avocado cellulase gene (cell) and E8 and bean chitinase, respectively (Figure 4, Cass *et al.*, 1990). The 44-bp region of similarity between cellulase and E8 does not overlap the consensus binding site of E4 and E8, however. The similarity between the cellulase cell gene and the bean chitinase CH5B gene was confined to a 13-bp region (Figure 4). It was demonstrated earlier that a deletion of 195 bp from the transcriptional start, which falls within the region of similarity, resulted in complete loss of ethylene inducibility of CH5B.

Finally, a comparison of cellulase and PG promoters did not reveal any similarity (Cass *et al.*, 1990). This is not so surprising, since biochemical evidence indicated that cellulase and PG have a different timing of appearance in avocado fruit (Awad and Young, 1979).

5. CONCLUSIONS AND PERSPECTIVES

One decade after the entire elucidation of the ethylene biosynthetic cycle, we have started to discover molecular aspects of its regulation. The cloning of three important enzymes of the pathway, including its key regulatory step ACC synthase, has provided us the necessary tools for elucidation of the factors governing expression of the respective genes. Moreover, the development of new techniques for rapid molecular analysis of a given mutation will provide access to genes encoding polypeptides in the ethylene signal transduction pathway. At this moment, at least seven different loci involved in this process have been identified in *A. thaliana*. Finally, with the recent identification of ethylene-responsive elements in a number of genes, we are also acquiring insight into the elements determining ethylene inducibility of a gene. It is clear that, taken together, these different approaches have enhanced our understanding of fundamental aspects of ethylene biology. A detailed analysis of the expression of different ACC synthase, EFE, ethylene receptor, and transducer genes will reveal the fine tuning system necessary to control the pleiotropic effects of this hormone, without turning on everything simultaneously. In addition, this knowledge might be the basis for a number of spinoffs in agricultural and horticultural biotechnology. The economic impact can be enormous, since substantial postharvest losses are due to bruising or overripening and, consequently, increased susceptibility to pathogens. Controlled endogenous ethylene production could bring a solution to this problem and ensure improved quality and storage life of many plant products. A first indication of the potential has been given by the results of antisense EFE transformants of tomato. We hope that future research will soon support these speculations.

ACKNOWLEDGMENTS. We thank Luc Van Wiemeersch, Renato Rodrigues-Pousada, Ann Djudzman, and Patrick Roose for their contributions to the ethylene project in our laboratory in the past 4 years. We acknowledge Dr. Allan Caplan for critical comments on the manuscript. The help of Martine De Cock for layout and of the art team for preparation of the figures is greatly appreciated. DVDS is a Senior Research Assistant of the National Fund for Scientific Research (Belgium).

6. REFERENCES

Abeles, F. B., 1967, Mechanism of action of abscission accelerators, *Plant Physiol.* **20**:442–454.

Abeles, F. B., 1973, *Ethylene in Plant Biology*, 302 pp., Academic Press, New York.

Abeles, F. B., 1984a, A comparative study of ethylene oxidation in *Vicia faba* and *Mycobacterium paraffinicum*, *J. Plant Growth Regul.* **3**:85–95.

Abeles, F. B., 1984b, Role of ethylene oxidation in the mechanism of C_2H_4 action, in *Ethylene: Biochemical, Physiological and Applied Aspects* (Y. Fuchs and E. Chalutz, eds.), pp. 75–86, Martinus Nijhoff/Dr. W. Junk Publishers, Den Haag.

Abeles, F. B., Dunn, L. J., Morgens, P., Callahan, A., Dinterman, R. E., and Schmidt, J., 1988, Induction of 33-kD and 60-kD peroxidases during ethylene-induced senescence of cucumber cotyledons, *Plant Physiol.* **87**:609–615.

Acaster, M. A., and Kende, H., 1983, Properties and partial purification of 1-aminocyclopropane-1-carboxylate synthase, *Plant Physiol.* **72**:139–145.

Adlington, R. M., Baldwin, J. E., and Rawlings, B. J., 1983, On the stereochemistry of ethylene biosynthesis, *J. Chem. Soc. Chem. Commun.* **6**:290–292.

Amrhein, A., Schneebeck, D., Skorupka, H., and Tophof, S., 1981, Identification of a major metabolite of the ethylene precursor 1-aminocyclopropane-1-carboxylic acid in higher plants, *Naturwissenschaften* **68**:619–620.

Apelbaum, A., Burgoon, A. C., Anderson, J. D, Solomos, T., and Lieberman, M., 1981, Some characteristics of the system converting 1-aminocyclopropane-1-carboxylic acid to ethylene, *Plant Physiol.* **67**:80–84.

Asada, K., and Takahashi, M., 1987, Production and scavenging of active oxygen in photosynthesis, in *Photoinhibition* (D. J. Kyle, C. B. Osmond, and C. J. Arntzen, eds.), pp. 227–287, Elsevier Science Publishers B.V., Amsterdam.

Awad, M., and Young, R. E., 1979, Postharvest variation in cellulase, polygalacturonase, and pectinmethylesterase in avocado (*Persea americana* Mill, cv. Fuerte) fruits in relation to respiration and ethylene production, *Plant Physiol.* **64**:306–308.

Baird, L. M., Reid, M. S., and Webster, B. D., 1984, Anatomical and physiological effects of silver thiosulphate on ethylene-induced abscission in *Coleus*, *J. Plant Growth Regul.* **3**:217–225.

Bangerth, F., 1986, Natural and synthetic plant growth substances in fruit growing, in *Plant Growth Substances 1985* (M. Bopp, ed.), pp. 387–390, Springer-Verlag, Heidelberg.

Baumgartner, B., Kende, H., and Matile, P., 1975, Ribonuclease in senescing morning glory. Purification and demonstration of *de novo* synthesis, *Plant Physiol.* **55**:734–737.

Bengochea, T., Dodds, J. H., Evans, D. E., Jerie, P. H., Niepel, B., Shari, A. R., and Hall, M. A., 1980, Studies on ethylene binding by cell-free preparations from cotyledons of *Phaseolus vulgaris* L.: Separation and characterisation, *Planta* **148**:397–406.

Bernatzky, R., and Tanskley, S. D., 1986, Toward a saturated linkage map in tomato based on isozymes and random cDNA sequences, *Genetics* **112**:887–898.

Beyer, E. M., 1972, Mechanism of C_2H_4 action. Biological activity of deuterated ethylene and evidence against isotopic exchange and *cis-trans*-isomerization, *Plant Physiol.* **49**:672–675.

Beyer, E. M., 1975, $^{14}C_2H_4$: its incorporation and metabolism by pea seedlings under aseptic conditions, *Plant Physiol.* **56**:273–278.

Beyer, E. M., 1976, A potent inhibitor of ethylene action in plants, *Plant Physiol.* **58**:268–271.

Beyer, E. M., 1979, [^{14}C] ethylene metabolism during leaf abscission in cotton, *Plant Physiol.* **64**:971–974.

Beyer, E. M., 1985, Ethylene metabolism, in *Ethylene in Plant Development* (J. A. Roberts and G. A. Tucker, eds.), pp. 125–137, Butterworths, London.

Beyer, E. M., and Morgan, P. W., 1971, Abscission: The role of ethylene modification of auxin transport, *Plant Physiol.* **48**:208–212.

Biale, J. B., and Young, R. E., 1981, Respiration and ripening in fruit—Retrospect and prospect, in *Recent Advances in the Biochemistry of Fruit and Vegetables* (J. Friend and M. J. C. Rhodes, eds.), pp. 1–39, Academic Press, London.

Bird, C. R., Smith, C. J. S., Ray, J. A., Moureau, P., Bevan, M. W., Bird, A. S., Hughes, S., Morris, P. C., Grierson, D., and Schuch, W., 1988, The tomato polygalacturonase gene and ripening-specific expression in transgenic plants, *Plant Mol. Biol.* **11**:651–662.

Bleecker, A. B., Kenyon, W. H., Somerville, S. C., and Kende, H., 1986, Use of monoclonal antibodies in the purification and characterization of 1-aminocyclopropane-1-carboxylate synthase, an enzyme in ethylene biosynthesis, *Proc. Natl. Acad. Sci. USA* **83**:7755–7759.

Bleecker, A. B., Estelle, M. A., Somerville, C., and Kende, H., 1988a, Insensitivity to ethylene conferred by a dominant mutation in *Arabidopsis thaliana*, *Science* **241**:1086–1089.

Bleecker, A. B., Robinson, G., and Kende, H., 1988b, Studies on the regulation of 1-aminocyclopropane-1-carboxylate synthase in tomato using monoclonal antibodies, *Planta* **173**:385–390.

Bleecker, A. B., Chang, C., and Meyerowitz, E. M., 1990, Molecular analysis of a gene involved in ethylene response. Abstract presented at the Fourth International Conference on *Arabidopsis* Research, Vienna (Österreich), p. 131.

Blomstrom, D. C., and Beyer, E. M., 1980, Plants metabolise ethylene to ethylene glycol, *Nature (Lond.)* **283**:66–68.

Boller, T., 1982, Ethylene-induced biochemical defenses against pathogens, in *Plant Growth Substances 1982* (P. F. Wareing, ed.), pp. 303–312, Academic Press, London.

Boller, T., and Kende, H., 1980, Regulation of wound ethylene synthesis in plants, *Nature (Lond.)* **286**:259–260.

Boller, T., and Vögeli, U., 1984, Vacuolar localization of ethylene-induced chitinase in bean leaves, *Plant Physiol.* **74**:442–444.

Boller, T., Herner, R. C., and Kende, H., 1979, Assay for and enzymatic formation of an ethylene precursor, 1-aminocyclopropane-1-carboxylic acid, *Planta* **145**:293–303.

Borochov, A., and Woodson, W. R., 1989, Physiology and biochemistry of flower petal senescence, *Hortic. Rev.* **11**:15–43.

Bousquet, J-F., and Thimann, K. V., 1984, Lipid peroxidation forms ethylene from 1-aminocyclopropane-1-carboxylic acid and may operate in leaf senescence, *Proc. Natl. Acad. Sci. USA* **81**:1724–1727.

Bowles, D. J., 1990, Defense-related proteins in higher plants, *Annu. Rev. Biochem.* **59**:873–907.

Brady, C. J., 1987, Fruit ripening, *Annu. Rev. Plant Physiol.* **38**:155–178.

Brady, C. J., MacAlpine, G., McGlasson, W. B., and Ueda, Y., 1982, Polygalacturonase in tomato fruits and the induction of ripening, *Aust. J. Plant Physiol.* **9**:171–178.

Broglie, K. E., Gaynor, J. J., and Broglie, R. M., 1986, Ethylene-regulated gene expression: Molecular cloning of the genes encoding an endochitinase from *Phaseolus vulgaris, Proc. Natl. Acad. Sci. USA* **83**:6820–6824.

Broglie, K. E., Biddle, P., Cressman, R., and Broglie, R., 1989, Functional analysis of DNA sequences responsible for ethylene regulation of a bean chitinase gene in transgenic tobacco, *Plant Cell* **1**:599–607.

Brown, J. H., Legge, R. L., Sisler, E. C., Baker, J. E., and Thompson, J. E., 1986, Ethylene binding to senescing carnation petals, *J. Exp. Bot.* **37**:526–534.

Bufler, G., Mor, Y., Reid, M. S., and Yang, S. F., 1980, Changes in 1-aminocyclopropane-1-carboxylic acid content of cut carnation flowers in relation to their senescence, *Planta* **150**:439–442.

Burg, S. P., and Burg, E. A., 1967, Molecular requirements for the biological activity of ethylene, *Plant Physiol.* **42**:144–152.

Cabrera, C. V., Alonso, M. C., Johnston, P., Phillips, R. G., and Lawrence, P. A., 1987, Phenocopies induced with antisense RNA identify the *wingless* gene, *Cell* **50**:659–663.

Cameron, A. C., Fenton, C. A. L., Yu, Y. B., Adams, D. O., and Yang, S. F., 1979, Increased production of ethylene by plant tissues treated with 1-aminocyclopropane-1-carboxylic acid, *HortScience* **14**:178–180.

Carlson, J. R., 1988, A new means of inducibly inactivating a cellular protein, *Mol. Cell. Biol.* **8**:2638–2646.

Carr, J. P., and Klessig, D. F., 1990, The pathogenesis-related proteins of plants, in *Genetic Engineering, Principles and Methods*, Vol. 11 (J. K. Setlow, ed.), pp. 65–109, Plenum Press, New York.

Cass, L. G., Kirven, K. A., and Christoffersen, R. E., 1990, Isolation and characterization of a cellulase gene family member expressed during avocado fruit ripening, *Mol. Gen. Genet.* **223**:76–86.

Chang, C., Bowman, J. L., DeJohn, A. W., Lander, E. S., and Meyerowitz, E. M., 1988, Restriction fragment length polymorphism linkage map for *Arabidopsis thaliana, Proc. Natl. Acad. Sci. USA* **85:**6856–6860.

Chappell, J., Hahlbrock, K., and Boller, T., 1984, Rapid induction of ethylene biosynthesis in cultured parsley cells by fungal elicitor and its relationship to the induction of phenylalanine ammonia-lyase, *Planta* **161:**475–480.

Child, R. D., Williams, A. A., Hoad, G. V., and Baines, C. R., 1984, The effects of aminoethoxyvinylglycine on maturity and postharvest changes in Cox's Orange Pippin apples, *J. Sci. Food Agric.* **35:**773–781.

Chou, P., and Fasman, G. D., 1978, Prediction of the secondary structure of proteins from their amino acid sequence, *Adv. Enzymol.* **47:**45–147.

Christoffersen, R. E., and Laties, G. G., 1982, Ethylene regulation of gene expression in carrots, *Proc. Natl. Acad. Sci. USA* **79:**4060–4063.

Christoffersen, R. E., Tucker, M. L., and Laties, G. G., 1984, Cellulase gene expression in ripening avocado fruit: The accumulation of cellulase mRNA and protein as demonstrated by cDNA hybridization and immunodetection, *Plant Mol. Biol.* **3:**385–391.

Cohen, E., and Kende, H., 1987, *In vivo* 1-aminocyclopropane-1-carboxylate synthase activity in internodes of deepwater rice, *Plant Physiol.* **84:**282–286.

Connern, C. P., Smith, A. R., Turner, R., and Hall, M. A., 1989, Putative ethylene binding protein(s) from abscission zones of *Phaseolus vulgaris*, in *Plant Cell Separation* (NATO ASI Series Vol. 35) (D. J. Osborne, and M. Jackson, eds.), section 7, part 32, Springer-Verlag, Berlin.

Cordes, S., Deikman, J., Margossian, L. J., and Fischer, R. L., 1989, Interaction of a developmentally regulated DNA-binding factor with sites flanking two different fruit-ripening genes from tomato, *Plant Cell* **1:**1025–1034.

Davies, K. M., and Grierson, D., 1989, Identification of cDNA clones for tomato (*Lycopersicon esculentum* Mill.) mRNAs that accumulate during fruit ripening and leaf senescence in response to ethylene, *Planta* **179:**73–80.

Davies, K. M., Hobson, G. E., and Grierson, D., 1988, Silver ions inhibit the ethylene-stimulated production of ripening-related mRNAs in tomato, *Plant Cell Environ.* **11:**729–738.

de Laat, A. M. M., and van Loon, L. C., 1982, Regulation of ethylene biosynthesis in virus-infected tobacco leaves. II. Time course of levels of intermediate and *in vivo* conversion rates, *Plant Physiol.* **69:**240–245.

Dean, C., Sjodin, C., Lawson, E., Lister, C., and Bancroft, I., 1990, Development of an efficient transposon tagging system in *Arabidopsis*. Abstract presented at the Fourth International Conference on *Arabidopsis* Research, Vienna, p. 8.

Deikman, J., and Fischer, R. L., 1988, Interaction of a DNA binding factor with the 5'-flanking region of an ethylene-responsive fruit ripening gene from tomato, *EMBO J.* **7:**3315–3320.

Dekeyser, R. A., Claes, B., De Rycke, R. M. V., Habets, M. E., Van Montagu, M., and Caplan, A. B., 1990, Transient gene expression in intact and organized rice tissues, *Plant Cell* **2:**591–602.

del Campillo, E., Durbin, M., and Lewis, L. N., 1988, Changes in two forms of membrane-associated cellulase during ethylene-induced abscission, *Plant Physiol.* **88:**904–909.

del Campillo, E., Reid, P. D., Sexton, R., and Lewis, L. N., 1990, Occurrence and localization of 9.5 cellulase in abscising and nonabscising tissues, *Plant Cell* **2:**245–254.

DellaPenna, D., Alexander, D. C., and Bennett, A. B., 1986, Molecular cloning of tomato fruit polygalacturonase: analysis of polygalacturonase mRNA levels during ripening, *Proc. Natl. Acad. Sci. USA* **83:**6420–6424.

DellaPenna, D., Alexander, D. C., and Bennett, A. B., 1986, Molecular cloning of tomato fruit polygalacturonase: analysis of polygalacturonase mRNA levels during ripening, *Proc. Natl. Acad. Sci. USA* **83:**6420–6424.

Dixon, R. A., and Lamb, C. J., 1990, Molecular communication in interactions between plants and microbial pathogens, *Annu. Rev. Plant Physiol. Plant Mol. Biol.* **41**:339–367.

Düring, K., Hippe, S., Kreuzaler, F., and Schell, J., 1990, Synthesis and self-assembly of a functional monoclonal antibody in transgenic *Nicotiana tabacum*, *Plant Mol. Biol.* **15**:281–293.

Ecker, J. R., and Davis, R. W., 1987, Plant defense genes are regulated by ethylene, *Proc. Natl. Acad. Sci. USA* **84**:5202–5206.

Evans, D. E., Dodds, J. H., Lloyd, P. C., Gwynn, I., and Hall, M. A., 1982, A study of the subcellular localisation of an ethylene binding site in developing cotyledons of *Phaseolus vulgaris* L. by high resolution autoradiography, *Planta* **154**:48–52.

Feldmann, K. A., Marks, M. D., Christianson, M. L., and Quatrano, R. S., 1989, A dwarf mutant of *Arabidopsis* generated by T-DNA insertion mutagenesis, *Science* **243**:1351–1354.

Felix, G., and Meins, F., 1986, Developmental and hormonal regulation of β-1,3-glucanase in tobacco, *Planta* **167**:206–211.

Felix, G., and Meins, F., 1987, Ethylene regulation of β-1,3-glucanase in tobacco, *Planta* **172**:386–392.

Finlayson, B. J., and Pitts, J. N., 1986, *Atmospheric Chemistry: Fundamentals and Experimental Techniques*, Wiley, New York.

Fuhrer, J., 1982, Ethylene biosynthesis and cadmium toxicity in leaf tissue of beans (*Phaseolus vulgaris* L.), *Plant Physiol.* **70**:162–167,

Fujino, D. W., Burger, D. W., Yang, S. F., and Bradford, K. J., 1988, Characterization of an ethylene overproducing mutant of tomato (*Lycopersicon esculentum* Mill. cultivar VFN8), *Plant Physiol.* **88**:774–779.

Fujino, D. W., Burger, D. W., and Bradford, K. J., 1989, Ineffectiveness of ethylene biosynthetic and action inhibitors in phenotypically reverting the *epinastic* mutant of tomato (*Lycopersicon esculentum* Mill.), *J. Plant Growth Regul.* **8**:53–61.

Gane, R., 1934, Production of ethylene by some ripening fruit, *Nature (Lond.)* **134**:1008.

Gepstein, S., and Thimann, K. V., 1981, The role of ethylene in the senescence of oat leaves, *Plant Physiol.* **68**:349–354.

Giovannoni, J. J., DellaPenna, D., Bennett, A. B., and Fischer, R. L., 1989, Expression of a chimeric polygalacturonase gene in transgenic *rin* (ripening inhibitor) tomato fruit resulting in polyuronide degradation but not fruit softening, *Plant Cell* **1**:53–63.

Giraudat, J., Hauge, B., Smalle, J., and Goodman, H. M., 1990, Progress towards the cloning of the abi-3 locus of *Arabidopsis*. Abstract presented at the Fourth International Conference on *Arabidopsis* Research, Vienna, p. 134.

Goeschl, J. D., Rappaport, L., and Pratt, H. K., 1966, Ethylene as a factor regulating the growth of pea epicotyls subjected to physical stress, *Plant Physiol.* **41**:877–884.

Goeschl, J. D., Pratt, H. K., and Bonner, B. A., 1967, An effect of light on the production of ethylene and the growth of the plumular portion of etiolated pea seedlings, *Plant Physiol.* **42**:1077–1080.

Gomez Lim, M. A., Kelly, P., Sexton, R., and Trewavas, A. J., 1987, Identification of chitinase mRNA in abscission zones from bean (*Phaseolus vulgaris* Red Kidney) during ethylene-induced abscission, *Plant Cell Environ.* **10**:741–746.

Goren, M., Mattoo, A. K., and Anderson, J. D., 1984, Ethylene binding during leaf development and senescence and its inhibition by silver nitrate, *J. Plant Physiol.* **117**:243–248.

Goren, R., and Sisler, E. C., 1986, Ethylene-binding characteristics in *Phaseolus, Citrus,* and *Ligustrum* plants, *Plant Growth Regul.* **4**:43–54.

Greenberg, J., Goren, R., and Riov, J., 1975, The role of cellulase and polygalacturonase in abscission of young and mature shamouti orange fruits, *Physiol. Plant.* **34**:1–7.

Grierson, D., Tucker, G. A., Keen, J., Ray, J., Bird, C. R., and Schuch, W., 1986, Sequencing and identification of a cDNA clone for tomato polygalacturonase, *Nucleic Acids Res.* **14**:8595–8603.

Grierson, D., Purton, M. E., Knapp, J. E., and Bathgate, B., 1987, Tomato ripening mutants, in *Developmental Mutants in Higher Plants* (H. Thomas and D. Grierson, eds.), pp. 73–94, Cambridge University Press, Cambridge.

Grill, E., and Somerville, C., 1991, Construction and characterization of a yeast artificial chromosome library of *Arabidopsis* which is suitable for chromosome walking, *Mol. Gen. Genet.* **226**:484–490.

Guy, M., and Kende, H., 1984, Conversion of 1-aminocyclopropane-1-carboxylic acid to ethylene by isolated vacuoles of *Pisum sativum* L, *Planta* **160**:281–287.

Guzmán, P., and Ecker, J. R., 1988, Development of large DNA methods for plants: Molecular cloning of large DNA segments of *Arabidopsis* and carrot into yeast, *Nucleic Acids Res.* **16**:11091–11105.

Guzmán, P. A., and Ecker, J. R., 1989, Genetic analysis of ethylene-mediated signal transduction in *Arabidopsis, J. Cell. Biochem.* **Suppl. 13D**:319 (M416).

Guzmán, P., and Ecker, J. R., 1990, Exploiting the triple response of *Arabidopsis* to identify ethylene-related mutants, *Plant Cell* **2**:513–523.

Hall, M. A., Bell, M. H., Connern, C. P., Raskin, I., Robertson, D., Sanders, I. O., Smith, A. R., Turner, R., Williams, R. A. N., and Wood, C. K., 1990, Ethylene receptors, in *Molecular Aspects of Hormonal Regulation of Plant Development* (M. Kutáček, M. C. Elliott, and I. Macháčková, eds.), pp. 233–240, SPB Academic Publishing, Den Haag.

Hamilton, A. J., Lycett, G. W., and Grierson, D., 1990, Antisense gene that inhibits synthesis of the hormone ethylene in transgenic plants, *Nature (Lond.)* **346**:284–287.

Hanley, K. M., Meir, S., and Bramlage, W. J., 1989, Activity of ageing carnation flower parts and the effects of 1-(malonylamino)cyclopropane-1-carboxylic acid-induced ethylene, *Plant Physiol.* **91**:1126–1130.

Henstrand, J. M., and Handa, A. K., 1989, Effect of ethylene action inhibitors upon wound-induced gene expression in tomato pericarp, *Plant Physiol.* **91**:157–162.

Hesse, T., Feldwisch, J., Balshüsemann, D., Bauw, G., Puype, M., Vandekerckhove, J., Löbler, M., Klämbt, D., Schell, J., and Palme, K., 1989, Molecular cloning and structural analysis of a gene from *Zea mays* (L.) coding for a putative receptor for the plant hormone auxin, *EMBO J.* **8**:2453–2461.

Hiatt, A., Cafferkey, R., and Bowdish, K., 1989, Production of antibodies in transgenic plants, *Nature (Lond.)* **342**:76–78.

Hicks, G. R., Rayle, D. L., and Lomax, T. L., 1989, The *diageotropica* mutant of tomato lacks high specific activity auxin binding sites, *Science* **245**:52–54.

Hobson, G. E., 1964, Polygalacturonase in normal and abnormal tomato fruit, *Biochem. J.* **92**:324–332.

Hobson, G. E., and Nichols, R., 1977, Enzyme changes during petal senescence in carnation, *Ann. Appl. Biol.* **85**:445–446.

Hobson, G. E., Nichols, R., Davies, J. N., and Atkey, P. T., 1984, The inhibition of tomato fruit ripening by silver, *J. Plant Physiol.* **116**:21–29.

Hoffman, N. E., and Yang, S. F., 1982, Enhancement of wound-induced ethylene synthesis by ethylene in preclimacteric cantaloupe, *Plant Physiol.* **69**:317–322.

Hoffman, N. E., Yang, S. F., Ichihara, A., and Sakamura, S., 1982a, Stereospecific conversion of 1-aminocyclopropanecarboxylic acid to ethylene by plant tissues. Conversion of stereoisomers of 1-amino-2-ethylcyclopropanecarboxylic acid to 1-butene, *Plant Physiol.* **70**:195–199.

Hoffman, N. E., Yang, S. F., and McKeon, T., 1982b, Identification of 1-(malonylamino)cyclopropane-1-carboxylic acid as a major conjugate of 1-aminocyclopropane-1-carboxylic acid, an ethylene precursor in higher plants, *Biochem. Biophys. Res. Commun.* **104**:765–770.

Holdsworth, M. J., Schuch, W., and Grierson, D., 1987a, Structure and expression of an ethylene-related mRNA from tomato, *Nucleic Acids Res.* **15**:731–739.

Holdsworth, M. J., Schuch, W., and Grierson, D., 1987b, Nucleotide sequence of an ethylene-related gene from tomato, *Nucleic Acids Res.* **15**:10600.

Holdsworth, M. J., Schuch, W., and Grierson, D., 1988, Organisation and expression of a wound/ripening-related small multigene family from tomato, *Plant Mol. Biol.* **11**:81–88.

Honma, M. A., Waddell, C. S., and Baker, B., 1990, Development of an *Ac/Ds* transposon tagging system in *Arabidopsis*. Abstract presented at the Fourth International Conference on *Arabidopsis* Research, Vienna, p. 12.

Horton, R., and Osborne, D. J., 1967, Senescence, abscission and cellulase activity in *Phaseolus vulgaris*, *Nature (Lond.)* **214**:1086–1088.

Hsieh, Y., and Sacalis, J., 1986, Levels of ACC in various floral portions during ageing of cut carnations, *J. Am. Soc. Hort. Sci.* **111**:942–944.

Hwang, I., Kohchi, T., Hauge, B. M., Goodman, H. M., Schmidt, R., Cnops, G., Dean, C., Gibson, S., Iba, K., Lemieux, B., Arondel, V., Danhoff, L., and Somerville, C., 1991, Identification and map position of YAC clones comprising one third of the *Arabidopsis* genome, *Plant J.* **1**:in press.

Iki, K., Sekiguchi, K., Kurata, K., Tada, T., Nakagawa, H., Ogura, N., and Takehana, H., 1978, Immunological properties of β-fructofuranosidase from ripening fruit, *Phytochemistry* **17**:311–312.

Imaseki, H., Nakagawa, N., and Nakajima, N., 1990, Wound-induced ACC synthase, an immunochemical comparison of the wound-induced and auxin-induced enzymes, in *Plant Growth Substances 1988* (R. P. Pharis and S. B. Rood, eds.), pp. 113–121, Springer-Verlag, Heidelberg.

Inohara, N., Shimomura, S., Fukui, T., and Futai, M., 1989, Auxin-binding protein located in the endoplasmic reticulum of maize shoots: Molecular cloning and complete primary structure, *Proc. Natl. Acad. Sci. USA* **86**:3564–3568.

Jackson, M. B., and Osborne, D. J., 1970, Ethylene, the natural regulator of leaf abscission, *Nature (Lond.)* **225**:1019–1022.

Jacobsen, J. V., and Chandler, P. M., 1987, Gibberellin and abscisic acid in germinating cereals, in *Plant Hormones and Their Role in Plant Growth and Development* (P. J. Davies, ed.), pp. 164–193, Martinus Nijfoff, Dordrecht.

John, P., 1983, The coupling of ethylene biosynthesis to a transmembrane, electrogenic proton flux, *FEBS Lett.* **152**:141–143.

Johnson, K. D., Höfte, H., and Chrispeels, M. J., 1990, An intrinsic tonoplast protein of protein storage vacuoles in seeds is structurally related to a bacterial solute transporter (GlpF), *Plant Cell* **2**:525–532.

Kacperska, A., and Kubacka-Zebalska, M., 1985, Is lipoxygenase involved in the formation of ethylene from ACC? *Physiol. Plant.* **64**:333–338.

Keefe, D., Hinz, U., and Meins, F. Jr., 1990, The effect of ethylene on the cell-type-specific and intracellular localization of β-1,3-glucanase and chitinase in tobacco leaves, *Planta* **182**:43–51.

Kelly, M. O., and Bradford, K. J., 1986, Insensitivity of the *diageotropica* tomato mutant to auxin, *Plant Physiol.* **82**:713–717.

Kende, H., 1989, Enzymes of ethylene biosynthesis, *Plant Physiol.* **91**:1–4.

Kende, H., and Boller, T., 1981, Wound ethylene and 1-aminocyclopropane-1-carboxylate synthase in ripening tomato fruit, *Planta* **151**:476–481.

Kende, H., and Hanson, A. D., 1976, Relationship between ethylene evolution and senescence in morning-glory flower tissue, *Plant Physiol.* **57**:523–527.

Kende, H., Bleecker, A. B., Kenyon, W. H., and Mayne, R. G., 1986, Enzymes of ethylene biosynthesis, in *Plant Growth Substances 1985* (M. Bopp, ed.), pp. 120–128, Springer-Verlag, Berlin.

Kionka, C., and Amrhein, N., 1984, The enzymatic malonylation of 1-aminocyclopropane-1-carboxylic acid in homogenate of mung-bean hypocotyls, *Planta* **162**:226–235.

Koncz, C., Mayerhofer, R., Koncz-Kalman, Z., Nawrath, C., Reiss, B., Redei, G. P., and Schell, J., 1990, Isolation of a gene encoding a novel chloroplast protein by T-DNA tagging in *Arabidopsis thaliana, EMBO J.* **9**:1337–1346.

Konze, J. R., and Kende, H., 1979, Interaction of methionine and selenomethionine with methionine adenosyltransferase and ethylene-generating systems, *Plant Physiol.* **63**:507–510.

Konze, J. R., and Kwiatkowski, G. M. K., 1981, Rapidly induced ethylene formation after wounding is controlled by the regulation of 1-aminocyclopropane-1-carboxylic acid synthesis, *Planta* **151**:327–330.

Koornneef, M., 1990, Linkage map of *Arabidopsis thaliana,* in *Genetic Maps, Plants* (S. J. O'Brien, ed.), pp. 6.95–6.97, Cold Spring Harbor Laboratory Press, Cold Spring Harbor, NY.

Koornneef, M., van Eden, J., Hanhart, C. J., Stam, P., Braaksma, F. J., and Feenstra, W. J., 1983, Linkage map of *Arabidopsis thaliana, J. Hered.* **74**:265–272.

Koornneef, M., Hanhart, C. J., and van der Veen, 1990, A genetic and physiological analysis of late flowering mutants in *Arabidopsis thaliana.* Abstract presented at the Fourth International Conference on *Arabidopsis* Research, Vienna, p. 114.

Kushad, M. M., and Poovaiah, B. W., 1984, Deferral of senescence and abscission by chemical inhibition of ethylene synthesis and action in bean explants, *Plant Physiol.* **76**:293–296.

Lagrimini, L. M., Burkhart, W., Moyer, M., and Rothstein, S., 1987, Molecular cloning of complementary DNA encoding the lignin-forming peroxidase from tobacco: Molecular analysis and tissue-specific expression, *Proc. Natl. Acad. Sci. USA* **84**:7542–7546.

Lagrimini, L. M., Bradford, S., and Rothstein, S., 1990, Peroxidase-induced wilting in transgenic tobacco plants, *Plant Cell* **2**:7–18.

Larrigaudière, C., Latché, A., Pech, J. C., and Triantaphilidès, C., 1990, Short-term effects of γ-irradiation on 1-aminocyclopropane-1-carboxylic acid metabolism in early climacteric cherry tomatoes, *Plant Physiol.* **92**:577–581.

Lawton, K. A., Huang, B., Goldsbrough, P. B., and Woodson, W. R., 1989, Molecular cloning and characterization of senescence-related genes from carnation flower petals, *Plant Physiol.* **90**:690–696.

Lawton, K. A., Raghothama, K. G., Goldsbrough, P. B., and Woodson, W. R., 1990, Regulation of senescence-related gene expression in carnation flower petals by ethylene, *Plant Physiol.* **93**:1370–1375.

Legge, R. L., Thompson, J. E., and Baker, J., 1982, Free radical-mediated formation of ethylene from 1-aminocyclopropane-1-carboxylic acid: A spin-trap study, *Plant Cell Physiol.* **23**:171–177.

Leslie, C. A., and Romani, R. J., 1986, Salicylic acid: a new inhibitor of ethylene biosynthesis, *Plant Cell Rep.* **5**:144–146.

Lewis, L. N., and Varner, J. E., 1970, Synthesis of cellulase during abscission of *Phaseolus vulgaris* leaf explants, *Plant Physiol.* **46**:194–199.

Lincoln, J. E., and Fischer, R. L., 1988a, Diverse mechanisms for the regulation of ethylene-inducible gene expression, *Mol. Gen. Genet.* **212**:71–75.

Lincoln, J. E., and Fischer, R. L., 1988b, Regulation of gene expression by ethylene in wild-type and *rin* tomato (*Lycopersicon esculentum*) fruit, *Plant Physiol.* **88**:370–374.

Lincoln, J. E., Cordes, S., Read, E., and Fischer, R. L., 1987, Regulation of gene expression by ethylene during *Lycopersicon esculentum* (tomato) fruit development, *Proc. Natl. Acad. Sci. USA* **84**:2793–2797.

Liu, Y., Hoffman, N. E., and Yang, S. F., 1985a, Ethylene-promoted malonylation of 1-aminocyclopropane-1-carboxylic acid participates in autoinhibition of ethylene synthesis in grapefruit flavedo discs, *Planta* **164**:565–568.

Liu, Y., Hoffman, N. E., and Yang, S. F., 1985b, Promotion by ethylene of the capability to convert 1-aminocyclopropane-1-carboxylic acid to ethylene in preclimacteric tomato and cantaloupe fruits, *Plant Physiol.* **77**:407–411.

Lizada, M. C. C., and Yang, S. F., 1979, A simple and sensitive assay for 1-aminocyclopropane-1-carboxylic acid, *Anal. Biochem.* **100**:140–145.

Lotan, T., and Fluhr, R., 1990, Xylanase, a novel elicitor of pathogenesis-related proteins in tobacco, uses a non-ethylene pathway for induction, *Plant Physiol.* **93**:811–817.

Lotan, T., Ori, N., and Fluhr, R., 1989, Pathogenesis-related proteins are developmentally regulated in tobacco flowers, *Plant Cell* **1**:881–887.

Lürssen, K., and Konze, J., 1985, Relationship between ethylene production and plant growth after application of ethylene releasing plant growth regulators, in *Ethylene and Plant Development* (J. A. Roberts and G. A. Tucker, eds.), pp. 363–372, Butterworths, London.

Lürssen, K., Naumann, K., and Schroder, R., 1979, 1-Aminocyclopropane-1-carboxylic acid—An intermediate of the ethylene biosynthesis in higher plants, *Z. Pflanzenphysiol.* **92**:285–294.

Lynch, D. V., Sridhara, S., and Thompson, J. E., 1985, Lipoxygenase-generated hydroperoxides account for the nonphysiological features of ethylene formation from 1-aminocyclopropane-1-carboxylic acid by microsomal membranes of carnations, *Planta* **164**:121–125.

Mansson, P.-E., Hsu, D., and Stalker, D., 1985, Characterization of fruit specific cDNAs from tomato, *Mol. Gen. Genet.* **200**:356–361.

Margossian, L. J., Federman, A. D., Giovannoni, J. J., and Fischer, R. L., 1988, Ethylene-regulated expression of a tomato fruit ripening gene encoding a proteinase inhibitor I with a glutamic residue at the reactive site, *Proc. Natl. Acad. Sci. USA* **85**:8012–8016.

Marks, D. M., and Feldmann, K. A., 1989, Trichome development in *Arabidopsis thaliana*. I. T-DNA tagging of the *GLABROUS1* gene, *Plant Cell* **1**:1043–1050.

Mauch, F., and Staehelin, L. A., 1989, Functional implications of the subcellular localization of ethylene-induced chitinase and β-1,3-glucanase in bean leaves, *Plant Cell* **1**:447–457.

Mauch, F., Hadwiger, L. A., and Boller, T., 1984, Ethylene: symptom, not signal for the induction of chitinase and β-1,3-glucanase in pea pods by pathogens and elicitors, *Plant Physiol.* **76**:607–611.

Maunders, M. J., Holdsworth, M. J., Slater, A., Knapp, J. E., Bird, C. R., Schuch, W., and Grierson, D., 1987, Ethylene stimulates the accumulation of ripening-related mRNAs in tomatoes, *Plant Cell Environ.* **10**:177–184.

Mayak, S., Vaadia, Y., and Dilley, D. R., 1977, Regulation of senescence in carnation (*Dianthus caryophyllus*) by ethylene, *Plant Physiol.* **59**:591–593.

Mayne, R. G., and Kende, H., 1986, Ethylene biosynthesis in isolated vacuoles of *Vicia faba* L.— Requirement for membrane integrity, *Planta* **167**:159–165.

McGlasson, W. B., 1978, Phytohormones and fruit ripening, in *Phytohormones and Related Compounds: A Comprehensive Treatise* (D. S. Letham, P. N. Goodwin, and T. J. Higgins, eds.), pp. 447–493, Elsevier, Amsterdam.

McGlasson, W. B., and Pratt, A. K., 1964, Effects of ethylene on cantaloupe fruits harvested at various ages, *Plant Physiol.* **39**:120–127.

McGlasson, W. B., Dostal, H. C., and Tigchelaar, E. C., 1975a, Comparison of propylene-induced responses of immature fruit of normal and mutant tomatoes, *Plant Physiol.* **85**:218–222.

McGlasson, W. B., Poovaiah, B. W., and Dostal, H. C., 1975b, Ethylene production and respiration in aging leaf segments and in disks of fruit tissue of normal and mutant tomatoes, *Plant Physiol.* **56**:547–549.

McKeon, T. A., and Yang, S. F., 1984, A comparison of the conversion of 1-aminocyclopropane-1-

carboxylic acid stereoisomers to 1-butene by pea epicotyls and by a cell-free system, *Planta* **160**:84–87.

McKeon, T. A., and Yang, S. F., 1987, Biosynthesis and metabolism of ethylene, in *Plant Hormones and Their Role in Plant Growth and Development* (P. J. Davies, ed.), pp. 94–112, Martinus Nijhoff, Dordrecht.

McKeon, T. A., Hoffman, N. E., and Yang, S. F., 1982, The effect of plant-hormone pretreatments on ethylene production and synthesis of 1-aminocyclopropane-1-carboxylic acid in water-stressed wheat leaves, *Planta* **155**:437–443.

McMurchie, E. J., McGlasson, W. B., and Eaks, I. L., 1972, Treatment of fruit with propylene gives information about the biogenesis of ethylene, *Nature (Lond.)* **237**:235–236.

McRae, D. G., Baker, J. E., and Thompson, J. E., 1982, Evidence for involvement of the superoxide radical in the conversion of 1-aminocyclopropane-1-carboxylic acid to ethylene in pea microsomal membranes, *Plant Cell Physiol.* **23**:375–383.

Mehlhorn, H., and Wellburn, A. R., 1987, Stress ethylene formation determines plant sensitivity to ozone, *Nature (Lond.)* **327**:417–418.

Mehta, A. M., Jordan, R. L., Anderson, J. D., and Mattoo, A. K., 1988, Identification of a unique isoform of 1-aminocyclopropane-1-carboxylic acid synthase by monoclonal antibody, *Proc. Natl. Acad. Sci. USA* **85**:8810–8814.

Meyerowitz, E. M., 1987, *Arabidopsis thaliana*, *Annu. Rev. Genet.* **21**:93–111.

Moore, T. C., 1989, Ethylene, in *Biochemistry and Physiology of Plant Hormones*, 2nd ed. (T. C. Moore, ed.), pp. 228–254, Springer-Verlag, New York.

Morgan, P. W., 1986, Ethylene as an indicator and regulator in the development of field crops, in *Plant Growth Substances 1985* (M. Bopp, ed.), pp. 375–390, Springer-Verlag, Heidelberg.

Morgens, P. H., Callahan, A. M., Dunn, L. J., and Abeles, F. B., 1990, Isolation and sequencing of cDNA clones encoding ethylene-induced putative peroxidases from cucumber cotyledons, *Plant Mol. Biol.* **14**:715–725.

Nakagawa, N., Nakajima, N., and Imaseki, H., 1988, Immunochemical difference of wound-induced 1-aminocyclopropane-1-carboxylate synthase from the auxin-induced enzyme, *Plant Cell Physiol.* **29**:1255–1259.

Nakajima, N., and Imaseki, H., 1986, Purification and properties of 1-aminocyclopropane-1-carboxylate synthase of mesocarp of *Cucurbita maxima* Duch. fruits, *Plant Cell Physiol.* **27**:969–980.

Nakajima, N., Nakagawa, N., and Imaseki, H., 1988, Molecular size of wound-induced 1-aminocyclopropane-1-carboxylate synthase from *Cucurbita maxima* Duch. and change of translatable mRNA of the enzyme after wounding, *Plant Cell Physiol.* **29**:989–998.

Nakajima, N., Mori, H., Yamazaki, K., and Imaseki, H., 1990, Molecular cloning and sequence of a complementary DNA encoding 1-aminocyclopropane-1-carboxylate synthase induced by tissue wounding, *Plant Cell Physiol.* **31**:1021–1029.

Nam, H-G., Giraudat, J., den Boer, B., Moonan, F., Loos, W. D. B., Hauge, B. M., and Goodman, H. M., 1989, Restriction fragment length polymorphism linkage map of *Arabidopsis thaliana*, *Plant Cell* **1**:699–705.

Napier, R. M., and Venis, M. A., 1990, Receptors for plant growth regulators: Recent advances, *J. Plant Growth Regul.* **9**:113–126.

Neljubow, D., 1901, Über die horizontale Nutation der Stengel von *Pisum sativum* und einiger anderen Pflanzen, *Beih. Bot. Zentralbl.* **10**:128–139.

Nelson, N., and Taiz, L., 1989, The evolution of H+-ATPases, *Trends Biochem. Sci.* **14**:113–116.

Ness, P. J., and Romani, R. J., 1980, Effects of aminoethoxyvinylglycine and counter effects of ethylene on ripening of "Bartlett" pear fruits, *Plant Physiol.* **65**:372–376.

Nichols, R., 1966, Ethylene production during senescence of flowers, *J. Hortic. Sci.* **41**:279–290.

Nichols, R., 1968, The response of carnations (*Dianthus caryophyllus*) to ethylene, *J. Hortic. Sci.* **43**:335–349.

Nichols, R., 1973, Senescence of cut carnation flowers: Respiration and sugar status, *J. Hortic. Sci.* **48:**111–121.

Nichols, S. E., and Laties, G. G., 1984, Ethylene-regulated gene transcription in carrot roots, *Plant Mol. Biol.* **3:**393–401.

Odawara, S., Watanabe, A., and Imaseki, H., 1977, Involvement of cellular membrane in regulation of ethylene production, *Plant Cell Physiol.* **18:**569–575.

Osborne, D. J., 1982, The ethylene regulation of cell growth in specific target tissues of plants, in *Plant Growth Substances, 1982* (P. F. Wareing, ed.), pp. 279–290, Academic Press, London.

Osborne, D. J., McManus, M. T., and Webb, J., 1985, Target cells for ethylene action, in *Ethylene and Plant Development* (J. A. Roberts and G. A. Tucker, eds.), pp. 197–212, Butterworths, London.

Pear, J. R., Ridge, N., Rasmussen, R., Rose, R. E., and Houck, C. M., 1989, Isolation and characterization of a fruit-specific cDNA and the corresponding genomic clone from tomato, *Plant Mol. Biol.* **13:**639–651.

Pegg, G. F., 1976, The involvement of ethylene in plant pathogenesis, in *Physiological Plant Pathology*. Encyclopedia of Plant Physiology, New Series, Vol. 4 (R. Heitefuss, and P. H. Williams, eds.), pp. 582–591, Springer-Verlag, Heidelberg.

Peiser, G., 1986, Levels of 1-aminocyclopropane-1-carboxylic acid (ACC) synthase activity ACC and ACC-conjugate in cut carnation flowers during senescence, *Acta Hortic.* **181:**99–104.

Peiser, G. D., Wang, T-T., Hoffman, N. E., Yang, S. F., Liu, H-W., and Walsch, C. T., 1984, Formation of cyanide from carbon 1 of 1-aminocyclopropane-1-carboxylic acid during its conversion to ethylene, *Proc. Natl. Acad. Sci. USA* **81:**3059–3063.

Peleman, J., Boerjan, W., Engler, G., Seurinck, J., Botterman, J., Alliotte, T., Van Montagu, M., and Inzé, D., 1989a, Strong cellular preference in the expression of a housekeeping gene of *Arabidopsis thaliana* encoding *S*-adenosylmethionine synthetase, *Plant Cell* **1:**81–93.

Peleman, J., Saito, K., Cottyn, B., Engler, G., Seurinck, J., Van Montagu, M., and Inzé, D., 1989b, Structure and expression of the *S*-adenosylmethionine synthetase gene family in *Arabidopsis thaliana*, *Gene* **84:**359–369.

Pesis, E., Fuchs, Y., and Zauberman, G., 1978, Cellulase activity and fruit softening in avocado, *Plant Physiol.* **61:**416–419.

Pickett, F. B., Smith, M. L., and Estelle, M. A., 1989, Recessive mutation at the *etr-2* locus of *Arabidopsis thaliana* confers resistance to some effects of ethylene exposure, *J. Cell. Biochem.*, **Suppl. 13D:**324 (M432).

Pirrung, M. C., 1983, Ethylene biosynthesis. 2. Stereochemistry of ripening, stress, and model reactions, *J. Am. Chem. Soc.* **105:**7207–7209.

Pirrung, M. C., 1986, Mechanism of a lipoxygenase model for ethylene biosynthesis, *Biochemistry* **25:**114–119.

Pirrung, M. C., and McGeehan, G. M., 1983, Ethylene biosynthesis. 1. A model for two reactive intermediates, *J. Org. Chem.* **48:**5143–5144.

Pollard, J. E., and Biggs, R. H., 1970, Role of cellulase in abscission of citrus fruits, *J. Am. Soc. Hortic. Sci.* **95:**667–673.

Privalle, L. S., and Graham, J. S., 1987, Radiolabeling of a wound-inducible pyridoxal phosphate-utilizing enzyme: Evidence for its identification as ACC synthase, *Arch. Biochem. Biophys.* **253:**333–340.

Rai, R. M., Tewari, J. D., and Pant, N., 1983, Physiological effect of growth regulators on ripening of cultivar early-Shanbury apples, *Prog. Hortic.* **15:**276–282.

Ramalingam, K., Lee, K-M., Woodward, R. W., Bleecker, A. B., and Kende, H., 1985, Stereochemical course of the reaction catalyzed by the pyridoxal phosphate-dependent enzyme 1-aminocyclopropane-1-carboxylate synthase, *Proc. Natl. Acad. Sci. USA* **82:**7820–7824.

Ray, J., Knapp, J., Grierson, D., Bird, C., and Schuch, W., 1988, Identification and sequence determination of a cDNA clone for tomato pectin esterase, *Eur. J. Biochem.* **174:**119–124.

Riov, J., 1974, A polygalacturonase from citrus leaf explants, *Plant Physiol.* **53:**12–16.

Riov, J., and Goren, R., 1979, Effect of ethylene on auxin transport and metabolism in midrib sections in relation to leaf abscission of woody plants, *Plant Cell Environ.* **2:**83–89.

Riov, J., and Yang, S. F., 1982a, Autoinhibition of ethylene production in citrus peel discs. Suppression of 1-aminocyclopropane-1-carboxylic acid synthesis, *Plant Physiol.* **69:**687–690.

Riov, J., and Yang, S. F., 1982b, Effects of exogenous ethylene on ethylene production in citrus leaf tissue, *Plant Physiol.* **70:**136–141.

Roberts, D. R., Walker, M. A., Thompson, J. E., and Dumbroff, E. B., 1984, The effects of inhibitors of polyamines and ethylene biosynthesis on senescence, ethylene production and polyamine levels in cut carnation flowers, *Plant Cell Physiol.* **25:**315–322.

Roberts, J. A., Tucker, G. A., and Maunders, M. J., 1985, Ethylene and foliar senescence, in *Ethylene and Plant Development* (J. A. Roberts and G. A. Tucker, eds.), pp. 267–275, Butterworths, London.

Roby, D., Toppan, A., and Esquerré-Tugayé, M.-T., 1986, Cell surfaces in plant-microorganism interactions. VI. Elicitors of ethylene from *Colletotrichum lagenarium* trigger chitinase activity in melon plants, *Plant Physiol.* **81:**228–233.

Roby, D., Broglie, K., Gaynor, J., and Broglie, R., 1991, Regulation of a chitinase gene promoter by ethylene and elicitors in electroporated bean protoplasts, *Plant Physiol.* (in press).

Rommens, C., Kneppers, T., Jongman, C., Nettekoven, M., Overduin, B., Nijkamp, J., and Hille, J., 1990, Transposon tagging strategies using *Ac* and *Ds* in tomato, *J. Cell. Biochem.* **Suppl. 14E:**289 (R145).

Rose, R. E., Houck, C. M., Monson, E. K., DeJesus, C. E., Sheehy, R. E., and Hiatt, W. R., 1988, The nucleotide sequence of the 5' flanking region of a tomato polygalacturonase gene, *Nucleic Acids Res.* **16:**7191.

Sacher, J. A., 1973, Senescence and postharvest physiology, *Annu. Rev. Plant Physiol.* **24:**197–224.

Saltveit, M. E., Bradford, K. J., and Dilley, D. R., 1978, Silver ion inhibits ethylene synthesis and action in ripening fruits, *J. Am. Soc. Hortic. Sci.* **103:**472–475.

Samac, D. A., Hironaka, C. M., Yallaly, P. E., and Shah, D. M., 1990, Isolation and characterization of the genes encoding basic and acidic chitinase in *Arabidopsis thaliana*, *Plant Physiol.* **93:**907–914.

Sanders, I. O., Smith, A. R., and Hall, M. A., 1989a, The measurement of ethylene binding and metabolism in plant tissue, *Planta* **179:**97–103.

Sanders, I. O., Smith, A. R., and Hall, M. A., 1989b, Ethylene metabolism in *Pisum sativum* L., *Planta* **179:**104–114.

Sato, T., and Theologis, A., 1989, Cloning the mRNA encoding 1-aminocyclopropane-1-carboxylate synthase, the key enzyme for ethylene biosynthesis in plants, *Proc. Natl. Acad. Sci. USA* **86:**6621–6625.

Satoh, S., and Yang, S. F., 1988, S-adenosylmethionine-dependent inactivation and radiolabeling of 1-aminocyclopropane-1-carboxylate synthase isolated from tomato fruits, *Plant Physiol.* **88:**109–114.

Scherer, S., and Davis, R. W., 1979, Replacement of chromosome segments with altered DNA sequences constructed *in vitro*, *Proc. Natl. Acad. Sci. USA* **76:**4951–4955.

Sexton, R., and Roberts, J. A., 1982, Cell biology of abscission, *Annu. Rev. Plant Physiol.* **33:**133–162.

Sexton, R., Durbin, M. L., Lewis, L. N., and Thomson, W. W., 1981, The immunochemical localization of 9.5 cellulase in abscission zones of bean (*Phaseolus vulgaris* c.v. Red Kidney), *Protoplasma* **109:**335–347.

Sexton, R., Lewis, L. N., Trewavas, A. J., and Kelly, P., 1985, Ethylene and abscission, in *Ethylene and Plant Development* (J. A. Roberts and G. A. Tucker, eds.), pp. 173–196, Butterworths, London.

Sheehy, R. E., Pearson, J., Brady, C. J., and Hiatt, W. R., 1987, Molecular characterization of tomato fruit polygalacturonase, *Mol. Gen. Genet.* **208:**30–36.

Sheehy, R. E., Kramer, M., and Hiatt, W. R., 1988, Reduction of polygalacturonase activity in tomato fruit by antisense RNA, *Proc. Natl. Acad. Sci. USA* **85:**8805–8809.

Sisler, E. C., 1977, Ethylene activity of some π-acceptor compounds, *Tobacco Sci.* **21:**43–45.

Sisler, E. C., 1979, Measurement of ethylene binding in plant tissue, *Plant Physiol.* **64:**538–542.

Sisler, E. C., 1982a, Ethylene-binding properties of a Triton X-100 extract of mung bean sprouts, *J. Plant Growth Regul.* **1:**211–218.

Sisler, E. C., 1982b, Ethylene binding in normal, *rin*, and *nor* mutant tomatoes, *J. Plant Growth Regul.* **1:**219–226.

Sisler, E. C., 1987, Purification of the ethylene-binding component from mung bean sprouts and seeds, in *Plant Hormone Receptors* (NATO ASI Series, Vol. H10) (D. Klämbt, ed.), pp. 297–301, Springer-Verlag, Berlin.

Sisler, E. C., 1990, Ethylene binding receptors—Is there more than one? in *Plant Growth Substances 1988* (R. O. Pharis and S. B. Rood, eds.), pp. 192–202, Springer-Verlag, Heidelberg.

Sisler, E. C., 1991, The ethylene-binding receptor in plants, in *The Plant Hormone Ethylene* (A. K. Mattoo and J. C. Suttle, eds.), CRC Press, Boca Raton, FL (in press).

Sisler, E. C., and Wood, C., 1987, Ethylene binding and evidence that binding *in vivo* and *in vitro* is to the physiological receptor, in *Plant Hormone Receptors* (NATO ASI Series, Vol. H10) (D. Klämbt, ed.), pp. 239–248, Springer-Verlag, Berlin.

Sisler, E. C., and Yang, S. F., 1984, Anti-ethylene effects of *cis*-2-butene and cycloolefins, *Phytochemistry* **23:**2765–2768.

Sisler, E. C., Goren, R., and Huberman, M., 1985, Effect of 2,5-norbornadiene on abscission and ethylene production in citrus leaf explants, *Physiol. Plant.* **63:**114–120.

Sisler, E. C., Reid, M., and Yang, S. F., 1986, Effect of antagonists of ethylene action on binding of ethylene in cut carnations, *Plant Growth Regul.* **4:**213–218.

Skriver, K., and Mundy, J., 1990, Gene expression in response to abscisic acid and osmotic stress, *Plant Cell* **2:**503–512.

Slater, A., Maunders, M. J., Edwards, K., Schuch, W., and Grierson, D., 1985, Isolation and characterisation of cDNA clones from tomato polygalacturonidase and other ripening-related proteins, *Plant Mol. Biol.* **5:**137–147.

Smith, C. J. S., Slater, A., and Grierson, D., 1986, Rapid appearance of an mRNA correlated with ethylene synthesis encoding a protein of molecular weight 35,000, *Planta* **168:**94–100.

Smith, C. J. S., Watson, C. F., Ray, J., Bird, C. R., Morris, P. C., Schuch, W., and Grierson, D., 1988, Antisense RNA inhibition of polygalacturonase gene expression in transgenic tomatoes, *Nature (Lond.)* **334:**724–726.

Spanu, P., and Boller, T., 1989, Ethylene biosynthesis in tomato plants infected by *Phytophthora infestans*, *J. Plant Physiol.* **134:**533–537.

Suttle, J. C., and Kende, H., 1980, Ethylene action and loss of membrane integrity during petal senescence in *Tradescantia*, *Plant Physiol.* **65:**1067–1072.

Tabor, C. W., and Tabor, H., 1984, Methionine adenosyltransferase (*S*-adenosylmethionine synthetase) and *S*-adenosylmethionine decarboxylase, *Adv. Enzymol.* **56:**251–282.

Tanksley, S. D., and Mutschler, M. A., 1990, Linkage map of the tomato (*Lycopersicon esculentum*) (2 = 24), in *Genetic Maps, Plants* (S. J. O'Brien, ed.), pp. 6.3–6.15, Cold Spring Harbor Laboratory Press, Cold Spring Harbor, NY.

Taylor, J. E., Grosskopf, D. G., McGaw, B. A., Horgan, R., and Scott, I. M., 1988, Apical localization of 1-aminocyclopropane-1-carboxylic acid and its conversion to ethylene in etio-lated pea seedlings, *Planta* **174:**112–114.

Theologis, A., 1986, Rapid gene regulation by auxin, *Annu. Rev. Plant Physiol.* **37:**407–438.

Theologis, A., Sato, T., and Huang, P.-L., 1990, Cloning the mRNA of ACC synthase, in *Plant Gene Transfer* (UCLA Symposia on Molecular and Cellular Biology, New Series, Vol. 129) (C. Lamb and R. Beachy, eds.), pp. 289–299, Wiley, New York.

Thompson, J. S., Harlow, R. L., and Whitney, J. F., 1983, Copper(I)-olefin complexes. Support for the proposed role of copper in the ethylene effect in plants, *J. Am. Chem. Soc.* **105:**3522–3527.

Tigchelaar, E. C., McGlasson, W. B., and Buescher, R. W., 1978, Genetic regulation of tomato fruit ripening, *HortScience* **13:**508–513.

Trewavas, A. J., 1982, Growth substance sensitivity, the limiting factor in plant development, *Physiol. Plant.* **55:**60–72.

Tsai, D.-S., Arteca, R. N., Bachman, J. M., and Phillips, A. T., 1988, Purification and characteriza-tion of 1-aminocyclopropane-1-carboxylate synthase from etiolated mung bean hypocotyls, *Arch. Biochem. Biophys.* **264:**632–640.

Tucker, G. A., and Brady, C. J., 1987, Silver ions interrupt tomato fruit ripening, *J. Plant Physiol.* **127:**165–169.

Tucker, G. A., and Grierson, D., 1982, Synthesis of polygalacturonase during tomato fruit ripening, *Planta* **155:**64–67.

Tucker, M. L., and Laties, G. G., 1984, Interrelationship of gene expression, polysome prevalence, and respiration during ripening of ethylene and/or cyanide-treated avocado fruit, *Plant Physiol.* **74:**307–315.

Tucker, M. L., Durbin, M. L., Clegg, M. T., and Lewis, L. N., 1987, Avocado cellulase: Nucleotide sequence of a putative full-length cDNA clone and evidence for a small gene family, *Plant Mol. Biol.* **9:**197–203.

Tucker, M. L., Sexton, R., del Campillo, E., and Lewis, L. N., 1988, Bean abscission cellulase. Characterization of a cDNA and regulation of gene expression by ethylene and auxin, *Plant Physiol.* **88:**1257–1262.

Ursin, V. M., and Bradford, K. J., 1989, Auxin and ethylene regulation of petiole epinasty in two developmental mutants of tomato, *diageotropica* and *Epinastic, Plant Physiol.* **90:**1341–1346.

Van Der Straeten, D., and Van Montagu, M., 1990, Ethylene biosynthesis and mode of action: molecular analysis of ACC synthase in tomato and isolation of ACC-insensitive mutants in *Arabidopsis thaliana,* in *Polyamines and Ethylene: Biochemistry, Physiology and Interactions* (Proceedings 1990 Penn State Symposium in Plant Physiology) (H. E. Flores, R. N. Arteca, and J. Shannon, eds.), pp. 36–49, American Society of Plant Physiologists, Rockville, MD.

Van Der Straeten, D., Djudzman, A., Van Wiemeersch, L., and Van Montagu, M., 1989a, A molecular genetic approach to the biosynthesis and mode of action of ethylene: Isolation of ACC-insensitive mutants of *Arabidopsis thaliana* (L.) Heynh, *Arch. Intern. Physiol. Biochim.* **97:**B188.

Van Der Straeten, D., Van Wiemeersch, L., Goodman, H. M., and Van Montagu, M., 1989b, Purification and partial characterization of 1-aminocyclopropane-1-carboxylate synthase from tomato pericarp, *Eur. J. Biochem.* **182:**639–647.

Van Der Straeten, D., Van Wiemeersch, L., Van Damme, J., Goodman, H., and Van Montagu, M., 1989c, Purification and amino-acid sequence analysis of 1-aminocyclopropane-1-carboxylic acid synthase from tomato pericarp, in *Biochemical and Physiological Aspects of Ethylene Production in Lower and Higher Plants* (H. Clijsters, M. De Proft, R. Marcelle, and M. Van Poucke, eds.), pp. 93–100, Kluwer Academic Publishers, Dordrecht.

Van Der Straeten, D., Van Wiemeersch, L., Goodman, H. M., and Van Montagu, M., 1990, Cloning and sequence of two different cDNAs encoding 1-aminocyclopropane-1-carboxylate synthase in tomato, *Proc. Natl. Acad. Sci. USA* **87:**4859–4863.

Veen, H., 1979, Effects of silver on ethylene synthesis and action in cut carnations, *Planta* **145**:467–470.

Venis, M., 1984, Cell-free ethylene-forming systems lack stereochemical fidelity, *Planta* **162**:85–88.

Venis, M., 1985, *Hormone Binding Sites in Plants,* Research Notes in Biosciences. Longman, New York.

Voelker, T. A., Herman, E. M., and Chrispeels, M. J., 1989, *In vitro* mutated phytohemagglutinin genes expressed in tobacco seeds: Role of glycans in protein targeting and stability, *Plant Cell* **1**:95–104.

Vögeli, U., Meins, F. Jr., and Boller, T., 1988, Co-ordinated regulation of chitinase and β-1,3-glucanase in bean leaves, *Planta* **174**:364–372.

von Heijne, G., 1983, Patterns of amino acids near signal-sequence cleavage sites, *Eur. J. Biochem.* **133**:17–21.

Walsh, C., Pascal, R. A., Jr., Johnston, M., Raines, R., Dikshit, D., Krantz, A., and Honma, M., 1981, Mechanistic studies on the pyridoxal phosphate enzyme 1-aminocyclopropane-1-carboxylate deaminase from *Pseudomonas* sp, *Biochemistry* **20**:7509–7519.

Wang, C. Y., and Adams, D. O., 1982, Chilling-induced ethylene production in cucumbers (*Cucumis sativus* L.), *Plant Physiol.* **69**:424–427.

Wang, H., and Woodson, W. R., 1989, Reversible inhibition of ethylene action and interruption of petal senescence in carnation flowers by norbornadiene, *Plant Physiol.* **89**:434–438.

Wang, T-T., and Yang, S. F., 1987, The physiological role of lipoxygenase in ethylene formation from 1-aminocyclopropane-1-carboxylic acid in oat leaves, *Planta* **170**:190–196.

Wilkins, T. A., Bednarek, S. Y., and Raikhel, N. V., 1990, Role of propeptide glycan in post-translational processing and transport of barley lectin to vacuoles in transgenic tobacco, *Plant Cell* **2**:301–313.

Wood, B. W., 1985, Effect of ethephon on IAA transport, IAA conjugation, and antidotal action of NAA in relation to leaf abscission of pecan, *J. Am. Soc. Hort. Sci.* **110**:340–343.

Woodson, W. R., 1987, Changes in protein and mRNA populations during the senescence of carnation petals, *Physiol. Plant.* **71**:495–502.

Woodson, W. R., and Lawton, K. A., 1988, Ethylene-induced gene expression in carnation petals. Relationship to autocatalytic ethylene production and senescence, *Plant Physiol.* **87**:498–503.

Wright, S. T. C., 1980, Effect of plant growth regulator treatments on the levels of ethylene emanating from excised turgid and wilted wheat leaves, *Planta* **148**:381–388.

Yang, S. F., 1987, The role of ethylene and ethylene synthesis in fruit ripening, in *Plant Senescence: Its Biochemistry and Physiology* (W. W. Thomson, E. A. Nothnagel, and R. C. Huffaker, eds.), pp. 89–97, American Society of Plant Physiologists, Rockville, MD.

Yang, S. F., and Hoffman, N. E., 1984, Ethylene biosynthesis and its regulation in higher plants, *Annu. Rev. Plant Physiol.* **35**:155–189.

Yanofsky, M. F., Ma, H., Bowman, J. L., Drews, G. N., Feldmann, K. A., and Meyerowitz, E. M., 1990, The protein encoded by the *Arabidopsis* homeotic gene *agamous* resembles transcription factors, *Nature (Lond.)* **346**:35–39.

Yip, W-K., Dong, J-G., Kenny, J. W., Thompson, G. A., and Yang, S. F., 1990, Characterization and sequencing of the active site of 1-aminocyclopropane-1-carboxylate synthase, *Proc. Natl. Acad. Sci. USA* **87**:7930–7934.

Yoshii, H., and Imaseki, H., 1981, Biosynthesis of auxin-induced ethylene, effects of indole-3-acetic acid, benzyladenine and aminocyclopropane-1-carboxylic acid (ACC) and ACC synthase, *Plant Cell Physiol.* **22**:369–379.

Yu, Y-B., and Yang, S. F., 1979, Auxin-induced ethylene production and its inhibition by aminoethoxyvinylglycine and cobalt ion, *Plant Physiol.* **64**:1074–1077.

Yu, Y-B., Adams, D. O., and Yang, S. F., 1979, 1-aminocyclopropane-1-carboxylate synthase, a key enzyme in ethylene biosynthesis, *Arch. Biochem. Biophys.* **198:**280–286.
Zobel, R. W., 1972, Genetics of the diageotropic mutant in the tomato, *J. Hered.* **63:**95–97.
Zobel, R. W., 1973, Some physiological characteristics of the ethylene requiring tomato mutant diageotropica, *Plant Physiol.* **52:**385–389.

Chapter 14

Gibberellin-Binding Proteins and Hormonal Regulation of Transcription in Cell Nuclei and Chloroplasts of Higher Plants

David I. Jokhadze

1. INTRODUCTION

In his famous work *The Power of Movement in Plants* (1880), on the basis of data obtained from his experiments, Darwin expressed the concept that vital processes in plants take place according to basic physical laws, as is the case with animal organisms. His well-known experiments on barley and other plant coleoptiles enabled him to understand that plants are able to receive light, gravitational, and other external stimuli by their special sensory sites, which transfer the stimuli to corresponding motor sites where responsive reactions, such as growth and movement, take place. Darwin suggested that the agency for transfer of stimuli through the plant should be a chemical substance. By means of a series of simple but shrewd experiments, it was demonstrated that this substance is produced at growth sites. The results of these observations enable us to make certain com-

David I. Jokhadze Institute of Plant Biochemistry, Georgian Academy of Sciences, 380059 Tbilisi USSR.

Subcellular Biochemistry, Volume 17: Plant Genetic Engineering, edited by B. B. Biswas and J. R. Harris. Plenum Press, New York, 1991.

parisons of the role of the apexes of basic plant organs with the coordinative role of the nervous system of lower animals. This is a well-known fact today, and the interaction among nerve cells occurs with the aid of special neuron mediators and neurohormones; nerve endings in muscles carry out their functional effect by release of their specific transmitter substance acetylcholine. It should be added that another neurohormone, serotonin, has a chemical structure close to indolil-3-acetic acid (IAA), a well-known phytohormone (auxin) essential to the function of plant vital processes.

The study of phytohormones as plant growth factors, which regulate absolutely all processes in plants and support the wholeness of the organism, began in the second half of the nineteenth century. Despite the fact that up to now great successes have been achieved in determining the chemistry, biochemistry, and physiological functions of these substances, interest in them has not declined. For the past two decades information has accumulated on the participation of phytohormones in regulation of the genetics of the plant cell, transcription in particular. In this review, I summarize information concerning the nature of participation of one of the main phytohormone groups, the gibberellins, in the transcriptional mechanisms of cell nuclei and chloroplasts and the gibberellin-binding proteins, which take part in hormone–genome interaction in higher plants.

2.   SOME GENERAL CHARACTERISTICS OF PHYTOHORMONES

Phytohormones are substances carrying out the regulation of cell metabolism: they are comparatively low-molecular-weight, endogenous organic substances and are present in small quantities in plant tissues. In their free state they are found at a concentration of about 10^{-10} to 10^{-5} M and at these concentrations perform their specific actions (Polevoi, 1982, 1986). As a rule, they are produced in one group of cells and tissues, but function in others. This enables phytohormones to carry out interactions in different parts of the plant organism. In particular, their functions include the regulation of physiological and morphogenetic programs, such as division and stretching of cells, root formation, flowering, and fruit ripening. At the same time, while fulfilling their functions, phytohormones interact with specific membrane components and also with specific regulatory proteins of the genetic apparatus. Attention should be paid to the fact that in these roles phytohormones differ from vitamins, many of which participate in the synthesis of a series of enzymes and are also part of their catalytic centers (Polevoi, 1982). There are, nevertheless, instances when phytohormones function in the very tissue where they have originated, this fact having been detected while cultivating calluses (Fox and Erion, 1977). It is interesting to note that the so-called histohormones, tissue regulators originating in the same tissues in which they perform their function, do exist in animals. They include, for example, some kinins, prostoglandins, histamine, and serotonin (Rozen, 1980). There is every reason to suppose that phytohormones can perform functions equivalent to both the true hormones and histohormones of animals (Kulaeva, 1982).

Five main types of phytohormones have been isolated to date. These are the auxins, gibberellins, abscissic acid, cytokinins, and ethylene (Elbersgame and Dervil, 1985) (see also Chapter 13). The natural active auxin, which exists in several forms, is β-indolyl-acetic-acid (IAA). Chemically, gibberellins are diterpens and in plants they form a large family of closely related compounds. Usually, the term "gibberellin" means gibberellic acid (GA_3). Cytokinins are derivatives of 6-aminopurine and exist in plants in the form of zeatin or its analogs and derivatives. Abscissic acid is an optically active sesquiterpenoid, and the two-carbon compound ethylene regulates certain physiological processes.

Recently information has appeared on the discovery of regulatory molecules in a new class of plants; these have been named oligosaccharines (Elbersgame and Dervil, 1985). They are located in the cell wall. Each oligosaccharine has the specific function of transferring a signal regulating a certain function, such as protection from illnesses or growth, and also the differentiation of cells during the process of growth.

The physiological function of all the phytohormone groups can be expressed by certain common features. Thus, all phytohormones are semifunctional (pleiotrops); i.e., each takes part in the regulation of not one, but many, physiological processes. In particular, they influence growth and division of cells, the processes of aging and adaptation, metabolite transport, respiration, and the synthesis of nucleic acids and proteins (Moore, 1979; Polevoi, 1982, 1986). We can say that their effect embraces all aspects of cellular metabolism. Nevertheless, each phytohormone group has specific features regarding its final effects. For example, IAA induces root growth, cytokinins affect the formation of stem buds, and gibberellins specifically influence the lengthening of stems, cuttings, and fibers (Polevoi, 1982, 1986). The final effect of these functions depends on both the hormone class and the object. In the regulation of an individual physiological process, not only a single, but several different phytohormones can participate. The final effect depends on the phytohormone concentrations (Kulaeva, 1982). Plants do not have a narrow specialization to phytohormones, but in different combinations and ratios they are used to regulate different physiological and morphogenetic processes.

According to numerous observations, auxins, cytokinins, and gibberellins activate in plants various growth and formation processes, while ethylene and abscissic acid often have an inhibitory influence on these processes. That is why gibberellins and cytokinins are often termed growth stimulators, while ethylene and abscissic acid are inhibitors of these processes. Nevertheless, such a division is rather arbitrary, as data on the occurrence of stimulatory and inhibitory effects of all phytohormone groups permit us to conclude that it is more logical to consider them substances regulating physiological processes in plants.

It should be noted, finally, that some metabolites, such as phenol compounds, participating in the regulation of transformation of phytohormones in a cell can have a great influence on the effect of functional phytohormones (Kefeli and Turetskaya, 1977).

3. GIBBERELLINS AND THE GENETIC SYSTEM OF PLANTS

Together with studies on the participation of phytohormones in the regulation of separate functions of plant organisms, a matter of no little interest is the revelation of the role of the genetic apparatus in these functional mechanisms, particularly in the key process of transcription.

The first reports on phytohormone effects on the initial expression of genome function, the synthesis of RNA, published in the early 1960s, relate to the then widely growing interest in the problems of molecular biology. Wollgiehn (1961) and later other authors (see Kulaeva's review, 1982) showed that cytokinins stimulate RNA synthesis in cut leaves; at the same time synthesis of all types of RNA is activated, including rRNA, tRNA, and fractions containing mRNA. As a result of investigations of mechanisms of such activation, it was revealed that cytokinins provoke an increase of activity of forms I and II of the genetic system's key enzyme, the DNA-dependent RNA polymerase (nucleosidetriphosphate-RNA nucleotide transferase, EC.2.7.7.6). Activation by cytokinins was also achieved by adding them to RNA polymerase, under the influence of isolated cell nuclei. As regards auxins, it was established that in all auxin-sensitive tissues under the influence of IAA and its synthetic analogs, incorporation of labeled precursors into RNA and proteins is activated. Special attention should be given to some studies (Jacobsen, 1977; Roy and Biswas, 1977; Biswas and Roy, 1978; Kulaeva, 1982) that showed that as a result of treating tissues with auxin, the transcription intensity increases in isolated nuclei and chromatin. At the same time, the activity of soluble RNA polymerase increases.

Many authors have revealed changes in genome expression, such as changes of the level of transcribed mRNA under the influence of auxin (Theologis and Roy, 1982; Walker and Key, 1982; Hagen et al., 1984; Van der Zaal et al., 1987), cytokinin (Chen et al., 1987), and ethylene (Boller et al., 1983).

Intensive studies of the participation of gibberellins in genome regulation started when Johry and Warner (1968) established that gibberellin acid (GA_3) produced a noticeable increase in RNA synthesis in cell nuclei of dwarf pea shoots. The gibberellin form GA_3 promoted lengthening of the plant stem. At the same time, the gibberellin form GA_8 did not show any effect, either on the growth of the plant or on RNA synthesis.

The increase of RNA synthesis under the influence of GA_3 was confirmed by experiments in *in vitro* systems (Duda and Cherry, 1971) as well as *in vivo* (Trewavas, 1981; Zwar and Jacobson, 1972). According to Wasilewska and Kleczkowski (1976), the treatment of etiolated maize shoots by GA_3 produces a significant increase in the formation of polyribosomes. The authors concluded that this effect was the result of the increase of RNA synthesis, related to the ribosome complexes and carrying polyadenylated sequences, characteristic of the majority of eukaryotic mRNAs, including plant organisms. Kopoor (1987) traced stimulation by the inclusion of [^{3}H]-uridine into the total RNA in pea shoots under the influence of GA_3 and cAMP. Experiments conducted with the barley aleuronic layer, the classical model system for investigation of hormonal induction of enzymes in plants, showed that under the influence of GA_3, induction of RNA formation, responsible for the synthesis of α-amylase, takes place (Verner, 1977; Muthukrishnan *et al.*, 1980, 1983). Confirmation of such a conclusion came from the cDNA to α-amylase mRNA, induced by GA_3, and cloning in a bacterial system (Jacobsen *et al.*, 1982). Induction of the transcription of an individual gene by gibberellin and synthesis of the corresponding enzyme were demonstrated for the first time by Chory and colleagues (1987). In the case of dwarf pea and maize, the mechanisms of stem lengthening under the influence of GA_3 were investigated and, simultaneously, the profile change of protein synthesis based on the change of mRNA and activation of gene transcription was revealed.

According to Mishra and Feltham (1975), transcription stimulation under the influence of different hormones, observed in animal organisms (see Rozen, 1980), can be caused in three ways: (1) A partial blocking of chromatin deproteination takes place, resulting in increase of the quantity of free loci for transcription. (2) Nuclear RNA polymerase is activated by changing its conformational features following allosteric interaction between the hormone and enzyme. (3) The quantity of enzyme and possibly its separate subunits and effectors increases by means of *de novo* synthesis.

It was shown in the plant RNA synthesis system, with the chromatin of a dwarf pea (McComb *et al.*, 1970) and also with chromatin of potato (Wielgat and Kahl, 1979), that under the influence of GA_3, activation of RNA polymerase bound to chromatin occurs. Later, Hironori and co-workers (1983a), while studying the influence of GA_3 on the RNA-polymerase system of top buds of dwarf pea (6-day shoots were sprinkled by the GA_3 solution and then the plants were kept in darkness for 48 hr), established that at the same time as the increase of activity of the two forms of RNA polymerase (I and II), an increase of chromatin template activity takes place, while the catalytic features of form II do not change. The same authors (Hironori *et al.*, 1983b) showed that the effects of GA_3 on the chromation matrix remain on the chromation components. In particular, GA_3 is loosely bound with chromation proteins from Ga_3-treated chromatin and enhances the *in vitro* template activity of LBP-depleted chromation. The authors suggest that these same proteins are responsible for enhancement of the *in vitro* template activity.

In our laboratory it was demonstrated that in cell nuclei, obtained from the leaves and roots of young peas, grown on a medium containing GA_3, the endogenous ability for RNA synthesis is considerably increased (Jokhadze and Goglidze, 1977, 1980). A similar picture was observed with the chloroplasts isolated from leaves (Tevzadze and Jokhadze, 1986), which do not have a genomic chromosomal structure. The latter circumstance made us conclude that under the influence of GA_3, the main synthetic events should take place with the transcribing enzyme RNA polymerase. As further investigations have shown, there occurs in leaf cell nuclei a significant redistribution of the quantity and activity of all three main forms of DNA-dependent RNA-polymerase—I, II, and III (Jokhadze *et al.*, 1990).

Electrophoretic analysis of RNA-polymerase forms revealed formation of two new bands of form I under the influence of GA_3 (Jokhadze *et al.*, 1990). On the basis of these experiments, it was concluded that by changing the quantity or activity of RNA-polymerase forms, phytohormones participate in the selective function of genes, as it is known that different forms of the enzyme are responsible for transcription of the corresponding genome types (Blair, 1988).

It is noteworthy that in cell nuclei, isolated from soybean epicotiles grown on medium with GA_3, the cAMP content is increased, as well as the lobe of poly-$(A)^+$ mRNA in the sum total of the RNA synthesized in such nuclei. Analysis of poly-$(A)^+$ mRNA and corresponding cDNA, obtained by means of reverse transcription, enabled us to conclude that under the influence of GA_3, transcriptional activation of some types of mRNA or appearance of the new types occurs (Jokhadze *et al.*, 1989). It is quite probable that GA_3 also affects the post-transcriptional stage and participates in polyadenylation of the newly synthesized mRNA, causing enhancement of the poly-A segment of mRNA. According to Vesely and Rochat (1980), GA_3 enhances guanylate cyclase activity in mammalians.

According to the experimental data of Vanushin and Kirnes (1988), GA and other phytohormones influence the methylation of cytosine of nuclear DNA and in this way participate in the regulatory mechanisms of genome transcription. Obviously, this mechanism is general for eukaryotes, but compared to hormonal systems in animals, phytohormones are perhaps less specialized and have a systemic action.

4. GIBBERELLIN-BINDING PROTEINS

There is much convincing evidence of the existence in plants of proteins with an ability to specifically bind to phytohormones and in this way to partici-

pate in the regulation processes taking place within a cell. Similar studies on
hormone-binding proteins in animals, during the last decade especially, have also
brought about great achievements (Romanov, 1989).

Together with the evidence for the existence of proteins with a high affinity
for phytohormones and having receptor function, DNA sites have been found in
promotor areas of genes regulated by phytohormones (Kuhlemeier *et al.*, 1987).
In addition, phytohormones specifically accumulate in tissue cells sensitive to
them (Brandes and Kende, 1968; Szivatava, 1987). In model systems *in vitro,*
where phytohormones show an effect, the presence of "protein factor" is necessary (Kulaeva, 1982; Venis, 1985; Libbenga *et al.*, 1986), and the discovery of
synthetic antihormones competitively inhibiting the influence of phytohormones
(Skoog *et al.*, 1973) is significant. Experiments on modeling of the site of
phytohormone binding to a hypothetical receptor, based on the data of physiological activity of phytohormones and their chemical analogs (Kulaeva, 1982;
Venis, 1985), provide further important data.

Hormone-binding proteins have been found in many representatives of the
vegetable kingdom, from mosses to the great variety of higher plant species
(Romanov, 1989). Many studies deal with proteins binding auxins, cytokinins,
and abscissic acid. For example, there exist many instances of the receptor
function of auxin-binding proteins. Cytokinin-binding proteins were isolated not
only from full leaf tissue, but also from separate organelles, i.e., chloroplasts
(Romanko *et al.*, 1986).

The first suggestion that protein fractions have an ability to bind gibberellins
appeared in 1974 (Stoddart *et al.*, 1974). These authors described the processes
of isolating, from dwarf pea homogenate, two protein fractions selectively binding gibberellin A_1. The protein fractions had molecular mass of 500 and 600 kDa
and did not bind with gibberellin A_8. In their detailed reviews Venis (1985),
Stoddart (1986), and Szivatava (1987) mention many studies dealing with gibberellin-binding proteins or fractions containing such proteins. In investigations
in which the epicotyls of cucumber and lettuce, maize and dwarf pea shoots,
protoplasts of barley and *Vigna,* and potato tubers were tested, the presence of
gibberellin-binding sites in protein fractions and also in cell wall was con-
clusively established.

Keith and his co-workers (1981, 1982) have isolated from potato hypocotyls
a soluble gibberellin-binding protein fraction with the molecular mass 80–100
kDa, which is probably a receptor. In their work the authors used gel filtration on
a Sephadex G-50 column and then elution from DEAE cellulose filters. The
isolated protein fraction bound chiefly with GA_4 ($kM = 7.10^{-8}$ M) at a pH
optimum 7.5; saturation was achieved in 30–40 min. This binding was reversible
and the affinity of protein with different gibberellin isoforms correlated well with
their activity in the biotest with the same object. However, with GA_9 and GA_{36},
which showed activity in the biotest, no binding with the protein fraction was

observed (Yalpaini and Szivatava, 1985). Nevertheless, by taking into account the fact that GA_{36} is an immediate precursor of the very active GA_4 (Hedden, 1983), its activity in biotest can be readily explained.

Yalpaini and Szivatava (1987) present data on the structural specialization and kinetics of gibberellin binding to cytosol proteins of cucumber epicotyls. These authors obtained a partially purified GA-binding protein by making use of ammonium sulfate precipitation and ion-exchange chromatography with hydroxymetylapatite. In the same laboratory, a study was made of Ga_4 binding with proteins by dwarf pea and high pea epicotyl cytosol fraction (Zin-Huang and Szivatava, 1987). Differences of some characteristic features among gibberellin-binding protein fractions obtained from these plant forms were examined.

Recently Keith and Rappoport (1987) reported on the existence in dwarf maize leaf vaginas of soluble protein fractions with a high affinity to gibberellin A_1, and from other authors (Stoddart et al., 1974; Lashbrook et al., 1987) we find analogous data from pea epicotyls. The molecular mass of these proteins was 40–100 kDa and they were characterized by two GA_1 binding pH optima: 7.0 and 10.0 for maize leaf vaginas, but pH 8.0 for pea epicotyls. The complex of gibberellin with this fraction showed a great stability, particularly in the case of pea epicotyls, where the semidecay period of the complex was 9 hr at 18°C and 38–38.5 hr at 0°C (Keith and Szivastava, 1980).

In Stoddart's laboratory (Stoddart, 1979), the example of gibberellin A_1 binding to the cell walls of lettuce was discovered. This was supported by electron microscopy, compositional analysis, density gradient centrifugation, and also electrophoretic analysis. It should be mentioned that the gibberellin-binding took place only with intact cells and a correlation was obvious between the quantity of bound hormone and the tissue growth rate. At decreased temperature or under the influence of protein biosynthesis inhibitors, a reduction of GA_1 binding with cell walls was noticeable. From these observations one can conclude that proteins having the ability to specifically bind with gibberellins are present not only in cell cytoplasm and organelles, but also in cell walls. Also, under the influence of these phytohormone groups, the plasticity of the polysaccharide complex incorporated into the cell wall matrix increases.

Experiments in which the influences of GA_3 on the endogenous transcriptional activity of cell nuclei and chromatin from bean leaves were compared showed that by taking into account DNA and under the influence of phytohormone in both cases, the level of RNA synthesis increases. Nevertheless, the stimulation is expressed more vividly in intact nuclei than in chromatin fractions. This enabled us to conclude that a specific gibberellin-binding factor should be present in bean nuclei, which, by means of interaction between GA_3 and genome, is participating in GA-dependent transcription stimulation and is apparently lost during isolation of chromatin from nuclei (Tevzadze et al., 1983). Furthermore, in the case of bean epicotyls, the existence of a protein fraction specifically binding with GA_3 was also shown. This fraction, having molecular mass of 80–100 kDa, is present in different parts of a cell, but is mainly located in a nucleus (Tevzadze and Jokhadze, 1989).

5. GIBBERELLIN-BINDING PROTEINS AND TRANSCRIPTION IN CELL NUCLEI AND CHLOROPLASTS

Against the background of data on the hormonal regulation of transcription in plants, there is not much dealing with hormone-binding proteins participating in this process, particularly the gibberellins. The importance of such sparse data increases because of the fact that it could throw certain light not only on the molecular mechanisms of gene function, but also on the specificities of the genetic systems of separate organelles of the plant cell, which possesses transcriptional apparatus in the nucleus, chloroplasts, and mitochondria. Specifically, knowledge of the degree of organelle autonomy, and the origin and evolution of an organelle in a cell, may be advanced. It should be pointed out that in chloroplasts and mitochondria, whose main function is the energy supply of a cell, up to 10% of genes of the whole cell are located (Sager, 1972) and may play the most important role in plant metabolism.

Experiments relating to the participation of the hormone-binding proteins in transcription in cell nuclei and chloroplasts have been performed mainly on cytokinins. Thus, Romanko and co-workers (1982) studied the influence of a cytokinin-binding protein fraction (CKP), isolated from barley leaves, on RNA synthesis *in vitro* by nuclei and chloroplasts, isolated from these leaves. In both cases, addition of CKP together with cytokinin stimulated transcription. It was also demonstrated that CKP in complex with cytokinin reveals a stimulating influence on chromatin matrix activity and RNA-polymerase activity (Selivankina *et al.*, 1985). The increase in RNA synthesis occurs only if the chromatin matrix or the enzyme from the same source (in the case of barley leaves) or both these components appear together. Romanko and co-workers (1986) have isolated CKP from barley leaf epicotyls and have investigated its effect on various transcriptional systems *in vitro*. They mentioned the high specificity of the isolated preparation with respect to chloroplast RNA synthesis and its lack of effect on RNA synthesis by nuclear RNA polymerase.

From analogous investigations on auxins, the experiments by Van der Linde and co-workers (1984) should be mentioned. These authors isolated the receptor for this phytohormone from tobacco callus contaminated with polyphenoles. Such a preparation, which the authors mentioned as being partially purified, stimulated transcription in isolated nuclei of tobacco callus. Stimulation occurs due to RNA polymerase II activation. Also, Baily and co-workers (1985) have shown the existence of auxin-binding proteins in the cytosol and in nuclei, which stimulate nuclear RNA synthesis.

Analogous works concerning the gibberellins are comparatively few. According to Tomi and co-workers (1983), a protein fraction exists in pea shoots, loosely bound to chromatin, that is capable of stimulating RNA-synthesizing matrix activity in the presence of GA_3. The authors consider that this fraction participates in 'the weakening of binding between DNA and histones in the chromatin. As mentioned above, in our laboratory the protein fraction with molecular mass 80–100 kDA has been isolated from nuclei of bean epicotyls that had an ability to specifically bind with GA_3. We have termed this fraction gibberellin-binding protein (GBP). As the determination of ^{14}C GA_3 binding with the various cell fractions has shown, GBP may be revealed in a number of cell compartments, but is mainly concentrated in the nucleus (Tevzadze and Jokhadze, 1989; Jokhadze et al., 1990). For the preparation and partial purification of GBP from isolated cell nuclei of bean epicotyls, a protocol of nucleoplasm gel filtration on a G-25 column and TSK-gel HW-55 was used. Experiments revealing the effect of GBP on the RNA synthesis in cell nuclei of bean epicotyls, separately and in combination with GA_3, have shown that addition of the preparation to the nuclei (preincubation) partially stimulates the endogenous RNA transcription 32–44% and addition together approximately 250%. Analogous experiments with chloroplasts have also shown that the addition of GA_3 as well as GBP separately stimulates endogenous ability for RNA synthesis (especially in the presence of GA_3). However, the joint effect is much greater; in this case transcription is stimulated by more than 350%.

In an RNA-polymerase system in vitro containing pea leaf RNA polymerase (form I) and homologous nuclear DNA or chloroplast RNA polymerase with the homologous or nuclear DNA of pea leaves (Jokhadze and Goglidze, 1980; Balashvili et al., 1985), the addition of GA_3 without GBP caused no increase or RNA synthesis and the addition of GBP separately stimulated the process approximately 35%. The last observation may apparently be explained by the fact that in this case CPB promotes the detection of newly formed RNA released from the chromatin matrix. The addition of GBP and GA_3 together in this system increases transcription approximately 400%. This finding enables us to conclude that the protein fraction GBP, obtained from bean epicotyl cell nuclei, has the ability to bind selectively to GA_3 and to participate in the functional interaction between GA_3 and the genome, not only in nuclei but also in chloroplasts.

6. CONCLUSIONS

In conclusion, it can be stated that further studies on GBP and analogous protein fractions for the other phytohormones may reveal the intimate mechanisms of their role in transcription in the cell nucleus and also in the other cell organelles. In the latter, where the genome is not organized as in the nucleus, new information about the functional role in the interaction between phytohormone and genome, as well as in the development of new approaches to artificial manipulation with the hereditary material, may be forthcoming.

From a survey of recent information on the phytohormone-binding proteins and their role in the regulation of transcription, we can conclude that such pieces of information on the various phytohormones are not identical. Much more is known about the auxins and less about abscissic acid. Few publications exist on the gibberellin-binding proteins. The existence of proteins with a high affinity for phytohormones, including the gibberellins, which reveal the regulatory DNA sites in the gene promotor areas controlled by phytohormone, indicates the existence of hormone receptors in plants equivalent to those in animals. GBPs can function as important components in the transcriptional mechanism of cell nuclei as well as of other cellular organelles possessing genomic systems. Whether there are different or specific GBPs for separate cell organelles and how this is reflected at the transcriptional level will be revealed by the results of future investigations.

7. REFERENCES

Baily, H. M., Backer, D. I., Libbenga, K. R., van der Linde, P. C., Mennes, A. M., and Elliott, M. C., 1985, Auxin-binding site in tobacco cells, *Biol. Plant.* **27**:105–109.

Balashvili, M. I., Goglidze, R. I., and Jokhadze, D. I., 1985, Isolation and partial characterization of chloroplast RNA polymerase from pea leaves, *(Sov.) Bull. Acad. Sci. Georgian SSR* **119**:617–620.

Biswas, B. B., and Roy, P., 1978, Plant growth substances as modulators of transcription, in *Subcellular Biochemistry* (D. B. Roodyn, ed.), pp. 187–219, Plenum Press, New York.

Blair, D. G. R., 1988, Eucaryotic RNA polymerases, *Comp. Biochem. Physicol.* **89**:647–670.

Boller, T. A., Gehry, A., Mauch, F., and Vogeli, M., 1983, Chitinase in bean leaves: Induction by ethylene, purification, properties, and possible function, *Planta* **157**:22–31.

Brandes, H., and Kende, 1968, Studies on cytokinin-controlled bud formation in moss protonemata, *Plant Physiol.* **43**:827–837.

Chen, C. M., Ertl, Y., Yang, M. S., and Chang, C. C., 1987, Cytokinin-induced changes in the population of translatable mRNA in excised pumpkin cotyledons, *Plant. Sci.* **52**:169–174.

Chory, J., Voytas, D. F., Olszevski, N. E., and Ausubel, F. M., 1987, Gibberellin-induced changes in the populations of the translatable mRNAs and accumulated polypeptides in dwarfs of maize and pea, *Plant Physiol.* **83**:15–23.

Darwin, C., 1880, The Power of Movement in Plants, J. Murray, London.

Dorffling, K., 1982, *Das hormonsystem der pflangen, Zeichnungen von Rudolf Brummer*, p. 303, Georg Thieme, Stuttgart, New York.

Duda, C. T., and Cherry, J. H., 1971, Chromatin and nuclear-directed ribonucleic acid synthesis in sugar beet root, *Plant Physiol.* **47**:261–268.

Elbersgame, P., and Dervil, A., 1985, Oligosaccharins, *Sci. Am.* **253**:16–23.

Fox, J. E., and Erion, J., 1977, Cytokinin-binding proteins in higher plants, in *Plant Growth Regulation* (P. E. Pilet, ed.), pp. 139–146, Springer-Verlag, Berlin, Heidelberg, New York.

Hagen, G., Kleinshmidt, A., and Guilfloyle, T. J., 1984, Auxin regulated gene expression in intact soybean hypocotil and excised hypocotil sections, *Planta* **162**:147–153.

Hedden, P., 1983, *The Biochemistry and Physiology of Gibberellines,* Vol. 1, pp. 99, Praeger, New York.

Hironori, T., Sasaki, Y., and Kamikubo, T., 1983a, Increase of RNA polymerase activity in pea buds treated with gibberellic acid A_3, *Plant Cell Physiol.* **24:**587–592.

Hironori, T., Sasaki, Y., and Kamikubo, T., 1983b, Stimulation of *in vitro* RNA synthesis by DNA-loosely bound proteins treated with gibberellin A_3, *Plant Cell Physiol.* **24:**1087–1092.

Jacobsen, H. I., 1977, Regulation of ribonucleic metabolism by plant hormones, *Annu. Rev. Plant Physiol.* **28:**537–564.

Jacobsen, T. V., Chandler, P. M., Higgins, T. J. V., and Zwar, J. A., 1982, Control of protein synthesis in barley aleurone layers by gibberellin, in *Plant Growth Substances* (P. E. Wareing, ed.), pp. 111–120, Academic Press, London, New York.

Johry, M. M., and Warner, J. E., 1968, Enhancement of RNA synthesis of isolated pea nuclei by gibberellic acid, *Proc. Natl. Acad. Sci. USA* **59:**269–276.

Jokhadze, D. I., and Goglidze, R. I., 1977, Comparative effect of gibberellic acid on RNA-polymerase activity of cell nuclei from pea leaves and roots, *(Sov.) Plant Physiol.* **24:**746–750.

Jokhadze, D. I., and Goglidze, R. I., 1980, Isolation and partial characterization of three forms of RNA polymerase from pea *(Pisum sativum)* leaves cell nuclei, *(Sov.) Bull. Acad. Sci. Georgian SSR* **99:**189–192.

Jokhadze, D. I., Mamulashvili, N. A., Lomidze, N. N., and Nazarova, L. A., 1989, Isolation and partial characterisation of poly-(A)+ mRNA soybean grown with Ga_3, *(Sov.) Ukr. Biochem. J.* **61:**107–111.

Jokhadze, D. I., Goglidze, R. I., and Tevzadze, N. N., 1990, Studies of gibberellic acid effect on the DNA-dependent RNA polymerase activity of higher plant cell organelles, 20th Meet., FEBS, Abstracts, Budapest, P-Th 480, p. 326.

Kefeli, V. N., and Turetskaya, R. X., 1977, Natural inhibitors of growth: The principal physiological aspects of action, in *Plant Growth and Natural Regulators,* pp. 234–245, Nauka Press, Moscow.

Keith, B., and Rappoport, L., 1987, *In vitro* [³H] gibberellic A_1 binding to soluble proteins from GA-sensitive and GA-insensitive dwarf maize mutants, in *Molecular Biology of Plant Growth Control,* pp. 289–308, Alan R. Liss, New York.

Keith, B., Foster, N. A., Bonettemaker, M., and Srivastava, L. M., 1981, *In vitro* gibberellin A_4 binding to extracts of cucumber hypocotyls, *Plant Physiol.* **68:**344–348.

Keith, B., Brown, S., and Srivastava, L. M., 1982, *In vitro* binding of gibberrelin A_4 to extracts of cucumber measured by using DEAE–cellulose filters, *Proc. Natl. Acad. Sci. USA* **79:**1515–1519.

Kopoor, H. C., 1987, Stimulation of RNA synthesis in cowpea *(Vigna sinensis)* seedlings by gibberellic acid and adenosine 3'5'-cyclic monophosphate, *Phytochemistry* **26:**31–36.

Kuhlemeier, C., Green, P. J., and Chua, N. H., 1987, Regulation of gene expression in higher plants, *Annu. Rev. Plant Physiol.* **38:**221–257.

Kulaeva, O. N., 1982, *Hormonal Regulation of Physiological Processes in Plants on the Level of RNA and Protein Synthesis* (A. L. Kursanov, ed.), pp. 383, Nauka Press, Moscow.

Lashbrook, C. C., Keith, B., and Rappaport, L., 1987, *In vitro* gibberellin A_1 binding to a soluble fraction dwarf pea epicotyles, in *Molecular Biology of Plant Growth Control,* pp. 299–303, Alan R. Liss, New York.

Libbenga, K. K., Maan, A. C., Linde, P. C. G., and van der Mennes, A. M., 1986, *Hormones, Receptors and Cellular Interactions in Plants,* p. 1, Cambridge University Press, Cambridge.

McComb, A. J., McComb, J. A., and Duda, C. T., 1970, Increaesd ribonucleic acid polymerase activity associated with chromatin from internodes of dwarf pea plants treated with gibberellic acid, *Plant Physiol.* **46:**221–223.

Mishra, R. K., and Feltham, Z. A. W., 1975, RNA polymerase stimulation: Effect of aldosterone and other adrenocorticoids on RNA turnover in rat kidney, *Can. J. Biochem.* **54:**70–78.

Moore, T. C., 1979, *Biochemistry and Physiology of Plant Hormones,* p. 274, Springer-Verlag, New York.

Muthukrishnan, S., Chandra, G. R., and Maxwell, E. S., 1980, Evidence for processing of α-amylase messenger RNA in barley aleurone cells, *Plant Physiol.* **65:**124–128.

Muthukrishnan, S., Chandra, G. R., and Maxwell, E. S., 1983, Hormonal control of α-amylase gene expression in barley: Studies using cDNA probe, *J. Biol. Chem.* **258:**2370–2375.

Polevoi, V. V., 1982, *Phytohormones,* p. 249, Plenum University of Leningrad, Leningrad.

Polevoi, V. V., 1986, *The Role of Auxin in the Regulational System in Plants,* Nauka, Leningrad.

Romanko, E. G., Selivankina, S. Y., and Ovcharov, A. K., 1982, Involvement of cytokinin-binding proteins from barley leaves in activation of cytokinin of RNA synthesis in isolated nuclei and chloroplasts, *(Sov.) Plant Physiol.* **36:**524–531.

Romanko, E. G., Selivankina, S. Y., Moshkov, J. E., and Novikova, G. V., 1986, Effect of cytokinin-binding proteins isolated from chloroplasts on the transcription, *(Sov.) Plant Physiol.* **33:**1078–1083.

Romanov, G. A., 1989, Hormone-binding proteins in plants and the problem of phytohormone receptor, *(Sov.) Plant Physiol.* **36:**166–177.

Roy, P., and Biswas, B. B., 1977, A receptor protein for indoleacetic acid from plant chromatin and its role in transcription, *Biochem. Biophys. Res. Commun.* **74:**1597–1606.

Rozen, V. B., 1980, *The Principles of Endocrinology,* p. 344, Vyssaya Shkola, Moscow.

Sager, R., 1972, *Cytoplasmic Genes and Organelles,* p. 423, Academic Press, New York, London.

Selivankina, S. Y., Romanko, E. G., Burkhanova, E. A., Karavajko, N. N., Zauakin, V. V., and Kulaeva, O. N., 1985, A comparison of effects of cytokinin–receptor complex on the template activity of chromatin and RNA polymerase activity in barley leaves, *(Sov.) Biochimia* **50:**47–52.

Skoog, F., Schmitz, R. Y., Bock, R. M., and Hecht, S. M., 1973, Cytokinin antagonists: Synthesis and physiological effects of 7-substituted 3-methylpyrazol-[4,3-d] pyrimidines, *Phytochemistry* **12:**25–37.

Szivatava, L. M., 1987, *Plant Hromone Receptors,* p. 199, Springer-Verlag, New York.

Stoddart, J. L., 1979, Interaction of [³H] gibberellin A₁ with subcellular fraction from lettuce (*Lactura sativa* L) hypocotils, *Planta* **146:**363–368.

Stoddart, J. L., 1986, *Hormones, Receptors and Cellular Interactions in Plants,* p. 91, Cambridge University Press, Cambridge.

Stoddart, J. L., Briedenbach, W., Nadeau, R., and Rappoport, L., 1974, Selective binding of [³H] gibberellin A₁, by protein fractions from dwarf pea epicotyls, *Proc. Natl. Acad. Sci. USA* **71:**3255.

Tevzadze, N. N., and Jokhadze, D. I., 1986, Stimulation of endogenous RNA polymerase activity of chloroplasts with gibberellic acid, *(Sov.) Bull. Acad. Sci. Georgian SSR* **122:**141–143.

Tevzadze, N. N., and Jokhadze, D. I., 1989, Gibberellin-binding proteins in the cell of kidney bean epicotyls, *(Sov.) Proc. Acad. Sci. Georgian SSR* **15:**389–394.

Tevzadze, N. N., Amzashvili, M. G., and Jokhadze, D. I., 1983, Comparative effect of gibberellic acid on the transcriptional activity of cell nuclei and chromatin in kidney bean leaves, *(Sov.) Plant Physiol.* **30:**404–405.

Theologis, A., and Roy, P. M., 1982, Early auxin-regulated plyadenylated mRNA sequences in pea stem tissue, *Proc. Natl. Acad. Sci. USA* **79:**418–422.

Tomi, H., Sasaki, J., and Katikava, T., 1983, Enhancement of template activity of chromatin in pea by gibberellic acid, *Plant Sci. Lett.* **30:**155–164.

Trewavas, A. J., 1981, How do plant growth substances work? *Plant Cell Environ.* **4:**203–228.

Van der Linde, P. C. G., Bouman, H., Mennes, A. M., and Libbenga, K. R., 1984, A soluble auxin-binding protein from cultured tobacco tissues stimulates RNA synthesis *in vitro, Planta* **160:**102–108.

Van der Zaal, E. J., Mennes, A. M., and Libbenga, K. R., 1987, Auxin-induced rapid changes in translatable mRNA in tobacco cell suspension, *Planta* **172**:514–519.

Vanushin, B. E., and Kirnes, M. O., 1988, Regulation of DNA by phytohormones as possible mechanism of cell differentiation regulation, in *Genomes of Plants*, pp. 122–126, Kiev.

Venis, M., 1985, Hormone binding sites in plants, p. 190, Longman, New York.

Verner, J. E., 1977, Hormonal control of protein synthesis, in *Nucleic Acids and Protein Synthesis in Plants* (L. Bogorad and J. H. Weill, eds.), pp. 293–307, Plenum Press, New York.

Vesely, D. Z., and Rochat, M. H., 1980, Gibberellic acid, a plant growth hormone, enhances mammalian guanylate cyclase activity, *Res. Commun. Chem. Pathol. Pharmacol.* **28**:123–132.

Walker, J. C., and Key, J. L., 1982, Isolation of cloned cDNAs to auxin-responsive poly(A) RNAs of elongating soybean hypocotyle, *Proc. Natl. Acad. Sci. USA* **77**:357–361.

Wasilewska, L., and Kleczkowski, K., 1976, Preferential stimulation of the plant mRNA synthesis by gibberellic acid, *Eur. J. Biochem.* **66**:405–412.

Wielgat, B., and Kahl, G., 1979, Gibberellic acid activates chromatin-bound DNA dependent RNA polymerase in wounded potato tuber tissue, *Plant Physiol.* **64**:867–871.

Wollgiehn, R., 1961, Untersuchungen uber der einfluss des kinetins auf den nucleinsaure und proteinstoffwechsel isolierter blatter, *Flora* **151**:411–437.

Yalpaini, N., and Szivatava, L. M., 1985, Competition for *in vitro* [³H] gibberellin A₄ binding in cucumber by gibberellins and their derivatives, *Plant Physiol.* **79**:963–967.

Yalpaini, N., and Szivatava, L. M., 1987, Partial purification of gibberellin binding protein from cucumber hypocotyles, in *Molecular Biology of Plant Growth Control*, pp. 309–314, Alan R. Liss, New York.

Zin-Huang, L., and Szivatava, L. M., 1987, *In vitro* binding of gibberellin A₄ in epycotyles of dwarf pea and tall pea, in *Molecular Biology of Plant Growth Control*, pp. 315–322, Alan R. Liss, New York.

Zwar, J. A., and Jacobson, T. V., 1972, A correlation between a ribonucleic acid fraction selectively labelled in the presence of gibberellic acid and amylase synthesis in barley alleurone layers, *Plant Physiol.* **49**:1000–1006.

Index